IUTAM SYMPOSIUM ON RHEOLOGY OF BODIES WITH DEFECTS

SOLID MECHANICS AND ITS APPLICATIONS
Volume 64

Series Editor: G.M.L. GLADWELL
Solid Mechanics Division, Faculty of Engineering
University of Waterloo
Waterloo, Ontario, Canada N2L 3G1

Aims and Scope of the Series

The fundamental questions arising in mechanics are: *Why?*, *How?*, and *How much?* The aim of this series is to provide lucid accounts written by authoritative researchers giving vision and insight in answering these questions on the subject of mechanics as it relates to solids.

The scope of the series covers the entire spectrum of solid mechanics. Thus it includes the foundation of mechanics; variational formulations; computational mechanics; statics, kinematics and dynamics of rigid and elastic bodies; vibrations of solids and structures; dynamical systems and chaos; the theories of elasticity, plasticity and viscoelasticity; composite materials; rods, beams, shells and membranes; structural control and stability; soils, rocks and geomechanics; fracture; tribology; experimental mechanics; biomechanics and machine design.

The median level of presentation is the first year graduate student. Some texts are monographs defining the current state of the field; others are accessible to final year undergraduates; but essentially the emphasis is on readability and clarity.

For a list of related mechanics titles, see final pages.

IUTAM Symposium on

Rheology of Bodies with Defects

Proceedings of the IUTAM Symposium
held in Beijing, China,
2–5 September 1997

Edited by

REN WANG

Department of Mechanics and Engineering Sciences,
Peking University,
Beijing, China

Springer-Science+Business Media, B.V.

A C.I.P. Catalogue record for this book is available from the Library of Congress.

ISBN 978-94-017-3829-3 ISBN 978-0-306-46937-4 (eBook)
DOI 10.1007/978-0-306-46937-4

Printed on acid-free paper

Originally published by Kluwer Academic Publishers in 1999.
Softcover reprint of the hardcover 1st edition 1999

IUTAM SYMPOSIUM ON RHEOLOGY OF BODIES WITH DEFECTS

September 2-5, 1997, Beijing, China

Sponsorship:

IUTAM (International Union of Theoretical and Applied Mechanics).

CSTAM (Chinese Society of Theoretical and Applied Mechanics).

NNSFC (National Natural Science Foundation of China).

Kluwer Academic Publishers.

Huazhong University, Department of Mechanics.

Peking University, Department of Mechanics and Engineering Science.

Xiangtan University, Institute of Rheological Mechanics.

Contributors at the IUTAM Symposium on Rheology of Bodies with Defects
September 2-5, 1997, Beijing, CHINA

Participants at the IUTAM Symposium on Rheology of Bodies with Defects
September 2-5, 1997, Beijing, CHINA

LIST OF PARTICIPANTS

Prof. Bai Yi-Long
LNM, Institute of Mechanics
Chinese Academy of Sciences,
Beijing, 100080,
P. R. CHINA

Dr. Alla V. Balueva
Institute for Problems in Mechanics
Russian Academy of Sciences
pr. Vernadskogo 101
Moscow 117526, RUSSIA.

Dr. Adam Bodnar
Institute of Structural Analysis
Cracow University of Technology
ul. Warszawska 24
Krakow,31-155,
POLAND

Mr. Cao Ping
College of Resource Environment and
Civil Engineering
Central South University of Technology
Changsha, Hunan , 410083
P.R.CHINA

Prof. Chen Zhida
Dept. of Appl. .Math. and Mech.
China University of Mining
Beijing Graduate School
Baijing, 100083, P. R. CHINA

Mr. Chen Jian Kang
Dept. of Mechanics & Engrg. Sci.
Peking University
Beijing, 100871
P. R. CHINA

Prof. Marcin Chrzanowski
Cracow University of Technology
Institute of Structural Analysis
ul. Warszawska 24
Krakow, 31-155 POLAND

Dr. Dai Lan-Hong
Dept. of Mechanics & Engrg. Sci.
Peking University
Beijing, 100871, P. R. CHINA

Prof. Duan Zhu-Ping
Lab. for Laser & Dynamic Behavior
of Materials
Institute of Mechanics C.A.S.
Beijing, 100080, P.R.CHINA

Prof. Gao Yu-Chen
Northern Jiaotong University
Beijing,100044
P. R. CHINA

Dr. Vadim Sergejovich Gudramovych
Institute of Technical Mechanics
National Academy of Science of Ukraine
15, Leshko Popel St.,
Dnepropetrovsk, 320600, UKRAINE

Ms. Guo Zhaohua
Institute of Rheological Mechanics
Xiangtan University
Hunan, Xiangtan 411105
P. R. CHINA

Mr. He Lin
Chinese Society of Theoretical &
Applied Mechanics
15 Zhong Guan Cun Road
Beijing, 100080 , P. R. CHINA

Prof. Huang Nai-Chien
Dept. Aerospace and Mechanical Engrg.
University of Notre Dame
373 Fitzpatrick Hall
Notre Dame, IN 46556-5637, U. S. A.

Dr. Francois Hild
LMT/ENS de Cachan/CNRS/
Univ. Paris VI
61 Avenue du President Wilson
F-94235 Cachan Cedex FRANCE

Prof. Huang Zhu-Ping
Dept. of Mechanics & Engrg.Sci.
Peking University
Beijing, 100871,
P. R. CHINA

Prof. Kuang Zhen-Bang
Dept. of Engineering Mechanics
Shanghai Jiaotong University
Shanghai 200240 P. R. CHINA

Dr. Li Hui-Lin
LNM, Institute of Mechanics C.A.S.
15, Zhong Guan Cun Rd.
Beijing, 100080, P. R. CHINA

Dr. Li Jun Shi
East China Airport Design and
Construction Corporation
2550 Hongqiao Road
Shanghai 200335, P.R.CHINA

Dr. Liu Chun-Tu
Institute of Mechanics, C.A.S
Beijing,100080,
P.R.CHINA

Dr. Li Zhao-Xia
Dept. of Math. and Mechanics,
Southeast university,
Nanjing, 210096, P. R. CHINA

Mr. Liu Yi
Dept. of Mechanics & Engrg.Sci..
Peking university,
Beijing, 100871, P. R. CHINA

Prof. Liu Zong De
Dept. of Mechanics & Engrg. Sci.
Peking University
Beijing, 100871, P. R. CHINA

Prof. Liu Zhong
Institute of Rheological Mechanics
Xiangtan University
Xiangtan, Hunan 411105, P. R. CHINA

Prof. Lu Zhi-Xin
Institute of Solid Mechanics
University of Aeronautics & Astronautics
Beijing, 100083
P. R. CHINA

Mr. Luo Wenbo
Institute of Rheological Mechanics
Xiangtan University
Xiangtan Hunan 411105,
P. R. CHINA

Prof. Luo Yingshe
Institute of Rheological Mechanics.
Xiangtan University
Xiangtan Hunan 411105
P. R. CHINA

Dr. Mamoru Mizuno
Dept. Micro System Engineering
Nagoya University
Furo-Cho, Chikusa-ku
Nagoya 404-01, JAPAN

Prof. Kazimierz Sobczyk
Center of Mechanics
Institute of Fundamental Tech. Research
Polish Academy of Sciences
Swietokrzyska 21,00-049, Warsaw
POLAND

Prof. Sun Jun
Dept. of Geotech. Engineering
Tong-ji University
Shanghai 200092,
P. R. CHINA

Prof. Piet Stroeven
Faculty of Civil Engineering
Delft University of Technology
Stevinweg 4, 2628 CN Delft
THE NETHERLANDS

Prof. Ion Suliciu
Institute of Mathematics of the Romanian Academy
P.O.Box 1-764
RO-70700 Bucharest, ROMANIA

Prof. Wang Li-Li
Mechanics & Materials Science Research Center
Ningbo University
Ningbo, Zhejiang 315211
P. R. CHINA

Prof. Wang Ren
Dept. Mechanics & Eng. Sciences
Peking University
Beijing, 100871,
P. R. CHINA

Prof. Meng Fen Xia
Dept. of Physics
Peking University
Beijing, 100871, P. R. CHINA

Prof. Yang Ting-Qing
Dept. of Mechanics
Huazhong University of Science & Technology
Wuhan 430074 , P. R. CHINA

Mr. Zhang-Xi
Dept. of Mechanical and Mechatronic Engineering
The University of Sydney
Sydney, NSW 2006, AUSTRALIA

Prof. Yu Shou Wen
Dept. of Engineering Mechanics
Tsinghua University
Beijing 100084,
P.R.CHINA

Prof. Yuan Long-Wei
Institute of Rheological Mechanics
Xiangtan University
Xiangtan, Hunan 411105
P. R. CHINA

Prof. Zheng Quan-Shui
Dept. of Engineering Mechanics
Tsinghua University
Beijing, 100084, P. R. CHINA

Mr. Zhang Ping
Institute of Rheological Mechanics
Xiangtan University
Xiangtan, Hunan 411105, P.R.CHINA

Prof. Zhang San-Ying
Institute of. Mechanics, C. A. S.
15 Zhong Guan Cun Road
Beijing, 100080, P. R. CHINA

Prof. Zhao Da Gang
Chinese Society of Theoreticl and Applied Mechanics
15 Zhong Guan Cun Road,
Beijing 100080, P. R. CHINA

Prof. Zhou Yi-Chun
Dept. of Physics
Xiangtan University
Xiangtan, Hunan 41105,
P.R.CHINA

CONTENTS

PREFACE

The IUTAM Symposium on Rheology of Bodies with Defects was held in Beijing in September,1997. It was aimed at the development of Rheology in Solid Mechanics.

Rheology is classified in Applied Mechanics Review under fluid mechanics, however, in its broadest content as was envisaged in its earlier days, it covers the whole spectrum of material behavior from elasticity, plasticity, and fluid mechanics to gas dynamics. It was thought of as a branch of continuum mechanics, but emphasized the physical aspects of different materials, and frequently proceeded from basic physical principles. As the temperature rises, the distinction between solid and fluid, and the distinction between their micro-mechanical movements, become blurred. The physical description of such materials and their movements must be based on the thermodynamic principles of state variable theory; the classical division between solid and fluid mechanics disappears.

Under the classification adopted by Applied Mechanics Reviews, the subjects dealt with in this symposium come closer to viscoelasticity and viscoplasticity, especially close to the subdivision of creep dealing with creep rupture. The symposium focused at building a bridge between macroscopic and microscopic research on damage and fracture behavior of defective bodies made of metal, polymer, composite and other viscoelastic materials. Two different approaches are presented at the symposium. The first is a continuum damage theory for time-dependent evolution of defects at the macro/meso/microscopic levels. The other is based on thermodynamics, and makes use of recent advances in non-equilibrium dynamics, non-equilibrium statistical physics and stochastic mechanics; it also tries to merge macro/microscopic levels. The first approach comes very close to that presented in the IUTAM Series on "Creep in Structures" and many other series on creep and fracture of materials. The symposium provided a forum for discussion and interaction between the two approaches. A similar trend can be seen in the IUTAM Symposium on "Creep in Structures"; as the emphasis changed from pure deformation and stability of structures toward the development of theories regarding the damage growth and propagation of creep cracks, it also tended to merge with physical aspects of material sciences.

The early papers in this proceeding deal mainly with the theoretical framework and related experimental work using both approaches. They are followed by more specific studies on different materials, including composites, cast iron, mortar, concrete, clay and rock-mass. Finally creep rupture analyses of structures are presented. By defects, we usually mean voids, micro-cracks, inclusions and damage in material, but when dealing with the stability analyses of structures, we also consider initial imperfections as defects; this is the subject of the last paper dealing with creep analysis in structures.

The symposium was held in Beijing, for the benefit of the Eastern countries and in the hope of bringing Western and Eastern scientists together. We had researchers from Australia, France, Japan, Netherlands, Romania, Poland, Ukraine, and U.S.A. It was a good international gathering with active discussion. It is with deep regret that several scientists from Israel, Japan, Mexico and U. S. A. were unable to come at the last moment.

The symposium was mainly organized by the Chinese Society of Theoretical and Applied Mechanics (CSTAM). Special thanks should be given to Mr. He Lin, Zhao Da-Gang and Ms. Tang Ya-Nan of the CSTAM office. The generous financial support of IUTAM is much appreciated; the supports from CSTAM, NNSFC (National Natural Science Foundation of China), Xiangtan University, Huazhong University, Peking University, and Kluwer Academic Press are also gratefully acknowledged.

All the papers were reviewed and revised after the symposium; this delayed the assembly of the manuscripts. However, I would like to take this opportunity to thank the authors for their unfailing and cooperative efforts in quickly revising and returning their manuscripts, and to thank the staffs of Kluwer Academic Press for publishing the proceeding on time. Helps by Ms. Zhang Chong-jing and Ming-zhen Wang in the process of editing of the volume is acknowledged. Special thanks is due to Prof. G. M. L. Gladwell, the general editor of this series, for his careful review and checking of all the manuscripts.

Ren Wang
Dept. of Mechanics & Engineering Sciences
Peking university, Beijing 100871, CHINA

THE RUPTURE THEORY OF RHEOLOGICAL MATERIALS WITH DEFECTS*

L.W. Yuan
Institute of Rheological Mechanics, Xiangtan University, Hunan 411105, P.R.C.

Abstract The rupture process of rheological material containing defects may be considered as a thermodynamically open, far-from-equilibrium system in which the self-organization of dissipative structures takes place.The self-similarity of dissipative structures on different scaling levels provides the opportunity for transition from micro- to meso- and macro-levels. In this paper, based on the experimental results of local temperature field and pyromagnetic effect formed in the evolution period of defects, the rupture process of rheological material containing defects have been studied from a unified standpoint of macro-/meso-/microscopic stratum system. It is proved that this rupture process is not only a pure mechanical process, the thermo-mechanical coupling effect and pyromagnetic effect are not negligible.

1. Introduction

The classical methods of solid mechanics assuming continuous deformation sometimes does not provide an adequate description of the rheological behavior and fracture process of a real material. Irreversible macro-deformation and the fracture of solids are predominated by material behavior in meso-and micro-scale.

The fact that continuum approximation is often unsatisfactory for a real material is now beyond doubt. In natural or man-made materials, a variety of micro-, meso- and macro-defects appears at the natural state or the production stage that may evolve during the material's service life.

The irreversible deformation and fracture of solids may be related to a class of processes in which the macroscopic effects are predominated by mesoscopic level behavior. For this reason, a reliable prediction of a rheological solid under external force should be based upon a clear understanding of the mechanics of the processes in meso-and micro-scales. It is evident that only the description of these processes within a single system, taking into account the interrelation of different processes in micro-, meso-and macro-scales, would provide an adequate theory of the deformation and fracture of rheological solids from the first and the second principles of non-equilibrium thermodynamics.

---

*Project 19632030 supported by National Natural Science Foundation of China

R. Wang (ed.), IUTAM Symposium on Rheology of Bodies with Defects, 1–19.

From this point of view, physical mechanism constructed on a general fundamental concept in micro-scale, may be extended to meso- and macro-scale, if one uses the experimental facts of statistical self-similarity and scaling invariance of cracking in solid materials. As a consequence, a deformed rheological solid may be considered as a thermodynamically open, far-from-equilibrium system in which the self-organization of dissipative structures takes place. For this reason, the development of rupture theory from the thermodynamics of solid physics, allowing an adequate treatment of the response of a deformed rheological solid to an external physico-mechanical action, is possible on the basis of the quantum-statistical approach. The self-similarity of dissipative structures on different scaling levels provides the opportunity for transition from micro- to meso- and macro-levels.

In this paper, the recent progresses in the rupture theories of rheological materials with defects have been systematically reviewed in different scaling levels. Its aim is to probe the following possibility: the failure process of rheological bodies with defects can be considered as the kinetic phase transitions related to the self-organization of corresponding dissipative structures. Here, the time, the temperature, the rheological character and the dissipation are emphasized, and great attention is paid to the concepts of stratum, synergism, statistics, pyromagnetism and magneto-conductivity.

2. The constitutive equation with ordered parameters

The rupture theory of material in the rheology of bodies with defects deems that[1,2] the relation between the evolutionary behavior of meso-defect and the rheological property of material depends on the disorder of defect distribution and the collective synergetic effect of the far-from-equilibrium state. This synergetic effect extends across the micro-, meso- and macro-levels, thus forming a stratum system. In this stratum system, the nonlinear rheological dynamic variables of the highest stratum control the variation of the variable of lower stratum. For a system under the action of external loading, the whole process of defect evolution to macroscopic rupture can be described by the ordered parameters. They are relatively slow-changing variables representing the rheological kinetic states in the evolutionary process of the subsystem of the macroscopic strata in the rheological body with defects.

2.1. THEORETICAL ANALYSES OF THE EXPERIMENTAL RESULTS

In the experiment on the local temperature field and its variation law formed by the evolutions of the macro-crack and the meso-defects[3-5],the following three fundamental phenomena are observed:

1) The global rheological state of system $|\Psi\rangle$ can be expressed as

$$|\Psi\rangle = \sum_n x_n(t) \mathrm{B}_n^+ |0\rangle \tag{2.1}$$

where $\mathrm{B}_n^+|0\rangle$ is the operator of excited state of the nth subsystem with $|0\rangle$ being the ground-state of meso-defect in the evolutionary process of defect, $x_n(t)$ are complex functions of coordinates in describing the collective synergetic motion of meso-defect which are ordered parameters. The free energy of the system can be written as

$$\mathcal{E}_f = \langle \Psi | \hat{\mathcal{E}} | \Psi \rangle \tag{2.2}$$

$\hat{\mathcal{E}}$ is an operator of energy, and is written as e_n for the subsystem.

2) There exists cognate interaction among the correlated subsystems and weak cognate interaction $J\left(\mathrm{B}_{n+1}^{+}\,\mathrm{B}_{n}^{-} + \mathrm{B}_{n}^{+}\,\mathrm{B}_{n+1}^{-}\right)$ among meso-defects, where the Hermitian conjugate B_n^- of B_n^+ is an operator of degenerated excitation, and

$$\left[\mathrm{B}_n^+, \mathrm{B}_n^-\right] = \delta_{nm} \tag{2.3}$$

δ_{nm} is the Kronecker delta, and J represents the intensity of weak cognate interaction among meso-defects.

3) The excitation of subsystem is nonlinearly coupled with the oscillation variable of the phonon of heat carrier q_n and we have $g\sum_n q_n B_n^+ B_n^-$, where g denotes the intensity of nonlinear coupling.

According to the above-mentioned phenomena 2) and 3), we have

$$\hat{\mathcal{E}} = \sum_n e_n B_n^+ B_n^- - J\sum_n \left(B_{n+1}^+ B_n^- + B_n^+ B_{n+1}^-\right) - g\sum_n q_n B_n^+ B_n^- + \sum_n \left(\frac{1}{2}P_n^2 + \frac{\omega_o^2}{2}Q_n^2\right) \tag{2.4}$$

Substituting Eq.(2.4) into Eq.(2.2), we obtain

$$\mathcal{E}_f = \sum_n e_n x_n^* x_n - J\sum_n x_n^*\left(x_{n+1} + x_n\right) - g\sum_n q_n x_n^* x_n + \sum_n \left(\frac{1}{2}P_n^2 + \frac{\omega_o^2}{2}Q_n^2\right) \tag{2.5}$$

where P_n and Q_n can be regarded respectively as normal coordinates and normal momenta of oscillations, ω_0 is the oscillation frequency, the superscript "* " denotes the corresponding conjugate quantity.

2.2. THE CONSTITUTIVE EQUATION WITH ORDERED PARAMETERS

First, we construct the Laplace function

$$L = \frac{1}{2}\bar{c}\sum_n x_n^* \dot{x}_n + \frac{1}{2}\bar{c}^*\sum_n x_n \dot{x}_n^* - \mathcal{E}_f \tag{2.6}$$

where $\bar{c}$ is a complex number and $\bar{c}^*$ is its conjugate quantity. Then the nonlinear rheological dynamic equation can be obtained from this equation as follows:

$$\frac{1}{2}(\bar{c} - \bar{c}^*)\dot{x}_n = \frac{\partial \mathcal{E}_f}{\partial x_n^*} \tag{2.7}$$

Let the complex number $\bar{c} = i\bar{b}$ and $\bar{b}$ is a real number. Without lose of generality when neglecting the terms $\bar{c}^*$ and $\frac{1}{2}$, we have

$$i\bar{b}\dot{x}_n = \frac{\partial \mathcal{E}_f}{\partial x_n^*} = e_n x_n - J(x_{n+1} + x_n) - g q_n x_n \tag{2.8}$$

For the evolution of meso-defects with heat transfer, the generalized momentum can be evaluated from Eq.(2.6) as follows:

$$\pi_n = \frac{\partial L}{\partial \dot{x}_n} = \frac{1}{2}\bar{c}x_n^* , \quad \pi_n^* = \frac{\partial L}{\partial \dot{x}_n^*} = \frac{1}{2}\bar{c}^* x_n \tag{2.9}$$

secondly, we construct the Hamilton quantity

$$H = F\left(\bar{x}_n, \bar{x}_n^*, Q_n, P_n\right) \tag{2.10}$$

where $$\bar{x}_n = \frac{x_n}{2} + \frac{\pi_n^*}{\bar{c}^*} , \quad \bar{x}_n^* = \frac{x_n^*}{2} + \frac{\pi_n}{\bar{c}} \tag{2.11}$$

$\bar{c}$ and $\bar{c}^*$ are determined by the self-consistent condition. From Eqs.(2.10), (2.11) and (2.8), we obtain

$$\bar{c}\dot{x}_n = \frac{\partial F}{\partial x_n^*} = e_n x_n - J(x_{n+1} + x_n) - g q_n x_n \tag{2.12}$$

Owing to the viscous effect in the rheological material, the evolutionary velocity of a meso-defect will be smaller than the heat wave velocity caused by collective transmission in the multi-phonons system, thus in consideration of $g\sum q_n B_n^+ B_n^-$ we have

$$q_n = g x_n^* x_n / \omega_o^2 \tag{2.13}$$

By substituting this equation into Eq.(2.12),we obtain

$$\bar{c}\dot{x}_n = e_n x_n - J(x_{n+1} + x_n) - \frac{g^2}{\omega_o^2}\left|x_n\right|^2 x_n \tag{2.14}$$

Under the condition of continuous approximation, the above equation can be written as

$$\bar{c}\frac{\partial x_n(x,t)}{\partial t} = -Jd^2 \frac{\partial^2 x_n(x,t)}{\partial x^2} + (e_n - 2J)x_n(x,t) - \frac{g^2}{\omega_o^2}\left|x_n\right|^2 x_n \tag{2.15}$$

where d is the statistically averaged distance between correlated meso-defects. Eq.(2.15) is just like the constitutive equation with ordered parameters for the evolutionary state in the rupture process of a rheological material containing defects.

If the nth meso-defect is evolving with velocity $\mathbf{v_n}$, and the center of this meso-defect is located at $x = v_n t$, then the ordered parameter of this mobile defect is

$$x_n(x,t) = \frac{\pi \varepsilon_s}{2} \tanh\left[(x - c_0 t)/r_n\right] \tag{2.16}$$

where ε_s is the excited energy of defect, r_n the statistically averaging radius of defects, c_0 is the constant which has a direct bearing on the stiffness and the density of rheological material.

2.3. THE SOLUTION OF CONSTITUTIVE EQUATION WITH ORDERED PARAMETER

So far as a rheological body with defects is concerned on the whole, the fundamental physical quantity is the statistical distribution $\boldsymbol{A}$ of meso-defect's density, from which the average values of all functions of coordinate and momentum can be computed. Therefore, we make a beginning from the Liouville's equation which gives a description of the evolution of $\boldsymbol{A}$ in the phase space. We write $\boldsymbol{A}(\boldsymbol{x}_n)$ as

$$\dot{A} = iL_o A \tag{2.17}$$

where L_o is the Liouville's operator, x_n may be regarded as the generalized coordinates of a point in phase space, and the relation between x_n and π_n still obeys Eq.(2.9). For the nonlinear system, the Liouville's operator is $L_0 = L_i + L_j$, where L_i and L_j are, respectively, the linear and the nonlinear part of this operator. Adopting the technique of Mori-Zwanzig's projective operator[6,7], by unfolding L_j in the light of A, and neglecting higher order terms, we obtain:

$$\frac{\partial A}{\partial t} = i\Omega A - D(t)A + i\Omega\int_0^t M(t,t_p)A_p(t-t_p)dt_p + iL_j A \tag{2.18}$$

where Ω is the frequency matrix, $D(t)$ the correlation function of defects which is dependent on time t. $M(t,t_p)$ the rheological memory function. $A_p(t-t_p)$ is the value of ordered parameter A at time $t-t_p$. t_p the past time. This equation shows the oscillation of ordered parameter of the nonlinear system. Because the defect evolution produces a local temperature field, as shown in Fig.1 and the infra-acoustic oscillation of collective transmission in the system for multi-phonons of heat carrier may be expressed as Eq.(2.13),we select the variable related to the oscillation to be the ordered parameter. Therefore, in the case $\bar{c} = i\bar{b}$, after some lengthy manipulation ,we finally obtain the solution of Eq.(2.15) as follows:

$$x_n(x,t) = \left(\frac{g^2}{2Jd^2\omega_o^2}\right)^{1/2}\left[\exp\left(\frac{g^2(x-x_o-\mathrm{v}_n t)}{4Jd^2\omega_o^2}\right) + \exp\left(\frac{-g^2(x-x_o-\mathrm{v}_n t)}{4Jd^2\omega_o^2}\right)\right]^{-1}$$

$$\cdot\exp\left[\frac{i\bar{b}\mathrm{v}_n(x-x_o)}{2Jd^2} - \frac{it}{\bar{b}}\left(e_n - 2J + \frac{\bar{b}^2\mathrm{v}_n^2}{4Jd} - \frac{g^4}{16Jd^2\omega_o^4}\right)\right] \tag{2.19}$$

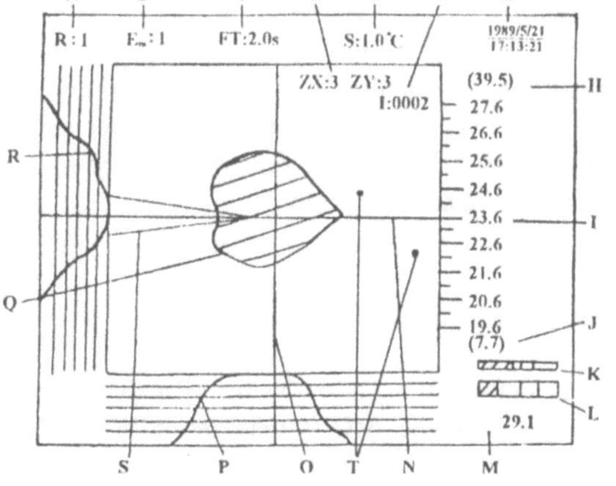

A-measuring range, B-emissivity,
C-scanning time, D-magnification multiple
E-sensitivity, F-frame number of tape record
G-measuring date and time
H-upper limit of measurable temperature
I-colored temperature values
J-lower limit measurable of temperature
K-objective distance
L-internal storaging page code of frame
M-temperature value of the cross point on the movable ocular micrometer
N-horizontal line, O-vertical line
P-distribution curve of vertical temperature
Q-thermal image of measured body
R-distribution curve of horizontal temperature
S-the main crack, T-the meso-defects

Figure 1. Display of thermal image for the local temperature fields near the crack tip and the neighborhoods of defects in the specimen under tension

This equation describes the meso-defect with evolutionary velocity v_n.
Moreover, the interaction between defects contributes to the respective temperature fields formed by various defects in the process of evolution and there finally occurs the collaboration. The collaborated condition of correlated defects leads to

$$\frac{2d}{(r_{T1}+r_{T2})\varsigma_c} \leq 1 \tag{2.20}$$

where ς_c is the parameter of critical collaboration in rheo-dynamics and is determined by the non-diagonalization characteristic of the nonlinear Liouville operator, r_{T1} and r_{T2} are the effectively projected radii on x-axis of two local temperature fields, respectively. In a certain range of ς_c (such as $\varsigma_c \approx 1$), there emerges a sensitivity to the incipient configuration of meso-defects, i.e., a chaotic behavior would appear. It is thus obvious that the evolution of meso-defect naturally introduces disorderness, which reflects the collective synergetic effect of nonlocal interaction among a large number of meso-defects.

2.4. BRIEF SUMMARY

We are led to two primary conclusions based on the above discussion as follows:
(1) The interior meso-defects with different sizes and types exhibit disordered distribution, and the collective synergetic effect is generated by the interaction among correlated meso-defects. The nonlinear coupling will prevail between the collective excitation and the oscillation, there is an important influence on the state of rheological motion of the defect evolution. Therefore, the disorderness of meso-defect distribution comes into action in a diversified form and the process of evolution is amplified. This disorderness may be expressed by the ordered parameters, thus can also be reflected on the constitutive equation of evolutionary state of the rheological material with defects.
(2) The rupture process of rheological material with defects is a process of nonlinear rheo-dynamics which is consistent with non-equilibrium statistical thermodynamics of the discontinuous system with a microstructure. The collaborative condition of correlated meso-defects depends on the nonlinear rheo-dynamic parameter ς_c, which is determined by the characteristic of non-diagonalization of the nonlinear Liouville's operator. The time sequence of energy transformation would be coupled with the thermo-mechanical character of material when the defect is evolving. The ordering degree of the power-flow distribution is reflected by the transformation direction of energy.

3. Meso-physical mechanism of the process of defect evolution

The dissipation of energy, due to plastic deformation, near the tip of a dynamically propagating crack and the surroundings of evolving defects may result in the heat generation that forms the local temperature fields[9-12] as shown in Fig.1. It is suspected that such temperature increase will strongly affect the nature of the deformation field of the process zone at crack tip and may result in changes of the dynamic fracture toughness of the material. Otherwise, because there exists the temperature gradient in the process zone at crack tip, it necessarily leads to the pyromagnetic effect in this zone [15]

3.1. THE THERMAL PROPERTY OF THE PROCESS ZONE AT CRACK TIP

As each stress and strain state is determined independently, unloading can be carried out in the same way as loading. In fact, the same procedure applies to any load-time history. The loss of energy, referred to as dissipation, can be calculated uniquely for each load or unload path. Let $\widetilde{W}$ denote the energy in unit volume of material within unit time. Then we define the energy dissipation density $\widetilde{D}$ and available energy density $\widetilde{A}$ in relation to $\widetilde{W}$ such that a differential change of energy state corresponds to

$$d\widetilde{W} = d\widetilde{D} + d\widetilde{A}, \qquad \widetilde{D} \geq 0; \qquad \frac{d\widetilde{D}}{dt} \geq 0 \tag{3.1}$$

As a result of volume and surface change of the process zone at crack tip, the temperature will alter accordingly. For an incremental change of stress σ and strain ε, thermal change can be obtained directly from[13]

$$\frac{\Delta\theta}{\theta} = -\frac{\Delta\sigma \ \Delta\varepsilon}{\Delta\widetilde{D}} \tag{3.2}$$

Here, θ can be calibrated experimentally in ^{o}K and it represents the non-equilibrium temperature.

The experimental results of the temperature field formed at the main macro-crack tip and the surroundings of meso-defects[1,2,4,5] demonstrate that the dissipation caused by temperature gradient $\boldsymbol{g}$ is the main part of energy dissipation, and is called the thermal dissipation D_{th}. While the others, such as the dissipation caused by defect evolution, rheological property of material, strain softening, crack propagation etc., are called the intrinsic dissipation D_{in} which are the non-thermal dissipation, we have

$$\Delta\widetilde{D} = \Delta D_{th} + \Delta D_{in} \tag{3.3}$$

Let us introduce the coefficient of energy dissipative capability C_θ which is defined as

$$C_\theta = \frac{d\widetilde{D}}{d\theta} \tag{3.4}$$

and the thermal conductivity coefficient k_c can be obtained from Fourier law:

$$k_c = C_\theta \left(\frac{\Delta V}{\Delta A}\right)_i \left(\frac{\Delta x_i}{\Delta t}\right), \qquad i = x, y, z \tag{3.5}$$

where ΔV and ΔA are , respectively, the increments of volume and surface of the process zone at crack tip. Because C_θ has directionality and is related to the rate, so according to the theory of the rate-dependence on temperature[16], we know the density function of dissipated energy in the process zone at crack tip $\widetilde{D}$ consists of D_{in} and D_{th}, and so

$$D_{in} = \sigma : \dot{\varepsilon} - \rho \frac{\partial \hat{\psi}}{\partial \boldsymbol{x}_n} \dot{\boldsymbol{x}} \geq 0$$

$$D_{th} = -\mathbf{h} \frac{\partial \ln \hat{\theta}}{\partial \theta} \nabla \theta \geq 0 \tag{3.6}$$

where $\hat{\psi}$ is the generalized Helmholtz free energy, **h** the thermal flux, the superscript " ^ " being the corresponding physical quantity of temperature rate-dependence.

Let G_d expresses the released energy for meso-defect evolution, F_g be the generalized force which is conjugated with the dissipation variable, then

$$G_d = -\rho \frac{\partial \mathcal{E}_f}{\partial Y}, \quad F_g = -\rho \frac{\partial \mathcal{E}_f}{\partial Z} \tag{3.7}$$

where $\mathcal{E}_f$ is the free energy as shown in Eq. (2.5), Y the dissipation variable affected by the rheological property of material, Z the statistical distribution variable of meso-defects. Thus, the relation between the rheological character of material, the evolution of internal meso-defect and the thermal dissipation can be written as

$$\sigma \dot{\varepsilon} + G_d \dot{Y} + F_g \dot{Z} - \frac{\mathbf{h}}{\theta} \mathbf{g} \geq 0 \tag{3.8}$$

Therefore, we can see that the dissipation has a direct bearing on the process, and the dissipation potential is inexistent.

Define

$$\hat{c}_s = -\hat{\theta} \frac{\partial^2 \hat{\psi}}{\partial \hat{\theta}^2}, \quad \hat{\Phi} = \hat{\theta} \frac{\partial \sigma}{\partial \hat{\theta}}$$

$$\hat{\chi} = -\frac{\partial \hat{\psi}}{\partial \boldsymbol{x}_n}, \quad \hat{\tau} = \frac{\partial \hat{\chi}}{\partial \hat{\theta}} \tag{3.9}$$

then we have the energy equation for the process zone at the crack-tip

$$\rho \hat{c}_s \dot{\hat{\theta}} + \nabla \mathbf{h} = \rho\, \mathbf{r} + \sigma : \dot{\varepsilon}_r + \hat{\Phi} : \dot{\varepsilon}_e + \rho\left(\hat{\chi} - \hat{\theta}\hat{\tau}\right) - \rho\left(\frac{\partial \hat{\psi}}{\partial \nabla \theta} - \hat{\theta} \frac{\partial^2 \hat{\psi}}{\partial \hat{\theta} \partial \nabla \theta}\right) \tag{3.10}$$

and

$$\sigma = \rho \frac{\partial \hat{\psi}}{\partial \varepsilon_e}, \quad \mathbf{h} = -\frac{\rho}{\partial \ln \hat{\theta} / \theta} \cdot \frac{\partial \hat{\psi}}{\partial \nabla \theta} \tag{3.11}$$

where c_s is the specific heat, **r** the intensity of heat source, ε_r the nonlinear strain field, and ε_e the linear stain field. The second, third and fourth terms on the right-hand side of Eq.(3.10) represent respectively, the dissipation caused by the nonlinear rheological deformation, the thermo-elasticity, and the oscillation of the ordered parameters, the 5th term is the dissipation caused by the dispersion of temperature.

3.2. THE ELECTROMAGNETIC CHARACTERS OF THE PROCESS ZONE AT CRACK TIP

In 1989, our experiments on polymer with defects and ductile alloy with crack under the action of tensile loading[1,15] discovered that the process zone at crack tip forms local heterogeneous temperature field. This zone has distinct electromagnetic character due to the existence of temperature gradient. The intensity of magnetization field of this process zone is dependent on the rheological property of materials, the rate of external loading and the temperature gradient in the process zone.

The experimental results of pyromagnetic effect reveal in the rupture process of rheological material with defects as shown in Fig.2. These experimental results show:

(1) The variations of magnetic induction field in the process zone at crack tip are small at the beginning of crack propagation. After the initial crack propagation, the intensity of magnetic induction field slowly increases with the growing load and the increase of temperature of the local temperature field. At about 2~3 sec. before fracture, the intensity of magnetic induction field grows rapidly with the increase of the area and the temperature of local temperature field near the process zone. After tensile failure of sample, the residual intensity on the fracture surface of the sample keeps up some time(the longest time is about 3 days)before it disappears.

(2) During crack propagation, the process zone at crack tip is simultaneously acted upon by the magnetic induction field **B** and the temperature gradient $\partial\theta/\partial x$ of the heat flow J_H (both are orthogonal to each other),thus in the direction which is perpendicular to the directions of **B** and $\partial\theta/\partial x$ .or in the direction which is perpendicular to the direction of J_H, there occurs either the effect of induced current **E** or the change of heat resistance ΔR as shown in Fig.3(a) and 3(b). These phenomena make clear that in the later stage of crack propagation, the local temperature field, the magnetic induction field and the electric induction field coexist in the process zone at crack tip, and have influence on each other.

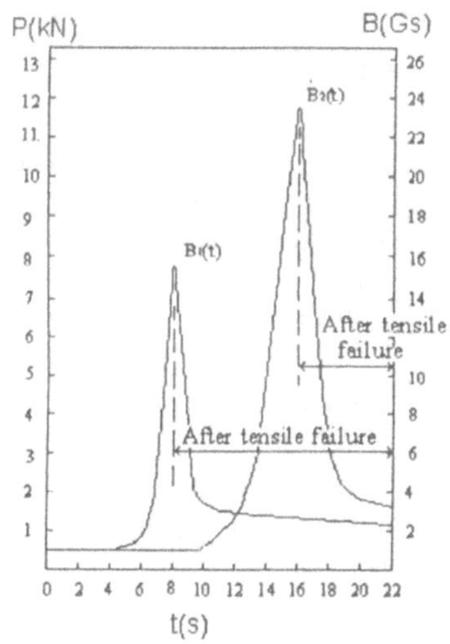

Figure 2. The changing law of magnetic induction fields $B_1(t)$ for galvanized steel plate with thickness 0.5mm and $B_2(t)$ for 45# steel sheet with thickness 1.2mm in the process zone at crack tip, under tensile loading P(t).

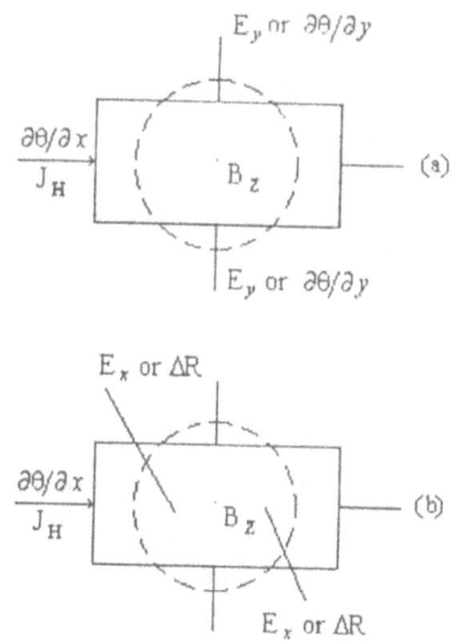

Figure 3. The effects of pyromagnetism and magneto-conduction in the process zone at crack tip
(a) The transverse effects
(b) The longitudinal effects

It is well known that the essential in the methodology of rheology in continuum mechanics is Galilei's invariant, but that for Maxwell equations in electromagnetics are Lorentz's invariants. Therefore, in order to write the characteristic equations for the process zone at crack tip, and let them be covariant with Galilei's transformation, we make use of non-dimensional analysis to obtain

$$\mathrm{curl}\mathbf{H} = \mathbf{i}, \quad \mathrm{div}\mathbf{B} = 0$$
$$\mathrm{curl}\mathbf{E} = -\frac{\partial \mathbf{B}}{\partial t}, \quad \mathrm{div}\mathbf{D} = 0 \tag{3.12}$$

where **H** expresses the magnetic field, **E** the electric field, **i** the electric current density, and **D** the electric induction field, here the following transformation laws have been used:

$$\mathbf{H}^0 = \mathbf{H}, \quad \mathbf{B}^0 = \mathbf{B}, \quad \mathbf{E}^0 = \mathbf{E} + \tilde{\mathrm{v}} \times \mathbf{B}$$
$$\mathbf{i}^0 = \mathbf{i}, \quad \mathbf{D}^0 = \mathbf{D} + \tilde{\mathrm{v}} \times \frac{\mathbf{H}}{c^2} \tag{3.13}$$

These transformation laws express the transform of quantity **X** in Galilei frame ι into quantity $\mathbf{X}^0$ in Lorentz frame ι^0, in which c is the light velocity, $\tilde{\mathrm{v}}$ the moving velocity of the frame ι^0 relative to the frame ι. Thus, the intensity of magnetic polarization **M** and the intensity of electric polarization **P** of the process zone at the crack-tip can be written respectively as

$$\mathbf{M} = \mathbf{B} - \mu_0 \mathbf{H}, \qquad \mathbf{P} = \mathbf{D} - \epsilon_0 \mathbf{E} \tag{3.14}$$

while

$$\mathbf{M}^0 = \mathbf{M}, \quad \mathbf{P} = \tilde{\mathrm{v}} \times \frac{\mathbf{M}}{\mu_0 c^2} \tag{3.15}$$

μ_0 and ϵ_0 are respectively the magneto-conductivity and the electro-conductivity of the process zone.

The boundary condition of the process zone at the crack-tip can be obtained from the local forms of global balance equations. Now we begin by writing the global forms of momentum balance equation and of energy balance equation as follows:

$$\frac{\mathrm{d}}{\mathrm{dt}}\int_V \rho \mathrm{v}'\mathrm{dV} = \int_V \rho f \mathrm{dV} + \int_A \sigma \cdot \mathbf{n}\mathrm{dA} \tag{3.16}$$

$$\frac{\mathrm{d}}{\mathrm{dt}}\int_V \rho\left(\frac{1}{2}c_s^2 + \mathrm{e_i}\right)\mathrm{dV} = \int_V \rho\left(f \cdot \mathrm{v}' + \mathbf{r}\right)d\mathrm{V} + \int_A \left(\sigma \cdot \mathrm{v}' + \mathbf{q}\right)\cdot \mathbf{n}\mathrm{dA} \tag{3.17}$$

where v' is the speed of transmission of the heat-carrying phonon, $\mathrm{e_i}$ the specific internal energy, **q** the energy flux vector, f the specific bulk force, **n** the unitary outward normal vector, for other symbols see Eq.(3.10). Note, **q** here is

$$\mathbf{q} = -\mathbf{E}^0 \times \mathbf{H}^0 + \mathbf{h} + \mathbf{s} \tag{3.18}$$

where **s** is the additional energy flux vector caused by the defect evolution in the process zone, and $-\mathbf{E}^0 \times \mathbf{H}^0$ expresses the electromagnetic energy flux vector. Define the strain field ε' and the stress field σ' as follows:

$$\varepsilon' = \varepsilon - \frac{\mu_0 \boldsymbol{b}^2}{2\rho} \qquad \sigma' = \sigma + \frac{1}{2}\mu_0 \boldsymbol{b}^2 \mathbf{I} - \mathbf{H} \otimes \mathbf{B} \tag{3.19}$$

where b is a characteristic quantity which is the magnetic inductance, **I** the electric current. Under the assumption of sufficient smoothness, the local forms of Eq.(3.16) and

Eq.(3.17) may be written as

$$\rho v' = \text{div}\, \sigma' - \mathbf{M}\cdot \text{grad}\, \mathbf{H} + \mu_0 \mathbf{i}\times \mathbf{H} + \rho f \quad (3.20)$$

$$\rho\dot{\varepsilon}' = \sigma : \text{grad}\, v' + \rho \mathbf{H}\dot{\widetilde{\mathbf{M}}} + \text{div}\,(\mathbf{h}+\mathbf{s}) + \mathbf{E}^0\cdot\mathbf{i} + \rho\mathbf{r} \quad (3.21)$$

In equation (3.20) the terms $\mathbf{M}\cdot\text{grad}\mathbf{H}$ and $\mu_0\mathbf{i}\times\mathbf{H}$ express, respectively, the magnetizing force and the Lorentz's force in the process zone. In equation (3.21) the term $\rho\mathbf{H}\dot{\widetilde{\mathbf{M}}}$ expresses the heat energy current of local temperature field near the process zone. In this case, the Clausius-Duhem inequality may be written as

$$\rho\theta\dot{\eta} \geq \text{div}\, \mathbf{h} - \frac{\mathbf{h}\cdot\text{grad}\,\theta}{\theta} + \rho\mathbf{r} \quad (3.22)$$

Thus, the operator of energy $\hat{\mathcal{E}}$ in Eq.(2.2) can also be written as

$$\hat{\mathcal{E}} = \mathcal{E}_i - \left(\theta\eta + \epsilon_0\, \widetilde{\mathbf{E}}\widetilde{\mathbf{P}} + \mu_0\, \widetilde{\mathbf{H}}\widetilde{\mathbf{M}}\right) \quad (3.23)$$

In the above-mentioned equations, $\widetilde{\mathbf{M}} = \mathbf{M}/\rho$ is the unitary magnetizing intensity, the superscript "~" expresses corresponding quantity for unitary density, $\mathcal{E}_i$ the internal energy of the process zone, η the specific entropy. It is thus clear that owing to the effects of thermodynamics, pyromagnetics and electromagnetics, the value of internal energy function of the process zone has been decreasing.

According to Eqs.(3.20)and(3.21), after tensile failure, on the fracture surface we have

$$|u\rho v'| = |\sigma|\cdot\mathbf{n}, \quad \left|u\rho\left(\frac{v'^2}{2} + E_i\right)\right| = |v'\cdot\sigma + E_f|\cdot\mathbf{n}$$

$$|u\rho H_{ij}| = |M_{ijk}|\cdot\mathbf{n}_k, \quad |u\rho\dot{\eta}| \geq \left|\frac{\mathbf{h}}{\theta}\right|\cdot\mathbf{n} \quad (3.24)$$

where u is the speed of transmission of the magnetic induction on the fracture surface. After a period of time(about 20-48 hours for polymer and 36-72 hours for ductile alloy), on the fracture surface we have

$$|\mathbf{H}|\times\kappa = 0, \quad |\mathbf{B}|\cdot\mathbf{n} = 0$$

$$|\mathbf{E} + v'\times\mathbf{B}|\times\kappa = 0, \quad |\mathbf{D}|\cdot\mathbf{n} = 0 \quad (3.25)$$

where κ is the unitary outward normal vector on the boundary of the process zone.

3.3. BRIEF SUMMARY

We are led to two primary conclusions based on the above analyses as follows:

(1) The interactions between electron and electron, electron and phonon, or phonon and phonon in the material give rise to the exchange of the momentum and of the energy between the correlated meso-defect and the molecular bond ,so that heat is transmitted in the form of waves. Because the heat wave transmission is influenced by the action of viscous damping, there will be formed a local heterogeneous temperature field at the surroundings of meso-defects and the crack-tip of mobile macro-crack.

(2) Because there exists a temperature gradient in the process zone at the crack-tip, it

necessarily leads to the pyromagnetic effect in this zone. The carrier for metal is electron or cavity, and for polymer is charged soliton or bipolaron, as is well known, the interaction between phonons and electrons comprises dynamic dissipative processes, thus the equations of conservation of linear momentum and energy ought to contain the contribution of electromagnetic effect.

4. Micro-physical mechanics of the rupture process

Traditionally, the analysis of processes that control rheological deformation and fracture of solids at the micro-level has been confined to consideration of models that take into account only paired interatomic bonds[17-19].But the rheological behavior of solids is determined by the dynamics of collective excitations induced by the external factors. Thus fracture processes are collective, far-from-equilibrium processes, whose kinetics are governed by the self-organization of dissipative structures that ensure an optimal, for specified loading, level of dissipation of energy[20]. Therefore, in developing a physical theory of the material rupture theory in the rheology of bodies with defects it is necessary to exhibit the mechanisms of microscopic processes limiting the defect dynamics, because it is this which determines the process of rheological deformation of a solid. Moreover, the rheological behavior of the process zone at crack tip during defect evolution is governed by the dynamics of collective excitations. The spectrum of structural excitations (defects) in the process zone can be determined correctly from the solution of the non-steady state equations of stochastic mechanics[21] with the potential that is formed by an ensemble of atoms, and this potential determines the "structural memory" of the process zone at the crack-tip. Therefore, the rupture process of rheological solid with defects can be predicted reliably only with a clear understanding of the nature and kinetics of the quantum process in a fractured solid.

4.1. THE INTERATOMIC PAIRWISE POTENTIAL

The usual approach to constructing the quantum fracture mechanics of irreversibly deformed solid is based on the use of an interatomic pairwise potential U(r) independent of the type of interatomic force. It has the shape shown in Fig.4 and is characterized by:

1) Minimum in $U(r_{ij})$, which corresponds to the equilibrium interatomic distance r_{ij} at zero absolute temperature, is determined by the competition of the forces of attraction and repulsion.

2) Position r_m is where the forces of interatomic interaction reach a

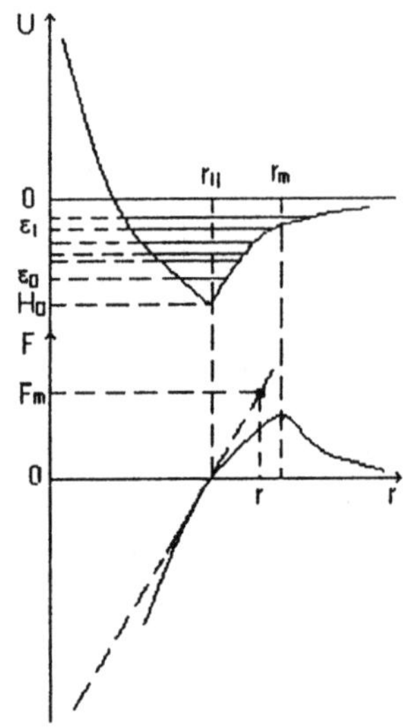

Figure 4. The shape of an interatomic pairwise potential and its energy spectrum

maximum, i.e., $\partial^2 U/\partial r^2\big|_{r_m} = 0$.

3) Energy spectrum of this potential represents a set of energy levels $\mathcal{E}_n$. At its ground state, an atom, owing to the oscillation is equal to zero , possesses a finite energy

$$\mathcal{E}_0 = \frac{\pi\hbar}{r_{ij}} \cdot \frac{\sqrt{U(r_{ij})}}{m} \tag{4.1}$$

reckoned from $U(r_{ij})=0$. Here $\hbar = 1.05 \times 10^{-34} J \times s$ is the Planck's constant and m is the mass of an atom

4) The amplitude of the zero atomic oscillation is equal to

$$\langle \Delta r \rangle = \frac{r_{ij} m C_a}{\sqrt{mU(r_{ij})}} \tag{4.2}$$

where

$$C_a = \frac{\pi\hbar}{m r_{ij}} \tag{4.3}$$

is the maximum velocity of a finite motion of atoms in the potential well U(r) at this energy level ,and

$$\Lambda_B = \frac{\langle \Delta r \rangle}{r_{ij}} = \frac{mC_a}{\sqrt{mU(r_{ij})}} \tag{4.4}$$

is the DeBoer's parameter.

5) Because of the Heisenberg's uncertainty principle, the energy levels of the atoms in the pairwise potential have finite widths .It is easy to show that the energy width of the ground level is equal to

$$\delta\,\mathcal{E}_0 = \frac{\pi^2\hbar^2}{2mr_{ij}} \tag{4.5}$$

We note that $\delta\,\mathcal{E}_0$ is equal to the kinetic energy of an atom for which the deBroglie wave length λ_D is equal to interatomic distance, i.e.,

$$\lambda_D = 2r_{ij} \tag{4.6}$$

This equality corresponds to the velocity of atomic motion $u = C_a$ of the atom.

6) Because of the asymmetrical shape of the potential U(r), the degeneracy of the levels n=1,2,....is elevated. So that the difference in the energies of the neighboring levels , $\Delta\mathcal{E}_n = \mathcal{E}_{n+1} - \mathcal{E}_n$ decreases with growing n. Each energy level is characterized by its stable atomic configuration and specific interatomic spacing $r_{ij}^{(n)}$. As a rule,

$$r_{ij}^{(n+1)} > r_{ij}^{(n)} \tag{4.7}$$

This condition is closely related to the local temperature field at the crack-tip of a mobile crack.

4.2. MICRO-PHYSICAL MECHANISM OF DEFECT EVOLUTION

It is thus evident that in the process zone of a mobile macro-crack there are the temperature gradient. The rheological stress is reduced with increment of the probability of thermal activation. So, for ordered crystalline rheological materials, the number of phonon (which can scatter the electron) increases with the rise of temperature, thus the resistivity is also enhanced .But for disordered amorphous rheological materials, they have higher **resistivity** and lower sensitivity of temperature .Therefore, the rheological behavior of a deformed media is governed by the dynamics of collective excitations, among which should contain the transmission of the heat-carrying phonon and the action of the electron-phonon coupling.

Based on these characters of potential U(r), it is sufficient for the main properties of the collective motion of excitated atoms to use the approximation of the potential relief V_{ij} in the following general form[22]:

$$V_{ij}(r) = V_{ij}^{0} f(r) \tag{4.8}$$

where f(r) is a periodic (quasiperiodic) function with period $2a_0$, and

$$V_{ij} = \begin{cases} U(r), & |r_i - r_j| \le a_0 \\ V_0 |r_i - r_j|^{-\alpha}, & |r_i - r_j| > a_0 \end{cases} \tag{4.9}$$

Here U(r) is a V-shaped potential which is defined by interatomic interaction, while $V_0 |r_i - r_j|^{-\alpha}$ takes into account the long-range interatomic correlation(for most crystals and glassy polymers), which determines the shear stiffness of material in the process zone, characterized by the shear modulus $G = \rho C_t^2$, where ρ is the density of the material and C_t is the velocity of transverse acoustic waves.

Models incorporating only pairwise interatomic bonds in solids are customarily used in the fracture analysis of kinetic processes[23].In this case the equations of the fracture macro-kinetics can be obtained by averaging the atomic fluctuations. Inclusion of the long-range interatomic forces responsible for the shear stability of the solid changes the situation drastically. Specifically, the effect of long-range forces results in a power dependence of the correlation fluctuations of atoms[24]:

$$\langle n(a) n(a - r) \rangle \sim r^{-\alpha} \tag{4.10}$$

which is characteristic of fractal structures. The exponent in Eq.(4.10) is determined by the fractal dimension d_f of the wave function ψ_i of the atoms forming the solid and dimension d of the enveloping space:

$$\alpha = d - d_f \tag{4.11}$$

The wave function ψ_i, satisfying the Schrödinger equation when both the short-and long-range interaction exist, are characterized by a hierarchy of spatial scales, which obviously determine the scales of the structural levels of the deformation and fracture of solids studied by Panin et al[25]. A strategy for solving problems involving many scales is given by an approach based on the renormalization group.

In the process zone at the crack-tip (or the surroundings of defects), there exists a hierachy of characteristic spatial scales of collective excitations in a deformed rheological solid (n=1,2,...) described by the ratios

$$R = \frac{\text{curl curl } \vec{u}}{\text{grad div } \vec{u}} = \frac{2(1-\nu_e)}{(1-2\nu_e)} \tag{4.12}$$

where $\vec{u}$ is the material vector velocity , ν_e is the effective coefficient of transverse deformation. In order to determine d_f for Eq.(4.11) the discussion may be shifted from the atomic level to the mesoscopic level. the exponent α can also be expressed in terms of Poisson's ratio ν. which specifies the change in volume during deformation of a rheological solid:

$$\alpha = 1 - 2\nu \tag{4.13}$$

Hence, using Eqs.(4.11)and (4.13),we obtain the relationship for fractal dimension of the wave functions of atoms in the form

$$d_f = 2(1+\nu) \tag{4.14}$$

Take into consideration the formation of electromagnetic effect and the evolution of local temperature field in the process zone. we can determine the Poisson's ratio ν by the mesoscopic structural value β_s (which is the statistical function of damage gradient and the number density of microdamages) and the constant of the electron-phonon coupling λ_c as

$$\nu = \frac{\beta_s + \gamma_n}{1-\gamma_n} = \frac{3K-2G}{2(3K+G)}$$

i.e.

$$G = \frac{3K(1-2\beta_s - 3\gamma_n)}{2(1+\beta_s)} \tag{4.15}$$

where

$$\gamma_n = \frac{1}{12}(2-3\beta_s)\lambda_c \cdot f(\theta)$$

$$f(\theta) = \int d\bar{p} \frac{\partial \widetilde{F}}{\partial \varepsilon_{\bar{p}}} \Big/ \int d\bar{p}\delta(\varepsilon_{\bar{p}} - \varepsilon_F) \tag{4.16}$$

$f(\theta)$ is a function determining the temperature dependence of the effective density of electron states. $\widetilde{F}$ the Fermi function. $\vec{p}$ the momentum of electron. $\varepsilon_{\bar{p}}$ the electron energy spectrum. ε_F the Fermi energy. $\delta(\ldots)$ the Delta function and K the bulk modulus. G the shear modulus. Based on our experimental results[1],the displacement distribution of evolving defect field and the corresponding strain distribution are shown in Fig.5 and Fig.6 respectively. The relations between the Poisson's ratio (or the shear modulus) and the variation of evolving defect location (x,y) are shown. respectively, in fig.7 and Fig.8. The generalized equations of the transport of mass. momentum and energy(including heat) in a deformed rheological solid with defects can be written in the form

$$\frac{\partial^{\zeta}\vec{u}}{\partial t^{\zeta}} = D_{\zeta} \frac{\partial^{\xi}\vec{u}}{\partial x^{\xi}} \tag{4.17}$$

where D_{ζ} is the effective diffusion coefficient (the thermal conductivity etc.); and $\partial^{\zeta}/\partial t^{\zeta}$, $\partial^{\xi}/\partial x^{\xi}$ are the fractional derivatives with respect to the time and the coordinate, respectively. The use of fractional derivatives permits a simplification of the mathematical form of the transport equations in far-from-equilibrium systems. To

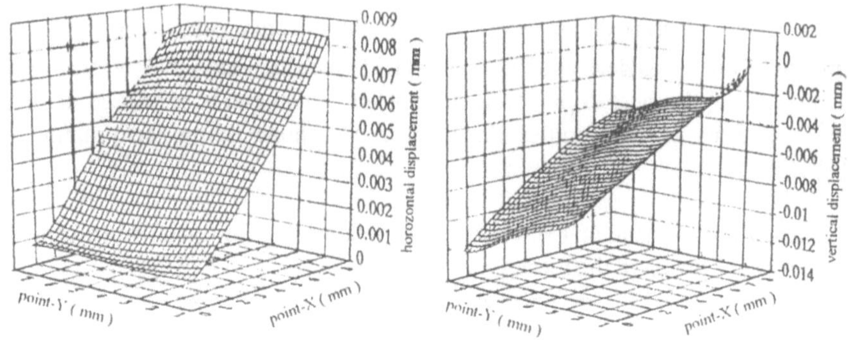

Figure 5. The displacement distribution of evolving defect field. (X,Y) are the distances from the crack tip

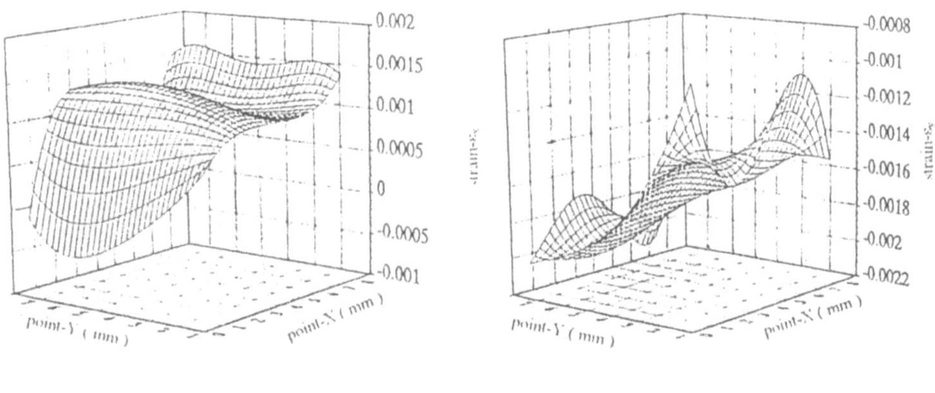

Figure 6. The strain distribution of evolving defect field. (X,Y) are the distances from the crack tip

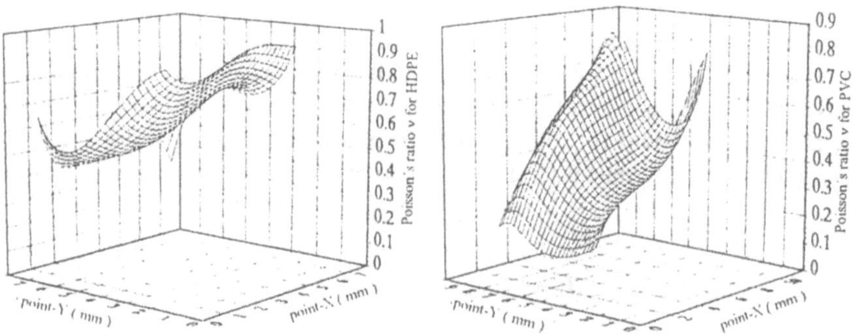

Figure 7. The relations between the Poison's ratio v and the variation of evolving defect location (X,Y)

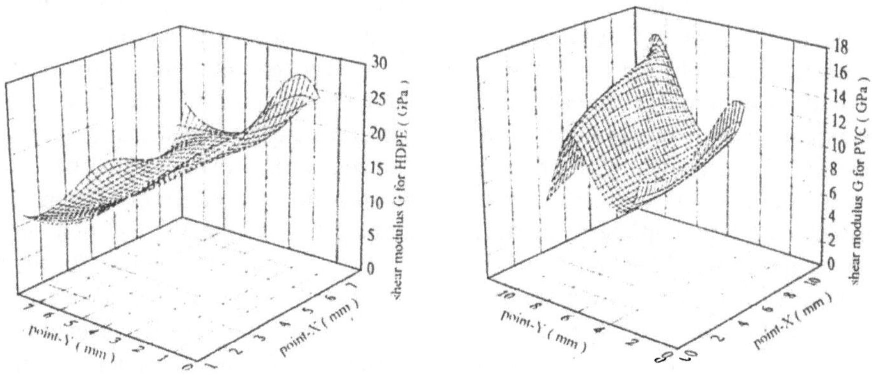

Figure 8. The relations between the shear modulus G and the variation of evolving defect location (X,Y)

elucidate the physical meaning of a transition to a space with a fractional dimension, it is convenient to write Eq.(4.17)in integral form:

$$G(t) = \int_0^t \widetilde{K}(t - \tau)u(\tau)d\tau \tag{4.18}$$

where $\widetilde{K}(t-\tau)$ is a memory function (the relaxation kernel. the creep kernel, etc.[26]). For creep of a rheological solid it is easy to obtain with the Laplace-Carson transform an expression for d_f in the form

$$d_f = \frac{6}{2+w}, \quad w = \frac{R(t)}{3K} \le \frac{2G}{3K} \tag{4.19}$$

where R(t) is the relaxation modulus. Thus we have

$$1 < \Phi_c = \frac{4-w}{2+w} \le 3 \tag{4.20}$$

This result having taken the rheological properties of material into account[27] is in good agreement with experiments of the local temperature field near the crack-tip.

5. Conclusions

As we have shown, the rheological materials with defects may be regarded as an open far-from-equilibrium system. Its main characters are the phenomena of rheology and dissipation which have manifested in the evoluting processes of various defects[28].The stratum hypothesis of micro-, meso-and macro-levels, indicates that the ordered parameters always describe the synergetic activities of the subsystems of the highest stratum. The constitutive equation with ordered parameters for the defect evolutionary state will be changed accordingly. The states of order ,disorder, chaos, anti-chaos and critical state of self-organization can, respectively, occur in different levels, and transform from one to another during the evolution period.

The formation of local temperature field near the crack tip or the surroundings of evolving defects demonstrates that the behavior depends on the rate of local energy transfer. In the creep process, there will be ample time for the material microstructure in the process zone to interact with the external load. Nucleation and growth of meso-defects in the process zone becomes unavoidable because the morphology of the microstructure would have undergone significant changes with time prior to global instability. This would be similar to the situation in phase transformation.

The rupture process of rheological material with defects can be predicted reliably only with a clear understanding of the nature and kinetics of the quantum processes in the process zone at the crack-tip. The nature of relations between processes of different scales constitute the central problem in fracture physics. In this case the equations of the macroscopic fracture kinetics can be obtained by averaging the atomic fluctuation. Specifically, the effect of long-range forces results in a power dependence of the correlation fluctuations of atoms as shown by Eq.(4.10) which is characteristic of fractal structure. In order to determine the fractal dimension d_f, the discussion must be shifted from the atomic level to the mesoscopic level, where the exponent α is expressed in terms of Poisson's ratio ν. which specifies the change in volume during deformation of a medium. When the process zone at the crack-tip forms in a rheological solid, the behavior of this solid becomes pronouncedly nonlinear and is determined by the self-organization of structural excitations on micro-, meso, and macro-scopic scales. The parameter that determines the kinetics of the dissipative processes in a deformable solid is the ratio of the distortion energy density to the dilation energy density in the process zone. When the critical state is reached, a non-equilibrium phase transition occurs with violation of the configuration order, causing the shear rigidity to decrease and the volume to increase. As a result, the local value of Poisson's ratio in the process zone is on the decrease. The instability of the critical state makes the critical phase to pass into a destructive phase by means of self-organization at the mesoscopic level. The formation of the destructive phase corresponds to the formation of a crack capable of self-similar growth, leading to macro-fracture of the solid.

Our experimental results also show that due to the evolution of meso-defects there exists a temperature gradient in the process zone at the crack-tip, it leads to the occurrence of the pyromagnetic effect. The carrier for metal is electron or cavity with electric charge -e or +e and spin ½, and for polymer is charged soliton which depends on the degeneracy of ground states or bipolaron with electric charge ±2e. The transports of these carriers give rise to the electromagnetic effect in the process zone at the mobile crack-tip. It is thus clear that the rupture process of rheological material with defects is not only a pure mechanical process, the thermo-mechanic-electromagnetic coupling effect definitely could not be neglected.

In summary, the rupture process of rheological material with defects is either a nonholonomic determinacy of the stratum system with nonlinear fluctuation caused by the electron-phonon transports, or a deterministic non-periodic process of rheological dissipation in which the global volume of the process zone at mobile crack tip changes with the area of interior surface of the defects. The determinacy and the randomness would be concomitant in the whole process. This paper attempts to combine the new achievements in nonlinear rheo-dynamics, non-equilibrium thermodynamics and statistical physics in the study of rupture theory of rheological material with defects from a unified standpoint of micro-/meso-/macroscopic stratum.

Acknowledgments

The author is grateful to professor R. Wang (Peking University) for useful discussions, to Prof. G. C. Sih and Prof. A.S. Balankin for sending me their reprints.

References

1. Yuan,L.W.(1994) *Rheology of Bodies with Defects* (in Chinese), Pub. House of Nat. Def.Ind, Beijing
2. Yuan, L.W, Swartz, E. S. and Hu, K.K.(1992) *The Phenomena of Rheology and Dissipation in the Process of Crack Propagation*, Kansas State Univ. Press, Kansas
3. Yuan, L. W, Zhi, R. B. and Li. Z. D. (1992) *Fundamentals of Rheological Fracture*(in Chinese), Publ. House of Nat. Def. Ind, Beijing
4. Luo, W.B. (1997) Experimental studies on the evolution of defect temperature field during deformation of ABS. In this proceeding.
5. Luo,W.B.(1997) Preliminary research on fractal character of the local temperature field near defects (in Chinese), *Nat. Sci. J. Xiangtan Univ.* **19**,32-34
6. Mori,H.(1965). *Progr. Theor. Phys.* **33**,423
7. Zwanzig,R.(1965), *Ann.Rev.Phys.Chem.* **16**,67
8. Duhamel,J. M. C. (1837) Second memoire sur les phenomenes thermo-mecaniques(in French), *J.de l'Ecode Polytechnique* **15**,1-57
9. Lindholm,U.S.(1990) Experimental basis for temperature-dependent viscoplastic constitutive equations, *Appl.Mech.Rev.* **43**(5),S338-344
10. Allen,D.H.(1991) Thermomechanical coupling in inelastic solids, *Appl. Mech.Rev.* **44**(8), 361-373
11. Noda,N.(1991) Thermal stresses in material with temperature- dependent properties, *Appl.Mech.Rev.* **44**(9), 383-397
12. Zehnder A.T. and Rosakis, A.J.(1992) A note on the use of high- speed infrared detectors for the measurement of temperature field at the vicinity of dynamically growing in 4340 steel, *J.Appl.Mech.* **59**,450-452
13. Sih, G.C.(1988) Thermomechanics of solids—nonequilibrium and irreversibility. *Theor. Appl. Fracture Mech.* **9**, 175-198
14. Yuan.L.W.(1997) Constitutive equation with ordered parameters for viscoelastic material containing defects, *Acta Mechanica Solida Sinica* **10**,11-27
15. Yan,J.C.(1990) Experimental research on the pyromagnetic effect in the process zone at mobile crack tip (in Chinese), M.S. thesis, Xiangtan University, Xiangtan, China.
16. Muller,I.(1972), *Arch.Rational Mech.Anal.* **46**,131
17. Rice,J.R. and Thomson,R.(1974) Ductile versus brittle behavior of crystals, *Philos Magazine* **29**,73-79
18. Anderson,T.L.(1991) *Fracture Mechanics: Fundamentals and Applications*, CRC Press, Boston
19. Cherepanov,G(1979) *Mechanics of Brittle Fracture*, Mc Graw Hill,N.Y.
20. Ivanova,V.S.,Balankin, A.S.,Bunin.L. and Oksogoev,A.(1994) *Synergetics and Fractals in Material Science* (in Russian), Nauka, Moscow
21. Yuan,L.W.(1995) Rheological fracture analysis of cable by considering the second order effects of random slack, *Acta Mechanica Solid Sinica* **8**,314-326
22. Balankin, A.S.(1995) Quantum-statistical approach to dynamic problems of solid mechanics, *Revista Mexicana de Fisica* **41**,147-180
23. Vladimirov, V. I.(1984) *Physical Nature of Metal Fracture* (in Russian), Metallurgiya, Moscow
24. Lifshitz,E.M. and Pitaevskii,L.P.(1978) *Statistical Physics*, Part 2 (in Russian), Nauka, Moscow
25. Panin, V.P, Likhachev, V.A. and Grinyaev, Yu. V. (1985) *Structural Levels of Deformation of Solids* (in Russian), Nauka, Moscow
26. Yuan,L.W.(1986) *Rheological Mechanics*(in Chinese), Science Publ. House, Beijing
27. Luo,W.B.(1995) Fractal analysis and experimental research on the local temperature field near defects and the interaction between defects (in Chinese), M.S. thesis, Xiangtan University, Xiangtan, China.
28. Yuan,L.W.(1989) On the rheological and dissipative phenomena in the process of crack propagation (in Chinese), *Advances in Mechanics* **19**, 20-35

RHEOLOGICAL BEHAVIOR AND FAILURE CHARACTERISTICS OF VISCOELASTIC SOLIDS WITH DEFECTS*

TING-QING YANG
(Huazhong University of Science & Technology, Wuhan 430074, China)

Abstract Some new advances in study on rheological behavior and failure process of viscoelastic solids with defects and some of our work are reviewed in this paper.

1 Introduction

Rheology of bodies with defects deals with the time-dependent damage evolution, the defect growth, deformation mechanism and fracture characteristics of materials. This topic of research involves the investigation of elasto-visco-plastic behavior and constitutive relation of bodies with defects, the rheological characteristic and the failure process of structures with defects (Yuan 1994, Yang 1996, 1996*a*). Some literatures in the last decades are found to be based on the early works of researchers, such as Williams M L, Knauss W G, Christensen R M, Schapery R A, McCartuey L N, Kaminskii A A, Graham G A C, who studied crack growth and fracture in viscoelastic solids. Significant development in study on rheology of bodies and structures with defects has been made in the past several years. In this paper, some recent advances in research on the rheological characteristics, deformation mechanisms and failure criteria of viscoelastic solids with defects are reviewed. Some works in author's research group are also reported.

---

* Project 19632030 supported by National Natural Science Foundation of China.

R. Wang (ed.), IUTAM Symposium on Rheology of Bodies with Defects, 21–32.

2 Viscoelastic Behavior and Fracture of Solid with Crack

Some textbooks such as Zhang (1994) have given an overall introduction on viscoelastic fracture mechanics (VEFM). In the following a brief outline of some recent development in the study on viscoelastic behavior and fracture characteristics of solids with crack is presented.

A series of **experiments in fracture mechanics** for viscoelastic materials such as metals, polymers and concretes have been studied. A number of papers have dealt with the experiments of VEFM (see Zyczkowski 1991). Hyde and Chambers (1991) have presented a summary of the pure mode-II and mixed-mode (with $K_I/K_{II}=1.6$) results for Jethete M152 - an aeroengine casing material. Huang et al (1991) studied the problem of plane stress crack growth in a rectangular linearly viscoelastic PMMA plate under a pair of equal and opposite forces P applied to the crack surface. They asserted that there exists a lower bound of the applied force P_L . When P is less than P_L there is no crack growth. There also exists a critical load P_C , whose value can be determined experimentally, for the transition from stable crack growth to an unstable crack growth. Under a given load, there exists a critical crack length a_c . Unstable crack growth would take place when the crack length $a(t)$ reached a_c. Based on some of Bazant's work, geometrically similar three-point bending concrete specimens were subjected to sudden changes in loading rate. The viscoelastic cohesive crack model with rate-dependent softening law is used to model the experimental results by Tandon et al (1995). Many experimental results dealing with polymers can be found in other papers, such as by Crook and Letton (1993), Yuan (1994), Liu and Li (1996).

Investigation concerning with concept and model of **fracture process zone** ahead of the crack tip has been developed. Knauss and Losi (1993) discussed crack propagation in a nonlinearly viscoelastic solid with relation to adhesive bond failure. Two zones of material behavior are presumed to govern the time-dependent growth: a linearly viscoelastic bulk domain and a narrow line zone representing unstable and nonlinearly viscoelastic response of the material. The displacement response in the bulk domain under traction, the average strain across the strip and the cohesive fracture energy are investigated. Assuming that there is a fracture process zone of Dugdale type in front of the crack tip, Huang (1991) analyzed the viscoelastic and viscoplastic deformations in the fracture process zone. The crack initiation and crack growth are considered to follow a fracture criterion of critical energy release rate. On the stress field at crack tip,

Atkinson and Eftaxiopoulos (1992) discussed the time evolution of the stress intensity factors at both tips of a finite crack lying near the interface of a viscoelastic anisotropic bimaterial. Eftaxiopoulos and Atkinson (1991) also studied the time evolution of the force on dislocation and near tip stress field of a semi-infinite crack, lying at an angle towards the welded interface of a bimaterial and just touching it.

A number of **fracture mechanics parameters** especially the integral parameters have been defined for application to VEFM in the last decade. A reveiw of the integral parameters in elastic-plastic fracture mechanics (EPFM) has been provided by Kim and Orange (1988). Five types of fracture parameters for creep crack growth are brought out in a paper by Brust and Majumder (1994). Kienzler and Hollstein (1991) showed the derivation of integral fracture concept from conservation laws of general continuum mechanics. In correspondence with J_k integral used in EPFM and introducing a creep potential, they gave the C^*_k integral of creep fracture mechanics. Experimental and numerical evaluations are given. An applicability to crack under steady-state creep conditions is demonstrated experimentally. Kuo et al (1992) have found that C_t is an effective parameter to characterize steady-state and transient creep crack growth rate. Kim and Van Stone (1995) investigated the ability of two other integrals as parameters for correlating time-dependent crack growth rates. The integrals are the rate forms of Kishimoto et al's integral $\hat{J}$ and Blackbum's integral J^*. Sensale and Creus (1993) have applied the transformed boundary element method (BEM) for investigating the viscoelastic fracture behavior. Lee et al (1995, 1995*a*) investigated the behavior of the cracked linear viscoelastic solid by using time-domain boundary element analysis. The $J(t)$ for the cracked linear viscoelastic body has been defined as the rate of change of the functional $M(t)$ for an incremental change in crack length. The computation of $J(t)$ has been performed from the time-domain BEM. Numerical computational procedure of $J(t)$ has been successfully applied in cracked linear viscoelastic plates.

A number of **other topics** in VEFM have been developed rapidly. There are many papers on the application of thermodynamic theory in viscoelastic fracture (Wang and Shen 1994), the boundary value problem for crack growth in viscoelastic media (Linkov 1994), the constitutive equation with ordered parameters for the meso-defect evolutionary state of viscoelastic material with defects (Yuan 1997), and the dynamic crack tip field (Wang Z-Q 1993, Li F-C 1993, Gao 1990, Busso et al 1995). More recently, Siegmund and Needleman (1997) presented a numerical study of dynamic crack growth in elastic-viscoplastic solids. The effects of the strain rate hardening characterization on crack initiation, crack growth and crack arrest were investigated.

3 Rheology for Viscoelastic Solids with Inclusion/Void

In recent years, the viscoelastic inclusion/void problem has attracted the attention of some scholars. Based on Eshelby's and Hill's works and by using Stieltjes convolutions to express the constitutive equations of viscoelastic solids, Laws (1980) has solved directly a viscoelastic inclusion problem under quasi-static condition , without the use of correspondence principle.

The influence of inclusion shape on the overall viscoelastic behavior of composites was studied by Wang and Weng (1992). The Eshelby-Mori-Tanaka method in elasticity is extended into the Laplace domain to examine the linearly viscoelastic behavior in two types of composites: a transversely isotropic one with aligned spheroidal inclusions and an isotropic one with randomly oriented inclusions. Wang and Weng (1993) also studied the nature of self-similarity and transient void growth in viscoelastic media with low concentration of voids by means of the elastic-viscoelastic correspondence principle.

Many researches concentrated on the microvoids evolution. Bahr et al (1991) discussed non-stationary nucleation of cavities at grain boundaries. The formation of creeping pores at grain boundaries as a result of stress concentration in slip processes is investigated by means of non-stationary nucleation theory. Van Der Giessen and Tvergaard (1991) have studied micro-cracking due to cavitation and grain boundary sliding in creeping polycrystals at elevated temperature. A plane strain cell model is used to analyse the development of intergranular creep fracture. Continuous nucleation of cavities and diffusive cavity growth are accounted for in the model, and also the effect of freely sliding grain boundaries is incorporated. In addition to elastic deformations, the grains are taken to deform by dislocation creep modelled by a power law creep. Li (1996) has discussed the evolution process of microvoids in the viscoelastic solid. Yu (1996) investigated the problem of steady-state creep crack growth by the nucleation and growth of creep cavities ahead of a crack . The link between crack velocity and rupture time in creep solids was explored. More recently, Steenbrink et al (1997) dealt with a study of voids in amorphous glassy polymers that exhibit elastic-viscoplasticity with rate dependent yield condition, intrinsic softening and progressive strain hardening at large strains. A detailed numerical study is carried out for plastic flow around voids and the resulting void growth, using the elastic-viscoplastic constitutive models for glassy polymers. Axisymmetric cells are presented by considering the viscoelastic deformation around initially spherical voids. The study on the cell model has shown that the axisymmetric deformation field around void for polymers with softening and subsequent strain hardening can be quite different from those found previously in metals under similar stress states.

4 The Use of Continuum Damage Mechanics

In the IUTAM fourth symposium on creep in structure (Cracow, Poland, 1990), several papers were concerned with the theory and application of continuum damage mechanics (CDM) in creep range. For instance: in the analysis of creep crack growth under irradiation condition, Murakami and Mizuno (1991) proposed that the constitutive equation might be demonstrated by analyzing the irradiation creep and irradiation creep damage; considering the interaction between creep and damage, the strain localization, Kruch and Chaboche (1991) presented a nonlocal damage approach for creep crack growth prediction; after summarizing the development of constitutive equation for creep CDM in metals, Hall and Hayhurst (1991) studied the size effects for notched and cracked specimens, and the global performance of a multi-material structure as in steam pipe weldments.

The use of CDM in viscoelastic solids leads to the improvement of mechanical modeling and the advance in the analysis of rheological process. It can be seen that the process zone model is improved when considering the CDM (Masuero and Greus 1993). Based on a series of research, Murakami and Liu (1995) discussed the mesh-dependence and its regularization in the local approach to creep fracture. By means of CDM and finite element method (FEM), the paper concluded: the most essential cause of the mesh-dependence in the local approach to fracture consists in the unlimited reduction of the crack region width by the refinement of the finite elements; to avoid this mesh-dependent width of the predicted crack, a nonlocal damage criterion is proposed; the damage localization and the related nonlocal damage theory are influential on the damage distribution in the process zone; stress sensitivity can be an important factor, and the damage localization can be effectively avoided by proper modification of the damage evolution equations. Recently, a viscoelastic constitutive model which describes time- and temperature-dependent deformation behavior of particulate composites with growing damage has been presented (Park and Schapery 1997). Ho (1996) has also discussed the characterization of damage by using a cumulative damage failure criterion on predamaged solid propellants.

Lately there have been lots of researches on nonlocal damage theory and approach of late years. Kruch et al (1991) introduced a nonlocal damage formulation, which consists in averaging the damage parameter D over the characteristic volume with a Gaussian function. So that the local approach with nonlocal damage can provide an excellent prediction of the structure lifetime with initial stress concentrations. To decrease the mesh-size effects as mentioned above, Murukami and Liu (1995) defined a nonlocal damage rate variable. Recently, considering the research work of Hall and Hayhurst (1991) and Bazant (1994) et al, a nonlocal continuum damage model for elasto/visco-plastic material was derived and used to describe fatigue crack growth behavior at high temperature by Qian et al (1996). A calculation method for nonlocal damage was introduced. By means of FEM , the analysis of stress, strain and nonlocal damage fields in a plate with a center crack was performed. The validity of the damage model proposed was also verified by comparing the FEM numerical simulation with experimental results.

More recently, Jie M and Yang (1997) have studied the slow extension of a mode-I crack in PMMA. A nonlinear differential equation has been derived to govern the crack propagation by using the Dugdale model. The craze length at the crack tip has been assumed to be constant throughout the crack propagation based on former experimental results. The viscous feature of the material in the craze zone has been taken into account via an empirical relationship between the craze stress and the crack speed shown by existing experiments. It has been found that the Dugdale model only works for crack extension when the crack speed is lower than a certain value, i.e. the so-called slow extension. An effective stress, instead of the stress intensity factor, has been introduced to describe the crack tip zone. Variations of the crack length and the crack speed with time have been computed and their dependence on the effective stress has been discussed. Jie M et al (1997a) have also studied the craze damage field near the crack tip of non-crystalline polymers. And, damage evolution and energy dissipation of polymers with craze have been discussed via a theoretical model. A mathematical relation has been derived between the damage of a single craze and that of a solid polymer with non-interacting parallel crazes (Jie M et al, 1997b). The final life of a single craze and of polymer with crazes is computed. It has been found that the energy dissipated in a craze is proportional to the far-field stress.

Based on a series of research, Hayhurst (1994) has investigated the use of CDM in creep analysis for design. The paper presented the results of CDM analysis on a range of components which included simple ligament situations, notch components, cracked members and butt welded steam pipes. It was shown in all cases that the lifetimes and the patterns of damage evolution can be accurately predicted from the uniaxial creep data, and from data which yield the multiaxial rupture criterion of the material. Othman et al (1993) have developed and successfully used a multi-axial constitutive equation with two damage state variables to describe behavior of nickel-based superalloys. The two damage state variables represent two physical mechanisms which take place in tertiary creep due to grain boundary cavity nucleation and growth, and to the multiplication of mobile dislocations, respectively.

5 Creep Crack Growth and Creep Failure

The development of study on creep fracture mechanics can be found in a lot of literature. Most of them deal with metals and many on polymers. Cock and Ashby (1982) dealt early with creep fracture by void growth in a long paper. The growth of void-like creep damage has been analyzed in more details. Many papers on creep crack growth and creep failure have been submitted to the IUTAM fourth Symposium on creep of structure (see Zyczkowski 1991): theoretical, experimental and numerical investigations of creep crack growth (by Kienzler R & Hollstein T); the assessment of creep-fatigue initiation and crack growth (by Priest R H & Miller D A); a constitutive model for creep and damage in composite materials (by Pyrz R); creep life estimates for defective structures (by Ainsworth R A); an assessment procedure for the high temperature response of structures (by Goodall I W et al); characterization of creep fatigue failure process based on stochastic behavior of small crack initiation and growth (by Kitamura T et al); fatigue crack propagation behavior under creep condition (by Ohji K and Kubo S), etc. Webster (1994) has made a review for the stress analysis procedures that are employed for estimating creep crack growth. A creep fracture mechanics parameter C^* has been defined for describing situations where creep strains dominate. Experimental, numerical and limit analysis procedures have been outlined for evaluating this parameter. Van der Burg & Van der Giessen (1994) employ the Delaunay network model to study the effect of randomness in the microstructure on the microcrack propagation behavior and the ultimate times to rupture. A simulation of microcrack propagation in creeping polycrystals due to diffusive grain boundary cavitation has been studied.

Recently the fracture behavior of creeping materials under biaxial loading by FEM has been investigated (Singh & Ramakrishnan 1995). A procedure for characterizing the stress strain rate field during fracture of creeping materials was developed. Hyde et al (1996) has analyzed the prediction of creep failure in aeroengine materials, nickel-base superalloy and titanium alloy, at high temperature under multi-axial stress states. From the uniaxial creep test results, a creep continuum damage model was established for each material. As ceramic matrix composites at high temperature exhibit time-dependent behavior due to fiber creep, even though the matrix remains elastic, Begley (1997) discussed the time dependent cracks in ceramic matrix composites. Using a bridging law developed before which describes the effects of fibers bridging a matrix crack and accounts for frictional sliding between the fibers and the matrix, the time-dependent behavior of bridged cracks for the composites is modeled.

Time-dependent creep-fatigue crack growth is an important consideration in the design and the estimation of remanent life of high temperature components. Grover and Saxena (1995) characterized the creep-fatigue behavior in a Cr-Mo steel, a steam header material, using the average value of the parameter C_t, , i.e. $(C_t)_{avg}$. The analytically estimated values of $(C_t)_{avg}$ were compared with the experimental values of $(C_t)_{avg}$. Sedmak A and Sedmak S (1995) introduced and applied a critical crack assessment procedure for high pressure steam turbine rotors.

More recently, Wang (1996) has reviewed the study on creep behavior, deformation mechanisms and creep failure of polymer and polymer composite. Author gave a survey to the researches and developments on VEFM in creep range, including creep crack growth, relevant fracture mechanics parameters, creep failure and life assessment in defective structure . Zhang and Yang (1997) reviewed the advances in research on cracks/defects evolution and creep failure of rock. Yang J-Y (1997) investigated the creep of the surrounding rocks of caverns in viscoelastic joint rock mass. The time-dependent displacement field of the surrounding rocks can be determined by FEM.

6 Concluding Remarks

Time-dependent mechanical behavior, deformation mechanisms and failure criteria of the bodies and structures with defects have become important research subjects in the past several years. Although the above overview is far from complete the papers presented, it can be seen all the same that the study on rheological behavior and failure characteristics of viscoelastic solids with defects has been developed rapidly and widely.

- Theoretical, experimental and numerical investigations are combined in studying the viscoelastic behavior of solids with defects. The numerical simulation is a very important approach.
- The topics of elastic-viscoplasticity, the creep of metals under high temperature and that of composites have attracted more attention of late, and may become more prosperous in the near future. CDM and micromechanics are very useful for the research.
- Mechanical behavior of materials with various defects (such as cracks, inclusions, voids etc) is strongly affected by their interaction. Such coupling and/or interaction are significant nonlinear subjects.
- The application of the theory on viscoelastic solids and structures with defects can be found in many aspects of engineering.

References

Atkinson C & Eftaxiopoulos D A. (1992) Crack tip stress intensities in viscoelastic anisotropic bimaterials and the use of t TING-QING YANG he *M*-integral, *Int J Fract*, 57:61-83

Bahr H-A et al. (1991) Non-stationary nucleation of cavities at grain boundaries in: Zyczkowski M (Ed). *Creep in Structures* , Springer-Verlag, pp.279-285

Bazant Z P. (1994) Nonlocal damage theory based on micromechanics of crack interactions, *J Engng Mech, ASCE*, 120:593-617

Begley M R. (1997) Time dependent cracks in ceramic matrix composites: crack growth initiation, *Int Solids Struct*, 34(9):1035-1051

Brust F W & Majumdar B S. (1994) Load history effects on creep crack growth, *Engng Fract Mech*, 49(6): 809-837

Busso E P et al. (1995) On the effect of loading conditions and geometry on time-dependent singular crack tip fields, *Engng Fract Mech*, 50(2):231-247

Cocks A C F & Ashby M F. (1982) On creep fracture by void growth, *Progress in Mat Science*, 27:189-244

Crook R A & Letton A. (1993) The viscoelastic behavior of notched glassy polymers, *Engng Fract Mech*, 44(2):167-173

Eftaxiopoulos D A & Atkinson C. (1991) Crack tip stress singularities and dislocation interactions with anisotropic viscoelastic bimaterial interfaces, *Int J Engng Sci* , 29(12):1569-1584

Gao Y C. (1990) Further study on strain singularity behavior of moving crack in elastic-viscoplastic materials,

Theor & Appl Fract Mech, 14:233-242

Grover P S & Saxena A. (1995) Characterization of creep-fatigue crack growth behavior in 2.25 Cr-1 Mo steel using $(C_t)_{avg}$, *Int J Fract*, 73:273-286

Hall F R & Hayhurst D R. (1991) Creep continuum damage mechanics-studies of size effects and weldments, in:Zyczkowski M (Ed). *Creep in Structures*, Springer-Verlag,, pp.313-332

Hall F R & Hayhurst D R. (1991) Modeling of grain size effects in creep crack growth using a non-local continuum damage approach, *Proc R Soc Lond*, A433:405-421

Hayhurst D R. (1994) The use of continuum damage mechanics in creep analysis for design, *J Strain Analysis*, 29(3):233-241

Ho S Y. (1996) High strain-rate impact studies of predamaged rocket propellants.I, *Combustion & Flame*, 104:524-534

Huang N C et al. (1991) Experiments on crack propagation in a viscoelastic material, *Int J Fract*, 48:299-309

Huang N C. (1991) Crack growth in linearly viscoelastic media, in:Zyczkowski M (Ed). *Creep in Structures*, Springer-Verlag, pp.323-330

Hyde T H & Chambers A C. (1991) Mixed-Mode creep and creep/fatigue crack growth, in:Zyczkowski M (Ed). *Creep in Structures*, Springer-Verlag, pp.331-338

Hyde T H et al. (1996) Prediction of creep failure in aeroengine materials under multi-axial stress states, *Int J Mech Sci*, 38(4):385-403

Jie M & Yang T-Q (1997). Slow extension of a crack in PMMA, Submitted to *Theor & Appl Fract Mech.*

Jie M et al. (1997a). Craze damage field near the crack tip of non-crytalline polymers, to appear in *Chinese J of Appl Mech*. *

Jie M et al. (1997b). Modelling of damage evolution and energy dissipation of polymers with craze, to appear in *Theor & Appl Fract Mech.*.

Kienzler R & Hollstein T. (1991) Theoretical, experimental and numerical investigations of creep crack growth, In: Zyczkowski (Ed). *Creep in Structures*, Springer-Verlag, pp.339-346

Kim K S & Orange T W. (1988) A review of path-independent integrals in elastic-plastic fracture, *Proc 18th Symp on Fracture Mechanics* ASTM STP, 945:713-729

Kim K S & Van Stone R H. (1995) Time-dependent crack growth characterization using rate integrals, *Int J Fract*, 70: 167- 181

Knauss W G & Losi G U. (1993) Crack propagation in a nonlinearly viscoelastic solid with relevance to adhesive bond failure, *J Appl Mech*,, 60:793-801

Kruch S, Chaboche J & Lesne P M. (1991) A new damage approach for creep crack growth prediction, In: Zyczkowski (Ed). *Creep in Structures*, Springer-Verlag, pp.355-362

Kuo An-Yu et al. (1992) An integral formulation of C_t for use in creep crack growth evalution, *Int J Fract*, 57:269-280

---

* In Chinese with English abstract

Laws N. (1980) Viscoelastic inclusion problem, *J Engng Mech Div*, Proc ASCE, 1980, 106:915-928

Lee S S & Westmann R A. (1995) Application of high-order quadrature rules to time-domain boundary element analysis of viscoelasticity, *Int J Numer Mech Engng*, 38:607-629

Lee S S & Kim Y J. (1995) Time-domain boundary element analysis of cracked linear viscoelastic solids, *Engng Fract Mech*, 51(4):585-590

Li F-C. (1993) Investigation on the asymptotic solution for the mode III moving crack-tip field, *Acta Mech Sinica*, 25(6):732-737 *

Li H-L. (1996) Statiatical Behavior of Microdamage's Evolution in Ductile Materials and its Stochastic Model, Ph. D Dissertation, Dept. of Mechanics, Peking University. *

Li Z-D. (1996) On the effect of thermo-mechanical coupling in the process of defect evolution, in:Jin R-G (Ed.) *Advances in Rheology*, Chem Ind Press, Beijing, pp.39-44 *

Linkov A M. (1994) Boundary value problem for crack growth in viscoelastic media, *Int J Fract*, 65:197-218

Liu Z & Li Z-D (1996) The observing experiment of the polymer fracture process and its damage fracture model, in:Jin R-G (Ed.) Advances in Rheology, Chem Ind Press, Beijing, pp.33-38 *

Masuero J R & Creus G J. (1993) Finite elements analysis of viscoelastic fracture, *Int J Fract*, 1993, 60:267-282

Murakami S & Liu Y. (1995) Mesh-Dependence in local approach to creep fracture, *Int J Damage Mech*, 4(3):230-250

Murakami S & Mizuno M. (1991) Mechanical modeling of irradiation creep and its application to the analysis of creep crack growth, In: Zyczkowski (Ed). *Creep in Structures*, Springer-Verlag, pp.237-252

Othman A M et al. (1993) Skeletal point stresses in circumferentially notched tension barsundergoingtertiary creep modeled with physically-based constitutive equations, *Proc Roy Soc Lond*, *ser A*, 441:343-358

Park S W & Schapery R A. (1997) A viscoelastic constitutive model for particulate composites with growing damage, *Int J Solids Struct*, 34(8):931-947

Qian Z-X et al. (1996) A nonlocal damage mechanics approach to high temperature fatigue crack growth, *Engng Fract Mech*, 53(4):535-543

Sedmak A & Sedmak S. (1995) Critical crack assessment procedure for high pressure steam turbine rotors, *Fatigue Fract. Engng Mater. Struct*, 18(9):923-934

Sensale B & Creus G J. (1993) Boundary element analysis of viscoelastic fracture, in: Brebbia C A (Eds). Boundary Element X V, Elsevier Sci Pub, pp.291-301

Siegmund T & Needleman A. (1997) A numerical study of dynamic crack growth in elastic-viscoplastic solids, *Int J Solids Struct*, 34(7):769-787

Singh R N & Ramakrishnan C V. (1995) Fracture behavior of creeping materials under biaxial loading by finite element method, *Engng Fract Mech*, 51(4):637-648

Steenbrink A C et al. (1997) Void growth in glassy polymers, *J Mech Phys Solids*, 45(3):405-437

Tandon S, Faber K T, Bazant Z P et al. (1995) Cohesive crack modeling of influence of sudden changes in loading rate on concrete fracture, *Engng Fract Mech*, 52(6): 987-997

Van der Burg M W D & Van der Giessen. (1994) Simulation of microcrack propagation in creeping polycrystals due to diffusive grain boundary cavitation, *Appl. Mech Rev*, 47(1.part II):S122-S131

Van der Giessen E & Tvergaard V. (1991) On microcracking due to cavitation and grain boundary sliding in creeping polycrystals, In: Zyczkowski (Ed). *Creep in Structures*, Springer-Verlag , pp.295-302

Wang Ren. (1996) A review on creep failure of polymer and polymer composite, in: Abe T & Tsuta T (Ed.), Proc 3rd Asia-pasific Symp on Advances in Engng Plasticity & its Applications, pp.43-52. Pergamon Press, 1996

Wang Y M & Weng G J. (1992) The influence of inclusion shape on the overall viscoelastic behavior of composites, *J Appl Mech,* 1992, 59:510-518

Wang Y M & Weng G J. (1993) Self-similar and transient void growth in viscoelastic media at low concentration, *Int J Fract,* 61:1-16

Wang X-M & Shen, Y-P. (1994) The application of thermodynamic theory in viscoelastic fracture, *Acta Mech Solida Sinica* 15:58-64 *

Wang Z-Q. (1993) Dynamic crack-tip field in an elastic-viscoplastic material, *Acta Mech Sinica*, 25(2):159-168 *

Webster G A. (1994) Fracture mechanics in creep range, *J Strain Anal*, 29(3):215-223

Yang J-Y. (1997) Research on creep characteristics of joint rockmass and viscoelastic FEM analysis for caverns, M.S. thesis, Dept. of Mechanics, HUST, Wuhan, China.*

Yang T-Q. (1996). A survey of rheology of bodies with defects, *Mechanics & Practice,* 18(3):13-17 **

Yang T-Q. (1996*a*) Study on rheology of bodies with defects and its application, in:Jin R-G (Ed.) *Advances in Rheology,* Chem Ind Press, Beijing, pp.11-14 *

Yu J. (1996) The link between crack velocity and rupture time in creeping solids, *Engng Fract Mech,* 53(2):213-230

Yuan L-W. (1994). *Rheology of Bodies with Defects*, Publishing House of National Defence Industry, Beijing **

Yuan L-W. (1997). Constitutive equation with ordered paramaters for viscoelastic material containing defects, *Acta Mech Solida Sinica,* 10(1):11-26 *

Zhang C-Y. (1994). *Viscoelastic Fracture Mechanics,* HUST Press, Wuhan **

Zhang X-C. & Yang, T-Q. (1997) Recent advances in the study on cracks/defects evolution and creep failure of rock, submitted to *Advances in Mechanics.* *

Zyczkowski M (Ed.). (1991) *Creep in Structures*, Springer-Verlag Berlin Heidelberg.

** In Chinese

SIMULATION OF SLOW KINETIC FRACTURE OF GAS EMISSIONABLE MATERIALS

A.V. BALUEVA
Institute for Problems in Mechanics of Russian Academy of Sciences
pr.Vernadskogo, 101, Moscow 117526, Russia

Abstract

Three-dimensional problem on a slow quasi-stationary crack growth in materials exhibiting specific properties of gas emission in bulk is considered. The crack occupies arbitrary domain in plane in initial moment t. The connected diffusion-elasticity 3D problem is reduced to two 2D boundary integro-differential equations which then are solved numerically.

1. Introduction

A number of materials used in modern engineering exhibit specific properties of gas emission in bulk under certain mechanical and/or physical influence or aging. Gas emission due to aging is typical for a number of polymers. Some metals and alloys applied in nuclear-power engineering become gas emissionable under radiation (Likhachev et al., 1982).

Gas emission in bulk can frequently cause crack or crack-like defects initiation and their slow propagation. Experiments show that the cracking resistance of the materials in gas presence is not a constant of the material, but is characterized by a function - the dependence of the crack growth velocity on the stress intensity factor (Barenblatt et al., 1966; Mishnaevsky 1994). In this case, the crack growth is observed long before the critical loading for materials in gas absence. Thus, crack kinetics analysis implies simultaneous consideration of gas diffusion into the crack and slow kinetic crack growth due to the action of inner gas pressure and other mechanical loads .

We suggest a numerical method for solving the 3D problem for a medium with cracks occupying a plane region. Problems of gas diffusion into crack and crack propagation are solved by reducing to 2D integro-differential equations in the crack domain. In model calculations, the

R. Wang (ed.), IUTAM Symposium on Rheology of Bodies with Defects, 33–45.

crack velocity v at each point of the crack contour is assumed to be dependent on the stress intensity factor K at this point. The kinetics calculation is performed step by step procedure. The algorithms applied here develop those suggested earlier (Balueva, 1993).

2. Statement of the Problem

We consider the slow quasi-steady growth of a tensile crack initiated at t=0 and occupying a domain G in the plane $x_3=0$. The velocity v at each crack contour point is assumed to be dependent on the stress intensity factor N (as is adopted in kinetic crack theories) and specified by a curve v(N) which is the material function. The crack is growing under the action of a gas produced by gas emission sources distributed in bulk. The crack is modeled by an ideal sink (far from equilibrium state). The crack velocity is assumed to be small as compared that in the transient period. Under this assumption, the flow into the crack can be found from the solution of the stationary diffusion problem for each t. Suppose that initially there are two diffusion sources of intensity W placed inside the body on the x_3-axis symmetrically at a distance ξ_3 from the crack .In view of the symmetry with respect to the crack plane, we can consider the problem in the half-space $x_3 \geq 0$.

The boundary value problem for the gas concentration $c(x_1, x_2)$ is following one:

$$\begin{aligned}
&\Delta c = -\frac{W}{D}\delta(x_1)\delta(x_2)\left[\delta(x_3-\xi_3)+\delta(x_3+\xi_3)\right], \\
&c|_{x_3=0} = 0, \quad (x_1,x_2)\in G; \\
&\left.\frac{\partial c}{\partial x_3}\right|_{x_3=0} = 0, \ (x_1,x_2)\notin G; \qquad (1.1)\\
&c|_{x_3=\infty} = 0,
\end{aligned}$$

where Δ is Laplacian, D is the coefficient of gas diffusion in the medium.

The diffusion flow density $q(x_1,x_2) = \partial c/\partial x_3|_{x3=0}$ for $(x_1,x_2) \in G$ is the unknown function in the problem. As usual, to construct an integral equation for q, we first consider the gas diffusion problem with sources in a medium without the crack:

$$\Delta c_o = -\frac{W}{D}\delta(x_1)\delta(x_2)[\delta(x_3-\xi_3)+\delta(x_3+\xi_3)] \tag{1.2}$$

$$\left.\frac{\partial c_o}{\partial x_3}\right|_{x_3=o} = 0, \quad \left.c_o\right|_{x_3=o} = 0$$

The solution to problem (1.2) is the function

$$c_o(x_1,x_2,x_3) = \frac{W}{4\pi D}\left(\frac{1}{R_1}+\frac{1}{R_2}\right), \quad \textit{where}$$
$$R_{1,2} = \sqrt{x_1^2+x_2^2+(x_3\pm\xi_3)^2} \tag{1.3}$$

and the gas concentration in the crack plane is given by

$$c_o(x_1,x_2,x_3) = \frac{W}{2\pi D}\frac{1}{\sqrt{x_1^2+x_2^2+\xi_3^2}} \tag{1.4}$$

Let us now write out the solution to the diffusion problem without sources but with the gas concentration inside the crack to be equal in magnitude and opposite in sign to that in the first problem:

$$\Delta c = 0, \quad \left.c\right|_{x_3=o} = -\frac{W}{2\pi D}\frac{1}{\sqrt{x_1^2+x_2^2+\xi_3^2}}, \; (x_1,x_2)\in G;$$

$$\left.\frac{\partial c_o}{\partial x_3}\right|_{x_3=o} = 0, \quad \left.c_o\right|_{x_3=o} = 0.. \tag{1.5}$$

The following integral equation is obtained for the diffusion flow density q from Eq.(1.5):

$$\Delta c = 0, \quad \left.c\right|_{x_3=o} = -\frac{W}{2\pi D}\frac{1}{\sqrt{x_1^2+x_2^2+\xi_3^2}}, \; (x_1,x_2)\in G; \tag{1.6}$$

Similarly, if a nonzero gas concentration $c^o : \left.c\right|_{x_3=o} = c^o,\ \textit{for}\ (x_1,x_2)\in G$ given inside the crack, then we arrive at the integral equation

$$\frac{1}{2\pi}\iint_G \frac{q(\xi_1,\xi_2)d\xi_1 d\xi_2}{\sqrt{(x_1-\xi_1)^2+(x_2-\xi_2)^2}} = -\frac{W}{2\pi D}\frac{1}{\sqrt{x_1^2+x_2^2+\xi_3^2}} \tag{1.7}$$

The following integral equation is valid if two sources symmetric with respect to the crack plane are placed at arbitrary points (a,b,ξ_3) and $(a,b,-\xi_3)$ in bulk:

$$\begin{aligned}\frac{1}{2\pi}\iint_G \frac{q(\xi_1,\xi_2)d\xi_1 d\xi_2}{\sqrt{(x_1-\xi_1)^2+(x_2-\xi_2)^2}} &= \\ = \frac{W}{2\pi D}\frac{1}{\sqrt{x_1^2+x_2^2+\xi_3^2}} - c^0(x_1,x_2).\end{aligned} \tag{1.8}$$

Using the superposition principle, we obtain the following equations for several point sources of gas diffusion inside the body or for those distributed with intensity density $W(x_1,x_2,x_3)$:

$$\begin{aligned}\iint_G \frac{q(\xi_1,\xi_2)d\xi_1 d\xi_2}{\sqrt{(x_1-\xi_1)^2+(x_2-\xi_2)^2}} &= \\ = \frac{1}{D}\sum_i \frac{W_i}{\sqrt{(x_1-a_i)^2+(x_2-a_i)^2+\xi_{3i}^2}},\end{aligned} \tag{1.9}$$

$$\begin{aligned}\iint_G \frac{q(\xi_1,\xi_2)d\xi_1 d\xi_2}{\sqrt{(x_1-\xi_1)^2+(x_2-\xi_2)^2}} &= \\ = \frac{1}{D}\iiint_T \frac{W(y_1,y_2,y_3)dy_1 dy_2 dy_3}{\sqrt{(x_1-y_1)^2+(x_2-y_2)^2+y_3^2}},\end{aligned} \tag{1.10}$$

where W_i are the intensities of the sources at the points $(a_i,b_i,\pm\xi_{3i})$ and T is the region of diffusion sources distribution with density $W(x_1,x_2,x_3)$. To search for the elastic fields induced by the gas diffusion into the crack, we consider the problem of a tensile crack with load p applied to its surfaces, where p is the gas pressure, which depends on the crack volume and mass of gas entered. The gas is assumed to be ideal; then the crack volume V, the mass of the gas M, and the pressure p are related by the Clapeyron equation $pV=MRT/\mu$. Here μ, R, and T are the molar mass of the gas, the gas constant per mole, and absolute temperature, respectively. Reducing the elasticity problem to boundary integral equations, we obtain the system

$$\frac{1}{2\pi}\iint_G \frac{q(\xi_1,\xi_2)d\xi_1\, d\xi_2}{\sqrt{(x_1-\xi_1)^2+(x_2-\xi_2)^2}} = -\frac{W}{2\pi D}\frac{1}{\sqrt{x_1^2+x_2^2+\xi_3^2}} \tag{1.11}$$

$$p(t) = -\frac{E}{4\pi(1-\nu^2)}\iint_{G(t)} \frac{u(\xi_1,\xi_2)d\xi_1\, d\xi_2}{\sqrt{(x_1-\xi_1)^2+(x_2-\xi_2)^2}}, \tag{1.12}$$

$$pV = nRT, \tag{1.13}$$

$$V(t) = \iint_{G(t)} u(\xi_1,\xi_2)d\xi_1\, d\xi_2, \tag{1.14}$$

$$Q = -\iint_{G(t)} q(\xi_1,\xi_2)d\xi_1\, d\xi_2, \tag{1.15}$$

$$n(t+\Delta t) = n(t) + Q\Delta t, \tag{1.16}$$

$$u(\xi,s,t) = \frac{4(1-\nu^2)}{E} N(s,t)\sqrt{\xi}, \tag{1.17}$$

$$v(s,t) = f(N(s,t)), \tag{1.18}$$

$$r(t+\Delta t,s) = r(t,s) + v(t,s)\Delta t, \qquad 1.19)$$

where integral equation (1.11) for $q(x_1,x_2)$ can be replaced by Eqs. (1.7)-(1.10) depending on the number of sources and their distribution. Equation (1.12) is the integro-differential equation for the crack surfaces displacement $u(x_1,x_2)$; further, n, $Q=\partial n/\partial t$, E, and ν are the number of gas moles in the crack, gas flow rate through the crack, Young's modulus, and Poisson's ratio of the medium, respectively. The term r is the radius vector of the crack contour and v in Eq.(1.19) is the crack contour point velocity in the direction of its outer normal. Equations (1.17) is used for the calculation the stress intensity factor N, where s is current point along the crack contour and ξ is the distance from the contour in direction of its inner normal (Slepyan, 1981).

Equations (1.17)-(1.19) provide the calculation of the stress intensity factor N and of the new crack contour.

The solution is performed stepwise (Balueva at al.,1992). The main computational difficulties of the first t-step are related to solving integro-differential equations (1.11)-(1.12) and in searching for a new crack contour via the calculated velocities v at the previous contour points (see Eq.(1.18)). The last computation is a separate calculational problem. A procedure for solving the elasticity problem for a normal tensile crack (Eq. (1.12)) has been developed (Goldstein at al., 1973). For this reason, we focus on a numerical method for solving the diffusion equation (1.11). In case of a circular crack region, we obtain an analytic solution.

3. Numerical Method for Solving the Diffusion Equation

Our method for solving the integro-differential equation is based on the variational-difference method (Balueva et al., 1985). Namely, after discretization, the values of q at the grid points are searched for as an expansion through a system of coordinate functions ψ_{p1p2},

$$q(x_1,x_2) = \sum_{p1p2} c_{p1p2}\psi_{p1p2}(x_1,x_2,h), \tag{2.1}$$

where ψ_{p1p2} is a bilinear spline function with a support in the four grid cells adjacent to the point (p_1h,p_2h) of the grid with the step h.

The coefficients c_{p1p2} coincide with the values of $q(x_1,x_2)$ at the grid points and are found by minimizing the corresponding quadratic functional:

$$\min\left\{ I(h) = \sum_{p_1p_2}\sum_{q_1q_2} a_{p_1q_1p_2q_2} c_{p_1p_2} c_{q_1q_2} + 2\sum_{p_1p_2} c_{p_1p_2} b_{p_1p_2} \right\} \tag{2.2}$$

$$a_{p_1p_2q_1q_2} = a_{|p_1-q_1|,|p_2-q_2|} = \\ = \frac{1}{(2\pi)^2}\int\int_{-\infty}^{\infty} \frac{1}{|\xi|} \psi_{p_1p_2}(\xi,h)\overline{\psi_{q_1q_2}(\xi,h)}\, d\xi, \tag{2.3}$$

$$\psi_{p_1p_2}(\xi,h) = h^2 e^{ih(p_1\xi_1+p_2\xi_2)} \frac{\sin^2\left(\frac{1}{2}h\xi_1\right)}{\left(\frac{1}{2}h\xi_1\right)^2} \frac{\sin^2\left(\frac{1}{2}h\xi_2\right)}{\left(\frac{1}{2}h\xi_2\right)^2}, \tag{2.4}$$

$$b_{p_1p_2} = \iint_{g^{2h}_{p_1p_2}} p(x_1,x_2)\psi(x_1,x_2)dx_1dx_2 \tag{2.5}$$

$$p(x_1,x_2) = \frac{W}{D}\frac{1}{\sqrt{x_1^2+x_2^2+\xi_3^2}} \tag{2.6}$$

The minimization is carried out by the gradient projection method with an automatic step choice according to the relation between the linear and the actual functional increments.

4. Axisymmetric Problem of Gas Diffusion into a Crack due Discrete Sources

The integral equation (1.11) in the case of a circular crack of radius a and a point source lying on the x_3-axis acquires the form

$$\int_0^{2\pi} d\varphi \int_0^a \frac{q(\rho)\rho d\rho}{\sqrt{r^2+\rho^2+2r\rho\cos\varphi}} = g(r), r \le a, \tag{3.1}$$

$$g(r) = \frac{W}{D}\frac{1}{\sqrt{r^2+\xi_3^2}}.$$

This integral equation (3.1) can be rewritten

$$\int_0^a q(\rho)K\left(\frac{2\sqrt{r\rho}}{r+\rho}\right)\frac{\rho\, d\rho}{r+\rho} = \frac{1}{4}g(r), \tag{3.2}$$

$$K(k) = \int_0^{2\pi}\frac{d\varphi}{\sqrt{1-k^2\sin^2\varphi}}, \tag{3.3}$$

where K(k) is the complete elliptic integral of the first kind.

Using properties of Bessel functions, we have

$$\frac{2}{\pi}K\left(\frac{2\sqrt{r\rho}}{r+\rho}\right)\frac{1}{r+\rho} = \int_0^{2\pi} J_0(ur)J_0(u\rho)\,du, \tag{3.4}$$

and Eq. (3.2) becomes

$$\int_0^\infty J_0(ur)\,du\int_0^a q(\rho)\rho\, J_0(u\rho)\,d\rho = \frac{1}{2\pi}g(r), \quad r\le a. \tag{3.5}$$

Denote

$$\int_0^a q(\rho)\rho\, J_0(u\rho)\,d\rho = \Phi(u). \tag{3.6}$$

Then Eq. (3.5) can be represented in the form

$$\int_0^\infty \Phi(u)\, J_0(ur)\,du = \frac{1}{2\pi}g(r), \quad 0\le r\le a. \tag{3.7}$$

Using properties of the Bessel transformation, from Eq. (3.6) we obtain

$$\int_0^\infty \Phi(u) u J_0(ur)\, du = \begin{cases} g(r), & \text{for } r \le a \\ 0, & \text{for } r > a \end{cases} \tag{3.8}$$

Thus, we arrive at the system of dual integral equations

$$\int_0^\infty \Phi(u) J_0(ur)\, du = \frac{1}{\pi} g(r), \quad 0 \le r \le a,$$

$$\int_0^\infty u\Phi(u) J_0(ur)\, du = 0, \quad r > a. \tag{3.9}$$

The solution to system (3.9) has the form

$$\Phi(u) = \frac{1}{\pi^2} \int_0^a \cos ut \left(\frac{d}{dt} \int_0^t \frac{y g(y)\, dy}{\sqrt{t^2 - y^2}} \right) dt, \tag{3.10}$$

or after substituting of g(y)

$$\Phi(u) = \frac{1}{\pi^2} \frac{W}{D} \int_0^a \cos ut \left(\frac{d}{dt} \int_0^t \frac{y\, dy}{\sqrt{y^2 + c^2} \sqrt{t^2 - y^2}} \right) dt,$$

then, from Eq.(3.6) we can find function q(r) by the formula

$$q(r) = \int_0^\infty \Phi(u) u\, J_0(ur)\, du. \tag{3.11}$$

Omitting cumbersome details of integration, let us write out the final expression for diffusion flux density through the crack

$$q(r) = \frac{1}{\pi} \frac{\xi_3}{\sqrt{(r^2 + \xi_3^{\,2})^3}} + \frac{1}{\pi^2} \frac{\xi_3}{(r^2 + \xi_3^{\,2})\sqrt{a^2 - r^2}} - \frac{1}{\pi^2} \frac{\xi_3}{\sqrt{(r^2 + \xi_3^{\,2})^3}} \arctan \frac{\sqrt{a^2 - r^2}}{r^2 + \xi_3^{\,2}} \tag{3.12}$$

As the crack radius $a \to \infty$, the solution has the asymptote

$$q(r) \to \frac{1}{2\pi} \frac{\xi_3}{\sqrt{(r^2 + \xi_3^{\,2})^3}}.$$

This formula coincides with the solution to the problem on a diffusion source in a half-space. On the crack contour, that is, as $r \to a$, we have the asymptote

$$q(r) \to \frac{1}{\pi^2} \frac{1}{(\xi_3 + a^2/\xi_3)\sqrt{a^2 - r^2}}.$$

Thus, this solution has a root singularity, which is actually observed in the problem of a gas diffusion into the crack for consideration with a source given at infinity (Balueva at al., 1994).

Comparison was made of numerical results with those obtained analytically. Good agreement is observed in the crack domain up to the last but one boundary node in the vicinity of the contour. The numerical solution becomes "bad" at the grid points adjacent to the boundary. This is due to the root singularity of the solution on the crack contour. An effective Boundary Refinement Method is applied for improvement of numerical solution near the boundary.

5. Crack Propagation due to Gas Diffusion from the Bulk Sources

Software is developed to calculate the crack propagation time and evolution of the crack shape and sizes under the action of gas diffusion from a unit source or sources distributed in bulk with a given density. These program is based on the described methods for solving the integro-differential equations (1.11) of the diffusion problem and eqs.(1.12) of the elasticity problem. A quasi-steady statement of the problem is used. System (1.11)-(1.19) is solved at each time step. The incubation period t_i before start of crack growth is calculated. This is a time before the crack opening under the gas diffusion action achieves the value for which the maximum stress intensity factor along the crack contour becomes greater than the fracture toughness threshold value K_{scc}.

Calculation of the growth time t_m is performed by the following scheme:

1) the gas pressure is calculated in the current crack region $G(t)$: $p^2(t) = n(t)RT / V_1(t)$, where $V_1(t)$ is the volume of the crack occupying the new region $G(t)$ for unit loading p=1, the gas mass n(t) being found at the previous step.

2) the stress intensity factor along the crack contour is calculated and used for calculation of the crack velocity v_1.

3) normal distances toward the crack contour passable by discrete contour points during this propagation are defined as $\Delta_i = v_i \Delta t$. Time interval Δt is calculated so that contour points spreading with maximum velocity v_1 pass a small distance Δu chosen experimentally.

4) a new contour shape is defined using a smoothing procedure over propagating and stationary points coordinates.

5) the diffusion problem is solved; the integral flow $q(x_1,x_2)$ through the crack surface, the total gas flow rate Q, and the new gas amount in the crack $n(t+\Delta t) = n(t) + Q\Delta t$ are defined.

6) the above procedure is repeated starting from step 1.

Model calculations were performed for a circular plane crack. Its kinetics was studied in the case of the gas diffusion from a unit bulk source. The incubation period t_i and time t_m of the crack growth from the initial to double radius were calculated. The dependence of the values t_i and t_m on the distance, ξ_3, from the diffusion source to the crack plane was studied. On diminishing the distance ξ_3, the gas flux to the crack increases and the gas pressure becomes greater, thus leading to the stress intensity factor growth along the contour. As a result, the velocity tends to its stationary value (on kinetic diagram), propagation and incubation times being practically independent of ξ_3. One more series of calculation was performed to study the dependence of life-time on the diffusion source intensity W. The greater the source intensity, the less is the life-time.

6. Acknowledgments

This work was supported by the Russian Foundation for Basic Research under grant N 96-01-01098. Author is indebted to Professor R.V.Goldstein of the Institute for Problems in Mechanics of RAS, Moscow, for his encouragement and many valuable discussions during the present investigations and also for helpful comments on the manuscript.

7. References

Likhachev, Yu.I., Pupko, V.Ya. and Popov, V.V. (1982) *Calculation Method for the Strength of Heat-Developing Nuclear Reactor Members*, Gergoatomizdat, Moscow.

Barenblatt, G.I., Entov, V.M., and Salganic R.L. (1966) On the kinetics of crack propagation. Fracture condition and long-term strength. *Mechanics of Solids*, **6**, 76-80.

Mishnaevsky, L. Jr (1994) A New Approach to the Determination of the Crack Velocity versus Crack Length Relation. *Int. J. Fatique Tract. Eng. Mater. Struct.*, **10**, 1205-1212.

Balueva, A.V. (1993) Three-Dimensional Problem on Crack Kinetics under Gas Diffusion. *Mechanics of Solids*, **6**, 123-131.

Slepyan, L.I. (1981) Mechanics of Cracks, Sudostroenie, Leningrad (1990) (in Russian).

Balueva, A.V. and Goldstein, R.V. (1992) Kinetic Propagation of Cracks in a Layer under Gas Diffusion, *Mechanics of Solids*, **2**, 114-124.

Goldstein, R.V., Klein, I.S. and Eskin, G.I. (1973) Variational-Difference Method for Solving Some Integral and Integro-Differential Equations of 3-D Elasticity, Preprint N 33, *Institute for Problems in Mechanics, USSR Academy of Sciences,* Moscow.

Balueva, A.V. and Zazovsky, A.F. (1985) Elastic-Hydrodynamic Problem on Fluid Inflow into a Crack in a Porous Medium. *Mechanics of Solids,* **5**, 157-166.

Balueva, A.V. and Dashevsky, I.N. (1994) Model of Internal Gas-filled Crack Growth in Materialas, Mechanics of Solids. 6, 113-118.

A NEW CREEP LAW AND ITS APPLICATION TO CRACK TIP FIELD ANALYSIS

Y. C. Gao
Northern Jiaotong University
100044, Beijing, China

Abstract

A crack is considered as a very thin failure zone but not an ideal discontinuous surface. Using a new constitutive relation that contains all of the three stages of creep, the deformation field near a Mode I growing crack tip is analyzed. The stress and strain remain finite at the crack tip. Asymptotic equations of the crack tip field are derived and solved numerically. Besides the local autonomy of crack tip field is discussed.

1. Introduction

Crack tip fields are related with fracture criterion. Most analysis of crack tip field is based on the concept of infinite failure strain, therefore the obtained solutions as [1-3] are singular. Indeed, such solutions can reflect the strain character in regions near but not too close to the crack tip. However, on physical grounds, finite strain is expected to remain at the crack tip. If the singular field is truncated when strain is large enough, the growing crack will become a very thin failure zone, at the boundary of which the stress and strain keep finite values. The margin of failure zone may form a sharp angle at its tip as given in [4] for plastic material. Also, the failure margin may form a smooth curve at its tip as given in [5] for creep material. Definitely, the shape of failure margin depends on the constitutive relation.

In the present paper the Mode I growing crack is analyzed using the new creep law given in [5]. The local autonomy of the near tip field is also discussed.

2. The Creep Law

The creep behavior of materials under constant stress can be divided into three stages. The first is called the initial stage, the second the steady creep stage and the third the unstable stage that correspond to failure. In [5], a formula that contains the above three stages was proposed, for uniaxial load it can be written as

R. Wang (ed.), IUTAM Symposium on Rheology of Bodies with Defects, 47–54.

$$\varepsilon = \varepsilon_e + \varepsilon_c \quad , \qquad \dot{\varepsilon}_c = C\varepsilon_c^{-\alpha}(\varepsilon_{cf} - \varepsilon_c)^{-\beta}\sigma^n \tag{1}$$

where ε is the total strain, ε_e the elastic strain, ε_c the creep strain, ε_{cf} the failure creep strain, σ the stress and C, α, β, n the material constants that may depends on the temperature. The superscript dot denotes derivative with respect to time t

For general stress state the creep law takes the form,

$$\dot{\varepsilon}_c = \frac{3}{2}Ce_c^{-\alpha}(e_{cf} - e_c)^{-\beta}s^{n-1}S \tag{2}$$

where ε_c is creep strain tensor, S the deviator of stress tensor, e_c and s are the effective creep strain and effective stress,

$$e_c = \int(\frac{2}{3}\dot{\varepsilon}_c{:}\dot{\varepsilon}_c)^{1/2}dt \quad , \qquad s = (\frac{3}{2}S{:}S)^{1/2} \tag{3}$$

It follows from (2) and (3) that

$$\dot{e}_c = Ce_c^{-\alpha}(e_{cf} - e_c)^{-\beta}s^n \tag{4}$$

Let

$$e = e_c / e_{cf} \tag{5}$$

Eq. (4) becomes

$$\dot{e} = Cs^n e_{cf}^{-(1+\alpha+\beta)}e^{-\alpha}(1-e)^{-\beta} \tag{6}$$

From (6) the failure time can be obtained,

$$t_f = \frac{1}{Cs^n}e_{cf}^{1+\alpha+\beta}B(\alpha+1,\beta+1) \tag{7}$$

where B is the Beta function defined as

$$B(\alpha+1,\beta+1) = \int_0^1 \xi^{-\alpha}(1-\xi)^{-\beta}d\xi \tag{8}$$

The relation between e and t/t_f is shown in Fig.1 for various α and β .

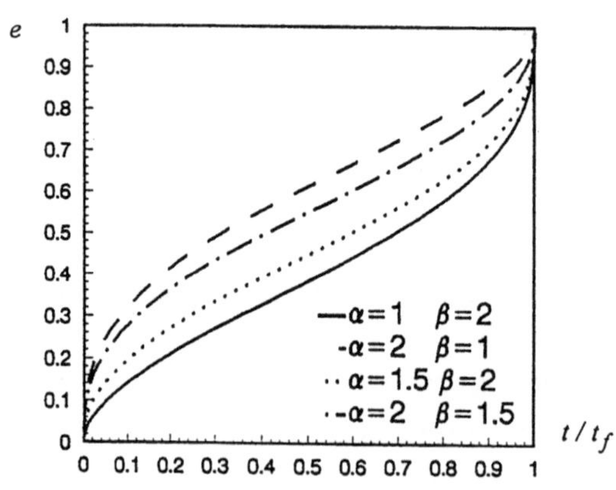

Fig. 1 Relation of creep strain and time

3. Steady Crack Growth

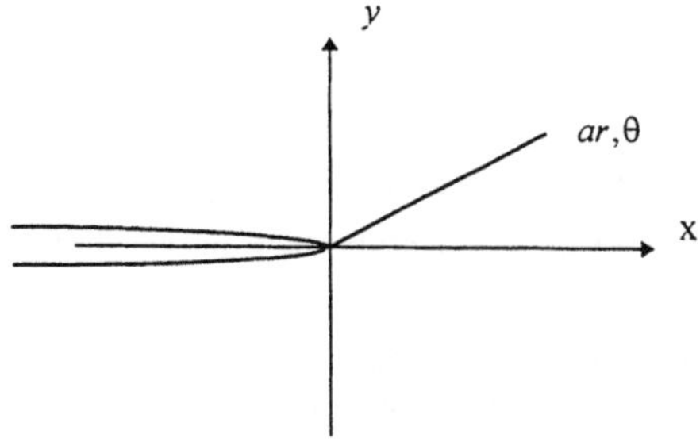

Fig.2 Failure margin and coordinate

Fig.2 shows a Mode I semi-infinite crack growing steadily in a creep material. The coordinates (x, y) are fixed on the moving crack tip.

Let u_x , u_y denote the displacements, for incompressible material a potential function Ω can be introduced such that

$$u_x = -\frac{\partial\Omega}{\partial y} \quad , \quad u_y = \frac{\partial\Omega}{\partial x} \tag{9}$$

the strains are given by

$$\varepsilon_{xx} = -\varepsilon_{yy} = -\frac{\partial^2\Omega}{\partial x\partial y} \quad , \quad \varepsilon_{xy} = \frac{1}{2}\left(\frac{\partial^2\Omega}{\partial x^2} - \frac{\partial^2\Omega}{\partial y^2}\right) \tag{10}$$

Considering the quasi-static problem, the stress function φ can be introduced

$$\sigma_{xx} = \frac{\partial^2\varphi}{\partial y^2} \quad , \quad \sigma_{yy} = \frac{\partial^2\varphi}{\partial x^2} \quad , \quad \sigma_{xy} = -\frac{\partial^2\varphi}{\partial x\partial y} \tag{11}$$

The rate operator for steady crack growth is

$$\frac{d}{dt}(\,) = (\,\dot{}\,) = -V\frac{\partial}{\partial x}(\,) \tag{12}$$

in which V is the velocity of crack tip. Then the rates of strain and stress are

$$\dot{\varepsilon}_{xx} = -\dot{\varepsilon}_{yy} = V\frac{\partial^3\Omega}{\partial x^2\partial y}, \quad \dot{\varepsilon}_{xy} = \frac{V}{2}\left(\frac{\partial^3\Omega}{\partial x\partial y^2} - \frac{\partial^3\Omega}{\partial x^3}\right) \tag{13}$$

$$\dot{\sigma}_{xx} = -V\frac{\partial^3\varphi}{\partial x\partial y^2}, \quad \dot{\sigma}_{yy} = -V\frac{\partial^3\varphi}{\partial x^3}, \quad \dot{\sigma}_{xy} = V\frac{\partial^3\varphi}{\partial x^2\partial y} \tag{14}$$

The constitutive relation can be written as

$$\begin{cases} \dfrac{\partial}{\partial x}(\varepsilon_{xx} - \varepsilon_{yy}) = \dfrac{1}{2G}\dfrac{\partial}{\partial x}(\sigma_{xx} - \sigma_{yy}) - \dfrac{3C}{2V}e_c^{-\alpha}(e_{ef} - e_c)^{-\beta}s^{n-1}(\sigma_{xx} - \sigma_{yy}) \\ \dfrac{\partial}{\partial x}\varepsilon_{xy} = \dfrac{1}{2G}\dfrac{\partial}{\partial x}\sigma_{xy} - \dfrac{3C}{2V}e_c^{-\alpha}(e_{ef} - e_c)^{-\beta}s^{n-1}\sigma_{xy} \end{cases} \tag{15}$$

in which G is elastic shear modules.

4. Asymptotic Equations

The failure boundary at crack tip may form either a smooth curve or an angle as shown in Fig.2, but the analysis can prove that only the smooth curve is reasonable shape, and that the stress state near the crack tip is in uniaxial tension for Mode I crack.

Far from the crack tip, stress and strain are relatively small. When the crack tip is approached, the effective creep strain would increase monotonically up to the value e_{cf} , but the effective stress may only increase up to a distance and then decrease because of creep relaxation.

Let s^* denote the effective stress at $x = y = 0$, considering the material to be incompressible, the total strain at $x = y = 0$ can be obtained according to (15),

$$\varepsilon|_{x=y=0} = (\frac{\sqrt{3}}{2}e_{cf} + \frac{1}{2G}\frac{s^*}{\sqrt{3}})(e_y \otimes e_y - e_x \otimes e_x) \tag{16}$$

where e_x and e_y are the unit vectors along x and y directions respectively, $\otimes$ is the dyadic symbol.

The stress strain field in the vicinity of crack tip can be expressed by an uniaxial tension plus a perturbation. For convenience, the dimensionless polar coordinate is introduced as shown in Fig.2

$$r = \frac{1}{a}(x^2 + y^2)^{1/2}, \qquad \theta = arctg\frac{y}{x} \tag{17}$$

where a is a character scale. According to Eq. (16) the displacement potential can be written as

$$\Omega = a^2 r^2 [\frac{1}{4}(\sqrt{3}e_{cf} + \frac{1}{\sqrt{3}}\frac{s^*}{G})\sin 2\theta + \frac{s^*}{2G}r^\delta g(\theta)] \tag{18}$$

in which δ is a positive exponent. Then

$$\begin{cases} u_r = -ar[\frac{1}{2}(\sqrt{3}e_{cf} + \frac{1}{\sqrt{3}}\frac{s^*}{G})\cos 2\theta + \frac{s^*}{2G}r^\delta g'] \\ u_\theta = ar[\frac{1}{2}(\sqrt{3}e_{cf} + \frac{1}{\sqrt{3}}\frac{s^*}{G})\sin 2\theta + (2+\delta)\frac{s^*}{2G}r^\delta g] \end{cases} \tag{19}$$

$$\begin{cases} \varepsilon_{rr} = -\varepsilon_{\theta\theta} = -\frac{1}{2}(\sqrt{3}e_{cf} + \frac{1}{\sqrt{3}}\frac{s^*}{G})\cos 2\theta - (1+\delta)\frac{s^*}{2G}r^\delta g' \\ \varepsilon_{r\theta} = \frac{1}{2}(\sqrt{3}e_{cf} + \frac{1}{\sqrt{3}}\frac{s^*}{G})\sin 2\theta - \frac{s^*}{4G}r^\delta[g'' - \delta(2+\delta)g] \end{cases} \tag{20}$$

$$\dot{\Omega} = Var[-\frac{1}{2}(\sqrt{3}e_{cf} + \frac{1}{\sqrt{3}}\frac{s^*}{G})\sin 2\theta + \frac{s^*}{2G}r^\delta \zeta] \tag{21}$$

Where

$$\zeta = g'\sin\theta - (2+\delta)g\cos\theta \tag{22}$$

then the velocity is

$$\begin{cases} v_r = V[\frac{1}{2}(\sqrt{3}e_{cf} + \frac{1}{\sqrt{3}}\frac{s^*}{G})\cos\theta - \frac{s^*}{2G}r^\delta\zeta'] \\ v_\theta = -V[\frac{1}{2}(\sqrt{3}e_{cf} + \frac{1}{\sqrt{3}}\frac{s^*}{G})\sin\theta - \frac{1+\delta}{2G}s^* r^\delta\zeta] \end{cases} \tag{23}$$

the strain rates are

$$\begin{cases} \dot{\varepsilon}_{rr} = -\dot{\varepsilon}_{\theta\theta} = -\frac{s^*}{2G}\cdot\frac{V\delta}{a}r^{\delta-1}\zeta' \\ \dot{\varepsilon}_{r\theta} = -\frac{s^*}{2G}\cdot\frac{V}{2a}r^{\delta-1}[\zeta'' + (1-\delta^2)\zeta] \end{cases} \tag{24}$$

Further, considering that the basic stress state at crack tip is uniaxial tension, the stress function is written as

$$\varphi = a^2 s^* r^2 [\frac{1}{\sqrt{3}}\cos^2\theta + r^\delta f(\theta)] \tag{25}$$

$$\begin{cases} \sigma_{rr} = s^*\left\{\frac{2}{\sqrt{3}}\sin^2\theta + r^\delta[f'' + (2+\delta)f]\right\} \\ \sigma_{\theta\theta} = s^*[\frac{2}{\sqrt{3}}\cos^2\theta + r^\delta(1+\delta)(2+\delta)f] \\ \sigma_{r\theta} = s^*[\frac{2}{\sqrt{3}}\cos\theta\sin\theta - r^\delta(1+\delta)f'] \end{cases} \tag{26}$$

$$\dot{\varphi} = aVs^* r^{1+\delta}\eta \tag{27}$$

where

$$\eta = f'\sin\theta - (2+\delta)f\cos\theta \tag{28}$$

the stress rates are

$$\begin{cases} \dot{\sigma}_{rr} - \dot{\sigma}_{\theta\theta} = \frac{s^*}{a}Vr^{\delta-1}[\eta'' + (1-\delta^2)\eta] \\ \dot{\sigma}_{r\theta} = -\frac{s^*}{a}V\delta r^{\delta-1}\eta' \end{cases} \tag{29}$$

The effective creep strain can be written as

$$e_c = e_{cf} - \frac{s^*}{G}r^\delta\psi \tag{30}$$

substituting (30) into (4) and matching the singularity it follows that

$$\delta = \frac{1}{1+\beta} \tag{31}$$

$$\psi'\sin\theta - \delta\psi\cos\theta + K\psi^{-\beta} = 0 \tag{32}$$

in which

$$K = \frac{ac}{V}s^{*n-\beta-1}G^{1+\beta}e_{cf}^{-\alpha} \tag{33}$$

Using (2),(24) and (26)-(30), it follows that

$$\begin{cases}\eta''+(1-\delta^2)\eta+2\delta\zeta'-2\sqrt{3}K\psi^{-\beta}\cos 2\theta=0\\ \zeta''+(1-\delta^2)\zeta-2\delta\eta'+2\sqrt{3}K\psi^{-\beta}\sin 2\theta=0\end{cases} \tag{34}$$

(22),(28),(32) and (34) are the basic equations of the near tip field.

5. Boundary Conditions and Solutions

For Eq. (22), since g is an odd function, so

$$g(0)=0 \tag{35}$$

(22) shows that ζ is also an odd function, further the regularity of (22) at $\theta=0$ requires that

$$\zeta'(0)=-(1+\delta)g'(0) \tag{36}$$

where $g'(0)$ is to be determined.

For Eq. (28), the regularity at $\theta=0$ requires that

$$\eta(0)=-(2+\delta)f(0) \tag{37}$$

where $f(0)$ is also to be determined.

For Eq. (32), the regularity at $\theta=0$ requires that

$$\psi(0)=(\frac{K}{\delta})^{\delta} \tag{38}$$

Besides, at $\theta=\pi/2$, the failure strain should be reached

$$\psi(\pi/2)=0 \tag{39}$$

For Eq. (34), the following conditions should be supplied,

$$\zeta(0)=0 \quad , \quad \eta'(0)=0 \tag{40}$$

At the failure margin $(\theta=\pi/2)$ the traction free conditions must be satisfied,

$$f(\frac{\pi}{2})=f'(\frac{\pi}{2})=0 \tag{41}$$

Eq. (22), (28), (32), and (34) can be solved under boundary conditions (35)-(41). The important fact is that the analytical solution to eq. (32) under conditions (38) and (39) is

$$\psi=(\frac{K}{\delta})^{\delta}(\cos\theta)^{\delta} \tag{42}$$

Then the Eqs. (22), (28) and (34) under conditions (35)-(37), (40) and (41) can be solved numerically. The undetermined values $f(0)$ and $g'(0)$ can be adjusted to fit condition (41). In order to reduce the calculation, the following nature of equations should be noted. If $\hat{f}$, $\hat{g}$, $\hat{\zeta}$, $\hat{\eta}$ are the solutions for $K=1$, then $k^{\delta}\hat{f}$, $k^{\delta}\hat{g}$, $k^{\delta}\hat{\zeta}$, $k^{\delta}\hat{\eta}$ will be the solutions for $K=k$. Hence, it suffices to calculate the solution for K=1. The calculation starts from $\theta=0.01$. The curves of function $f(\theta)$ for $K=1$ are shown in Fig.3 for various β. The adjusted values of $f(0)$ and $g'(0)$ are listed in Table 1.

TABLE 1. The values of $f(0)$ and $g'(0)$

β	10	6	2	0.5
$f(0)$	0.613	0.715	0.688	0.280
$g'(0)$	-0.20	-0.30	-0.51	-0.786

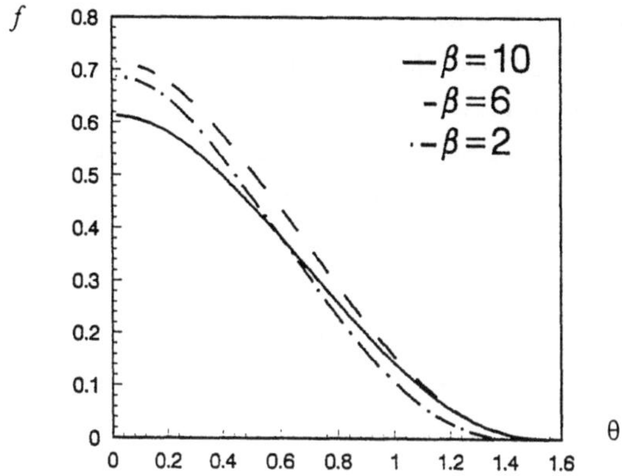

Fig.3 Function of $f(\theta)$

6. Local Autonomy

The singular solution obtained in [1] was questioned in [6] for the character of "local autonomy", i.e. the crack tip field does not have a free parameter. But [6] contains some algebraic mistakes, that was pointed out in [3]. The solutions obtained in [2] and [3] do also possess the "local autonomy" feature. In [5] it is stated that the local autonomy is just the proper feature for a growing crack tip field in rate sensitive materials. Now, we can understand this problem in a simpler way. Actually, for rate sensitive materials, the near tip field does not need a parameter to fit the far field because the far field itself is already fixed by a given V (the velocity of crack tip). In order to explain this we can consider the growing crack in creep material, since V is contained in the constitutive relation, the solution will depend on the specified value of V. Once the value of V is given, the field is fixed, so it does not need to be fitted. On the contrary, for the crack growing in plastic material, the V does not enter the basic equations and boundary conditions, therefore when V is given the solution of far field is not fixed. For this case the near tip solution has to have a free parameter to fit the various far fields.

If the far field material is plastic one but the near tip field material is considered as rate sensitive, the crack tip velocity V cannot be determined by the state of far field, in this case, V is the free parameter that enable the near tip field to fit the far field. In sum, the so called "local autonomy" is not a real problem. It should be pointed out that in the present paper, the near tip solution contains an unknown parameter s^* that will be determined by the given far field, but it is not for fitting various far field.

7. Concluding Remarks and Discussion

· Considering that the failure strain is finite, a growing crack is regarded as a thin failure zone, therefore the singular solution to the crack tip field should be truncated when the failure strain is reached.

· According to the new creep law given in [5], the near tip field of Mode I growing crack is obtained. The results show that the stress and strain state near the crack tip are in uniaxial tension. At the failure margin, the effective failure creep strain is reached but the effective stress s^* is an unknown .

· Near the crack tip, the rates of stress and strain possess the singularity of $r^{\delta-1}$ ($\delta < 1$).

· As s^* is unknown, the near tip field does not contain a parameter to fit the far field because the far field is fixed already.

· The so called "local autonomy" is not a real problem for the near tip field.

· The possibility of $s^* = 0$ still remains open, that will be another case of solution for $\beta + 1 - n > 0$.

Acknowledgment

This work is supported by the National Science Foundation of China.

References

[1] C. Y. Hui and H. Riedel, An analysis of antiplane shear crack growth in a rate sensitive elastic-plastic material, Int. J. Fract. 17, 409-425 (1981).

[2] Y. C. Gao, Further study on strain singularity behavior of moving cracks in elastic-visoplastic materials, J. Theor. Appl. Fract. Mech. 14, 233-242 (1990).

[3] T. J. Delph, The steadily propagating viscoplastic crack with elastic unloading, Int. J. Fract. 68, 183-191 (1994).

[4] Y. C. Gao and G. Rousselier, Near tip quasi-static crack growth behavior in strain hardening and softening material. J. Theor. Appl. Fract. Mech. 20, 149-155 (1994).

[5] Y. C. Gao and Z. Q. Wang, Stress and strain field near tip of Mode III growing crack in materials with creep behavior, J. Theor. Appl. Fract. Mech., 25, 113-126 (1996).

[6] W. Yang and L. B. Freund, The asymptotic stress and strain field near the tip of a growing crack under creep conditions, Int. J. Fract., 30 157-174 (1986).

DAMAGE FIELD EQUATION AND CRITERION FOR DAMAGE LOCALIZATION

Y.L.BAI [1] , M.F.XIA [1,2], F.J.KE [1,3] and H.L.LI [1]

1 Laboratory for Non-linear Mechanics of Continuous Media, Institute of Mechanics, Chinese Academy of Sciences, Beijing 100080, China

2 Department of Physics, Peking University, Beijing 100871, China

3 Department of Applied Mathematics and Physics, Beijing University of Aeronautics and Astronautics, Beijing 100083, China

Abstract

For heterogeneous materials with distributed microcracks or microvoids, damage evolution should be described in terms of a system of damage field and continuum mechanics equations. It was found that the dynamic function of damage $f = f(D, \sigma)$, i. e. the intrinsic damage evolution rate and the macroscopic formulation of the nucleation, growth and coalescence of microdamages, plays a key role in the evolution. The population of microdamages has a tendency to form localized damage, namely a precursor to failure. The increase of the relative gradient of damage signifies the occurrence of damage localization. Under quasistatic small deformation in one dimensional strain state, this leads to the following criteria $f_D > f/D + \dot{\theta}$ and $f_D > f/D$ in Eulerian and Lagrangian co-ordinates respectively, where $\dot{\theta}$ is dilatation rate. Whereas under the same assumptions the criterion for maximum stress is $f = \dot{\theta}$. Clearly, damage localization is a distinct feature of solids. It is relevant to the attainment of maximum stress via the dynamic function of damage f.

1. Introduction

Failure of Solids is a highly rate-dependent and non-linear process(Curran et al., 1985). Sometimes failure is noted as a process of reduction of dimension, like a roughly two-dimensional fracture surface formed in a three-dimensional body of solids. This is also

R. Wang (ed.), IUTAM Symposium on Rheology of Bodies with Defects, 55–66.

known as localized failure, though distributed microdamages developed. Fragmentation may be another kind of failure in solids, where distributed flaws appear to dominate the formation of fragments. In creep, it was also found two types of failure, bulk and localized damage predominance (Chan and Page, 1993). Usually, localized failure occurs suddenly, hence becomes more dangerous. According to the statistical evolution of microdamage, this paper intends to reveal what intrinsic factor in solids governs the localized failure.

We have established a fundamental equation of microdamage in the light of statistical mesoscopic damage mechanics (Bai et al., 1991, 1997). This is the evolution equation of microdamage in phase space

$$\frac{\partial n}{\partial t}+\sum_{i=1}^{I}\frac{\partial\left(n\cdot P_i\right)}{\partial\ p_i}=n_N-n_A\ . \tag{1}$$

where n is the number density of microdamage in phase space and t is generalized time. $P_i = \dot{p}_i$ are the rates of variables p_i. n_N and n_A are nucleation and annihilation rate densities of microdamage respectively. Now we examine a phase space {c, **x**}, i.e. $p_1 = c$ and $p_2 = \mathbf{x}$, where c is current size of microdamage and **x** is spatial coordinates. In this formulation, it is assumed that microdamage can be properly described by its current sizes c.

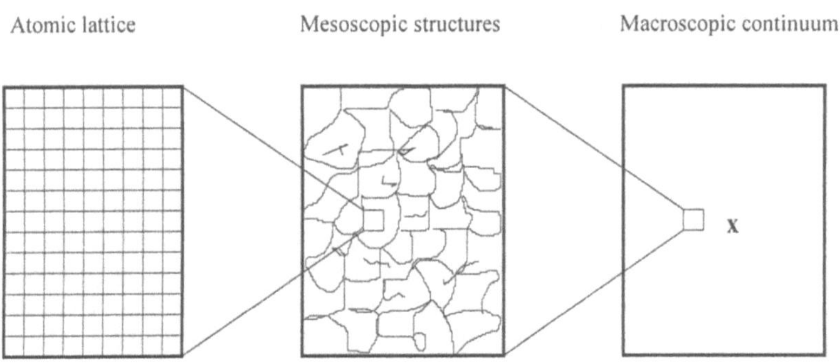

Figure 1 Schematic of length scales

We investigate such an element of solid that is large enough to contain a number of microdamages with different sizes but is small enough to be handled as a point macroscopically, Fig.1 (Pantelides, 1992). Actually, this is a revised version of continuum to include microstructual variables. Suppose that τ denote the average failure volume of a microdamage with current size c. Continuum damage should be defined by

$$D\ =\ \int_0^{\infty} n\left(t\,,c\,,\mathbf{x}\,\right)\tau d\,c\ \ . \tag{2}$$

It describes the fraction of damage in unit current volume. After multiplying one-dimensional form of (1) by τ and integrating it, we obtain the evolution equation of damage field

$$\frac{\partial D}{\partial t}+\nabla\cdot(D\mathbf{v})=f \tag{3}$$

$$f = \int_0^\infty (n_N - n_A)\tau\, dc + \int_0^\infty nA\,\tau' dc \tag{4}$$

where $\tau' = \dfrac{\partial \tau}{\partial c}$ and **v** is particle velocity, A is growth rate of microdamage. f is defined as the dynamic function of damage(DFD), which is uniquely dependent on mesoscopic dynamics of microdamage. In this formulation, macroscopic damage is described by the field variable $D = D(t, \mathbf{x})$, which is deduced from statistical description of microdamage (2) and governed by evolution equation (3). This damage field equation (3) should be associated with other mechanical equations, such as continuum, momentum and energy equations,(Xia et al., 1995).

As comparison, we recall continuum damage mechanics(Kachanov, 1986, Lemaitre and Plumtree, 1979, Chaboche, 1988). The core of damage mechanics is its evolution law

$$\dot{D} = \frac{K}{(1-D)^\gamma}\ , \tag{5}$$

where γ is an undetermined parameter and K is a stress-dependent function $K(\sigma)$. The nominal stress in damaged solid σ is assumed to be related to the stress in the matrix σ_s,

$$\sigma = \sigma_s\,(1 - D). \tag{6}$$

Actually, the continuum damage D in (5) should be a collective response of microdamages, as described by (2).

2. One-dimensional Damage Field Equation

As a case study, we examine one-dimensional strain state. This means that all velocity components and spatial derivatives are zero, except velocity component v and derivative $\dfrac{\partial}{\partial y}$. For simplicity, we ignore the energy equation in this paper. Then, we have a system of mechanical equations in one dimensional strain state,

$$\frac{\partial D}{\partial t} + v\frac{\partial D}{\partial y} + D\frac{\partial v}{\partial y} = f \tag{7}$$

$$\frac{\partial V}{\partial t} + v\frac{\partial V}{\partial y} - V\frac{\partial v}{\partial y} = 0 \tag{8}$$

$$\frac{\partial v}{\partial t} + v\frac{\partial v}{\partial y} - V\frac{\partial \sigma}{\partial y} = 0 \tag{9}$$

where V is specific volume and v is particle velocity. Additionally, this is an Eulerian formulation.

In continuum damage mechanics, damage evolution is usually assumed as an internal variable in the constitutive equation. Instead of damage field equation (7), it assumes a damage evolution law,

$$\dot{D} = F(D,\sigma) \tag{10}$$

Kachanov's damage evolution law (5) is an example. In order to compare the difference between the two formulations (7) and (10), we should examine other forms of damage field equation.

3. Eulerian, Lagrangian and Hybrid Formulations of Damage Field Equation

The transformation from Eulerian coordinates (t, y) to Lagrangian ones (T, Y) are

$$T = t \tag{11}$$

$$Y = y - vt \tag{12}$$

So, the system of equations (7-9) in Lagrangian coordinates is

$$\frac{\partial D}{\partial T} + D\frac{V_o}{V}\frac{\partial v}{\partial Y} = f \quad , \tag{13}$$

$$\frac{\partial V}{\partial T} - V_o\frac{\partial v}{\partial Y} = 0 \quad , \tag{14}$$

$$\frac{\partial v}{\partial T} - V_o\frac{\partial \sigma}{\partial Y} = 0 \quad , \tag{15}$$

where V_0 is the initial specific volume. Actually, the damage evolution rate should be defined as Lagrangian derivative, $\dot{D} = \frac{\partial D}{\partial T}$. Comparing (10) and (13), we can find that the difference is the term of dilatation rate, $\dot{\theta} = \frac{V_o}{V}\frac{\partial v}{\partial Y}$, where θ is natural strain. The dynamic function of damage f is defined in Eulerian coordinates and depends on the mesoscopic mechanisms of microdamage evolution. Whereas, the damage evolution rate F in (10) should include the dilatation rate in it.

One more significant point is that damage D is defined in Eulerian coordinates, either in statistical formulation (2) or in continuum damage mechanics like (10). Therefore, the system of equations (13-15) is a hybrid form. For clarification, we define a Lagrangian damage $\mathsf{D} = D\,V/V_0$ and corresponding dynamic function $\mathsf{f} = f\,V/V_0$ respectively. They are all defined in a unit initial volume. Table 1 summarizes the three forms of damage field equation.

Table 1. Three forms of damage field equation

	Damage	Derivative	Evolution equation	No. of Eqs.
Eulerian (Natural)	D	$\frac{\partial}{\partial t}$	$\frac{\partial D}{\partial t} + v\frac{\partial D}{\partial y} + D\frac{\partial v}{\partial y} = f$	(7)
Lagrangian (Engineering)	D	$\frac{\partial}{\partial T}$	$\frac{\partial \mathsf{D}}{\partial T} = \mathsf{f}$	(16)
Hybrid (Practical)	D	$\frac{\partial}{\partial T}$	$\frac{\partial D}{\partial T} = f - D\dot{\theta}$	(13)

4. Criterion for Damage Localization

One of the major tasks of damage mechanics is to predict the transition from accumulation of damage to catastrophic failure. This is usually attributed to the occurrence of damage localization.

For easy understanding, we firstly examine a pilot problem in Lagrangian formulation. Its inhomogeneity of damage can be defined by $(\frac{\partial D}{\partial Y})/D$. Then we focus on a transition of the rate of the damage inhomogeneity being positive, namely

$$\frac{\partial}{\partial T}[(\frac{\partial D}{\partial Y})/D] \geq 0. \tag{17}$$

This implies an increase of relative inhomogeneity of damage. (17) is equivalent to

$$(\frac{\partial(\frac{\partial D}{\partial Y})}{\partial T})/(\frac{\partial D}{\partial Y}) \geq (\frac{\partial D}{\partial T})/D. \tag{18}$$

Clearly, this means that the relative increase of damage gradient becomes greater than the relative increase of damage itself. After substitution of Eq.(16) into (18) and considering $f = f(\sigma, D)$, one can derive

$$f_D + f_\sigma \cdot \frac{\frac{\partial \sigma}{\partial Y}}{\frac{\partial D}{\partial Y}} \geq \frac{f}{D} \tag{19}$$

where $f_D = \frac{\partial f}{\partial D}$ and $f_\sigma = \frac{\partial f}{\partial \sigma}$. Now, we consider quasi-static deformation. That is to say, the inertial term in Eq.(15) can be ignored. Under this assumption, the momentum equation (15) leads to vanishing stress gradient. In another word, stress becomes uniform. In this case, (19) becomes

$$f_D \geq \frac{f}{D} . \tag{20}$$

This is an intrinsic condition for transition, depending on DFD f only. More importantly, the compound damage f_D (see Davison and Stevens, 1974), i.e. the tangent of dynamic function of damage, is the motivation of the transition.

When turning to dynamic case, we can rewrite (19) as

$$f_D + f_\sigma \gamma \frac{\sigma L_D}{D L_\sigma} \geq \frac{f}{D}, \tag{21}$$

where $L_D = \left|\frac{1}{D}\frac{\partial D}{\partial Y}\right|^{-1}$ and $L_\sigma = \left|\frac{1}{\sigma}\frac{\partial \sigma}{\partial Y}\right|^{-1}$ are the characteristic length scales of spatial variations of damage and stress, respectively. $\gamma = 1$ when high damage region corresponding to high stress region or $\gamma = -1$ when high damage region corresponding to low stress region. From (21), we can conclude that inertia term, namely stress gradient, will promote the transition, if high damage and stress regions coincide, i.e. $\gamma = 1$. Moreover, the effects of the two characteristic length scales are opposite.

Actually, the above transition in Lagrangian coordinates does not exactly mean real damage localization, because damage localization occurs in current configuration. In this configuration, a material element can expand during deformation. So, we should turn to the evolution of damage D in Eulerian coordinates. Since we have known the inertial effect qualitatively, in the following we shall examine quasistatic approximation only.

In order to do this, we look at a second pilot problem in hybrid form. Similar to (18), we examine

$$\left(\frac{\partial(\frac{\partial D}{\partial Y})}{\partial T}\right) / \left(\frac{\partial D}{\partial Y}\right) \geq \left(\frac{\partial D}{\partial T}\right) / D \ . \tag{22}$$

Similar manipulations lead to a criterion as

$$f_D - \left(D \frac{\partial \dot{\theta}}{\partial Y}\right) / \left(\frac{\partial D}{\partial Y}\right) \geq f/D, \tag{23}$$

One more term, damage multiplied by the gradient of dilatation rate appears on the left side of (23). For small strain and damage, the term $(D\frac{\partial \dot{\theta}}{\partial Y})/(\frac{\partial D}{\partial Y})$ becomes smaller than others. This is estimated as follows,

$$D \frac{\partial \dot{\theta}}{\partial Y} \sim D \frac{\partial(\frac{\partial D}{\partial Y})}{\partial T} << \frac{\partial(\frac{\partial D}{\partial Y})}{\partial T} \ . \tag{24}$$

Therefore, an approximate condition is

$$f_D \geq f/D \ . \tag{25}$$

Again, like (20), the compound damage f_D plays a key role in (25), except the dynamic function of damage f rather than f. Fig.2 shows the criterion (25) for damage localization schematically. To our surprise, criterion (25) looks like the well-known Chapman-Jouguet condition for detonation front (Taylor and Tankin, 1958).

Now, it is easy to turn to Eulerian formulation. The condition for damage localization is

$$\left(\frac{\partial(\frac{\partial D}{\partial y})}{\partial t}\right) / \left(\frac{\partial D}{\partial y}\right) \geq \left(\frac{\partial D}{\partial t}\right) / D. \tag{26}$$

Once more, similar but a little troublesome manipulations lead to the following inequality

$$f_D - \dot{\theta} \geq f/D - f_\sigma \frac{\frac{\partial \sigma}{\partial y}}{\frac{\partial D}{\partial y}} - \frac{v}{D}\frac{\partial D}{\partial y} + D\frac{\frac{\partial^2 v}{\partial y^2}}{\frac{\partial D}{\partial y}} + v\frac{\frac{\partial^2 D}{\partial y^2}}{\frac{\partial D}{\partial y}} \ . \tag{27}$$

The second term on the right hand side of (27) vanishes under quasistatic deformation. For small deformation in Eulerian coordinates,

$$v \frac{\partial}{\partial y} Z << \frac{\partial}{\partial t} Z \ . \tag{28}$$

where Z is a displacement-type variable. Then, similar to (24), the last three terms in (27) become negligibly small. So, a criterion for damage localization is obtained as

$$f_D - \dot{\theta} \geq f/D. \tag{29}$$

Very clearly, as long as we subtract the dilatation rate $\dot{\theta}$ from the compound damage f_D, the criterion for damage localization is equivalent to (25).

Now, to summarize this section, we notice that three rates are involved in damage field evolution: compound damage f_D, the intercept of DFD f/D and dilatation rate $\dot{\theta}$. The compound damage f_D is the motivation of damage localization. Since the dynamic function of damage is usually concave(Hayhurst et al., 1984), especially at the late stage of damage evolution owing to coalescence of microdamages, there is a tendency to damage localization. We leave the discussion on the intercept of DFD f/D to Section 6. As for the effect of dilatation rate $\dot{\theta}$ on damage localization in Eulerian formulation, it is obvious that $\dot{\theta} > 0$ retards damage localization due to its geometric expansion. So, in the following sections, we shall examine hybrid formulation and discuss criterion (25) only.

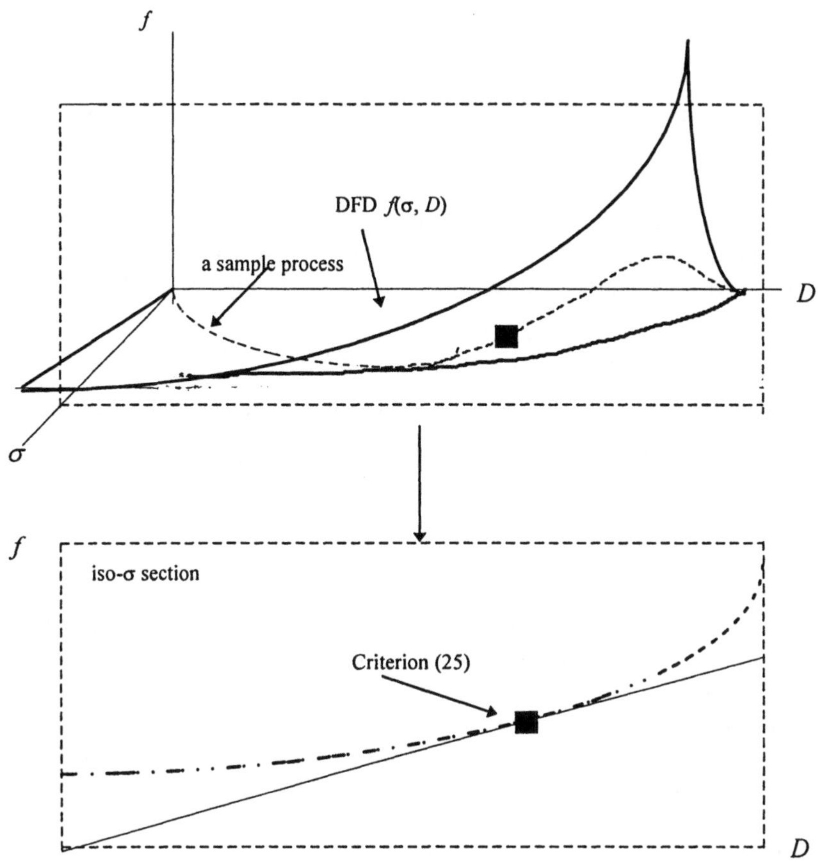

Figure 2 Schematic of the lower bound for damage localization (25)

5. Criterion Based on Mean Variables

One may notice that the discussion made in Section 4 concerns variables in damage and deformation field. Actually, it is quite inconvenient and even impossible to know values of all variables at every moment and every point in the field and then to determine if damage localization occurs. In practice, we may prefer to use average values of variables to do this. Now, we examine when this is appropriate.

Damage D can be decomposed into two parts: the average $\overline{D}$ and fluctuation $\tilde{D}$,

$$D = \overline{D} + \tilde{D} \tag{30}$$

For small fluctuation, Eq. (13) can be divided into two equations, one for the average $\overline{D}(T)$ and the other for the fluctuation $\tilde{D}(T, Y)$,

$$\frac{1}{\overline{D}}\frac{d\overline{D}}{dT} = \frac{f(\overline{\sigma},\overline{D})}{\overline{D}} - \dot{\overline{\theta}}\ , \tag{31}$$

$$\frac{1}{\tilde{D}}\frac{\partial\tilde{D}}{\partial T} = f_{\overline{D}}(\overline{\sigma},\overline{D}) + f_{\overline{\sigma}}(\overline{\sigma},\overline{D})\frac{\tilde{\sigma}}{\tilde{D}} - \dot{\overline{\theta}} - \overline{D}\frac{\dot{\tilde{\theta}}}{\tilde{D}}. \tag{32}$$

Similar to (17) and (18), the increase of relative fluctuation of damage can occur, when

$$\left(\frac{\partial\left(\frac{\tilde{D}}{\overline{D}}\right)}{\partial T}\right)/\left(\frac{\tilde{D}}{\overline{D}}\right) = \frac{1}{\tilde{D}}\frac{\partial\tilde{D}}{\partial T} - \frac{1}{\overline{D}}\frac{d\overline{D}}{dT} \geq 0\ . \tag{33}$$

Substitution of Eqs.(31-32) into the inequality (33) gives a criterion for the further increase of relative fluctuation of damage under the approximation of small damage,

$$f_{\overline{D}}(\overline{\sigma},\overline{D}) - \frac{f(\overline{\sigma},\overline{D})}{\overline{D}} + f_{\overline{\sigma}}(\overline{\sigma},\overline{D})\frac{\tilde{\sigma}}{\tilde{D}} - \overline{D}\frac{\dot{\tilde{\theta}}}{\tilde{D}} \geq 0\ . \tag{34}$$

Also, the third term vanishes for quasistatic deformation and the fourth one, like the term $D\frac{\partial\dot{\theta}}{\partial Y}$ in (23), becomes negligible for small deformation. Then, the criterion (34) becomes

$$f_{\overline{D}}(\overline{\sigma},\overline{D}) \geq \frac{f(\overline{\sigma},\overline{D})}{\overline{D}}. \tag{35}$$

The significance of (35) is that we can use the average values of variables and criterion (35) to predict the occurrence of damage localization. This is much easier than using the field variables in (25). Hence (35) is truly operational in practice. Additionally, the assumption of small fluctuation does not set limitation to the application of (35), because the problem of damage localization itself is whether small fluctuation of damage can evolve to localization or not. In the following, we shall ignore the bars over average values.

6. Other Criteria for Damage Evolution

Actually, in the concerned problem of damage evolution, there are four rate variables. Three are intrinsic: DFD f, compound damage f_D and the intercept of DFD f/D. And one external is strain rate $\dot{\theta}$. Their relative greatness can govern different modes of damage evolution. Following previous procedures, we derived corresponding conditions for these under quasistatic and small deformation. Table 2 summarizes the results.

From Table 2, it is very clear that modes 2 and 3 are equivalent. From modes 1 and 3, it can be seen that when strain rate $\dot{\theta} > 0$, this external variable retards both increases of homogeneous damage and damage gradient and balances the intercept of DFD and compound damage in the two modes, respectively. Noticeably, mode 4, the increase of relative damage gradient, signifies that the increase of damage gradient(mode 3) overtakes that of homogeneous damage(mode 1). Therefore, mode 4 becomes independent of strain rate $\dot{\theta}$ but dependent on two intrinsic rates f_D and f/D only.

Table 2. Conditions for different modes of damage evolution

	Mode of damage evolution	Definition	Criterion	No.
1	Increase of homogeneous damage	$\frac{\partial D}{\partial T} \geq 0$	$f/D \geq \dot{\theta}$	(36)
2	Increase of small disturbances $\tilde{D} \sim \exp(\alpha T)$	$\alpha > 0$	$f_D > \dot{\theta}$	(37)
3	Increase of damage gradient	$\frac{\partial(\frac{\partial D}{\partial Y})}{\partial T} \geq 0$	$f_D \geq \dot{\theta}$	(38)
4	Increase of relative damage gradient	$\frac{\partial(\frac{\partial D}{\partial Y}/D)}{\partial T} \geq 0$	$f_D \geq f/D$	(25) or (35)

Although loading condition decides the moment damage localization occurs, the lower bound for damage localization is essentially governed by intrinsic property of materials. Now, one may wonder, why DFD f does not appear in all these modes and what role does DFD f play. We leave this point to next section.

7. Criterion for Maximum Stress

Maximum stress attained under external loading decides the strength of materials. An obvious question is if there is any connection between maximum stress and damage localization. Under one-dimensional strain state, there is no difference between Cauchy and engineering stresses and no macroscopically geometric softening. Hence, softening comes from inside materials only.

The condition for attainment of maximum stress is

$$d\sigma = 0 . \tag{39}$$

Substitution of (6) into (39) gives

$$d\sigma = (1 - D)\, d\sigma_s - \sigma_s\, dD = 0 \tag{40}$$

or

$$\frac{d\sigma_s}{\sigma_s} = \frac{dD}{(1-D)} \quad . \tag{41}$$

The left hand side term is strain hardening of matrix and the right is damage induced softening. As an example, let us examine linear elastic matrix.. In this case, criterion (41) can be expressed by

$$(1-D)^2\, d\theta = [1-2D+2\,(1-D)\,\theta\,]\, dD\,, \tag{42}$$

where $\theta = exp\theta - 1 = V/V_0 - 1$ is engineering strain. For small deformation, substituting damage evolution equation into (42) and ignoring all small terms (0-th order approximation), one can deduce an approximate criterion for maximum stress,

$$f = \dot{\theta}\,. \tag{43}$$

This approximate condition implies that at this moment load supporting matrix no longer deforms further and all the increase of deformation $\dot{\theta}\,dt$ is due to damage increase $f\,dt$. One may recall, we have wondered what role does DFD f play in modes of damage evolution. Now we find that the relative importance of DFD f and strain rate $\dot{\theta}$ dominates the attainment of maximum stress. This is a transition very different from all aboves on damage evolution.

8. Comparison

We have found that damage localization and attainment of maximum stress are relevant implicitly. To illustrate the interrelation, we examine a sample DFD

$$f = (A + B\,D^m)\,\sigma \tag{44}$$

In this case, the critical condition for damage localization (29) leads to a D^m order equation

$$D^m - \frac{\dot{\theta}}{(m-1)B\sigma}D - \frac{A}{(m-1)B} = 0\;. \tag{45}$$

The lower bound for damage localization (25) or (35) becomes

$$D \geq \left(\frac{A}{B(m-1)}\right)^{\frac{1}{m}}. \tag{46}$$

Whereas the criterion for maximum stress (43) becomes

$$D = \left(\frac{\dot{\theta}}{B\sigma} - \frac{A}{B}\right)^{\frac{1}{m}}. \tag{47}$$

In the diagram of D vs. σ (Fig. 3), one can see that the failure occurring at low strength is

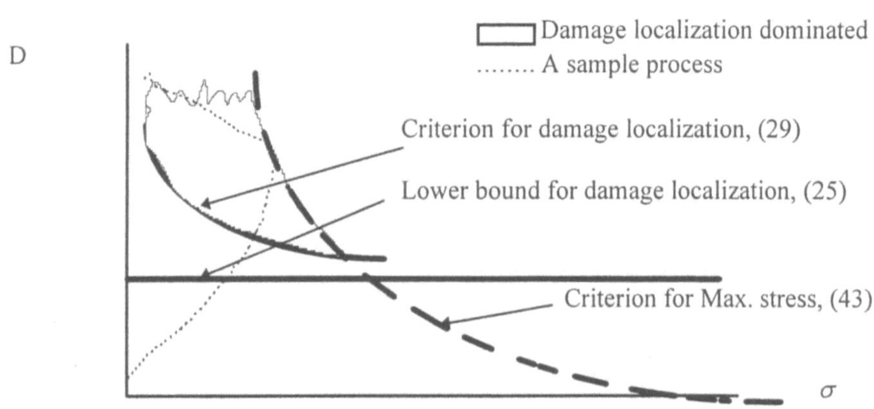

Figure 3 Comparison of damage localization and attainment of Maximum stress

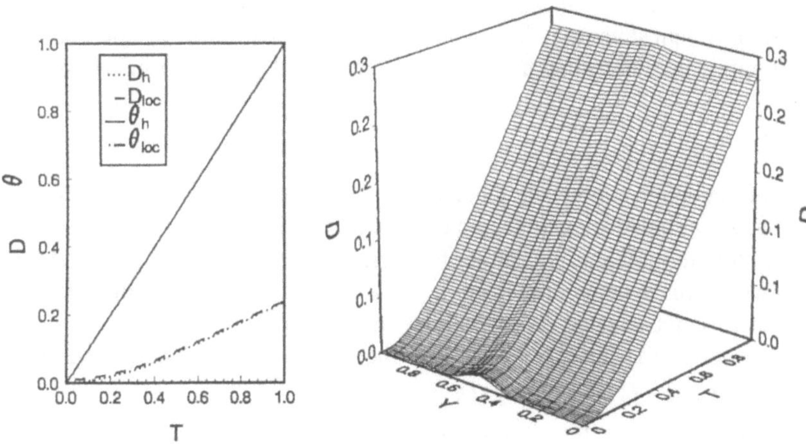

Figure 4 Homogenous damage evolution

(One-dimensional strain state, quasistatic; constant dragging velocity v_0; DFD $f=(A+BD^m)\sigma$; linear elastic matrix; dimensionless parameters (by $\dot{\theta}_m = v_0/L$) A=1, B=0, m=2,

initial damage distribution $D_0(Y) = 0.01\ \exp\left[-100\cdot\left(\frac{Y}{L}-\frac{1}{2}\right)^2\right]$)

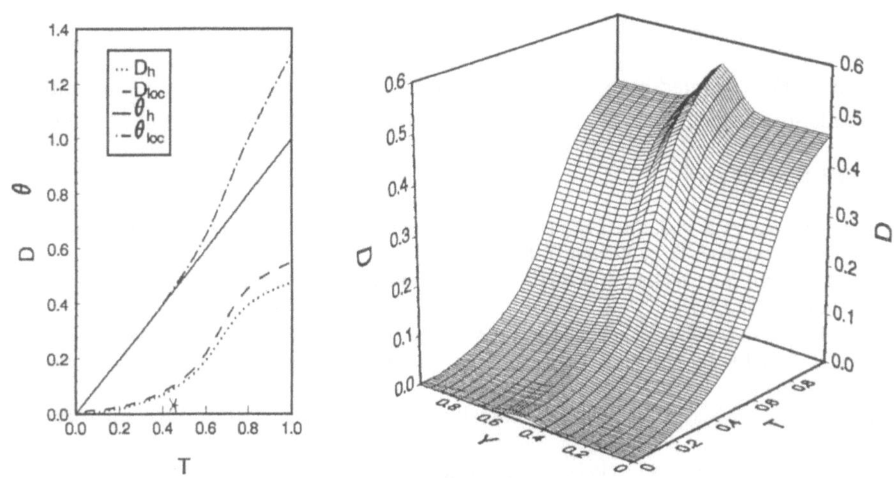

Figure 5 Damage localization

(One-dimensional strain state, quasistatic; constant dragging velocity v_0; DFD $f=(A+BD^m)\sigma$; linear elastic matrix; dimensionless parameters (by $\dot{\theta}_m = v_0/L$) A=1, B=100, m=2,

initial damage distribution $D_0(Y) = 0.01\ \exp\left[-100\cdot\left(\frac{Y}{L}-\frac{1}{2}\right)^2\right]$

× + are the times predicted by criteria (25 and 29), respectively.)

mainly due to damage localization, hence most dangerous. Depending on loading conditions, maximum stress may be relevant to damage localization or not. Figs. 4 and 5 provide examples demonstrating homogenous damage and damage localization.

9. Summary

1. A formulation concerning inhomogeneous damage field evolution is constructed based on statistical mesoscopic damage mechanics.
2. In the formulation, dynamic function of damage (DFD) $f(\sigma, D)$, linking mesoscopic dynamics (nucleation, growth and coalescence) of microdamage and macroscopic damage evolution, plays a key role.
3. A criterion for damage localization under quasistatic and small deformation is obtained as $f_D - \dot{\theta} \geq f/D$.
4. Positive dilatation rate $\dot{\theta}$ spreads damage geometrically and then retards damage localization.
5. Lower bound for damage localization is $f_D \geq f/D$. This is uniquely dependent on DFD and then an intrisic property of materials.
6. As comparison, an approximate criterion for attainment of maximum stress under the same assumption is also derived as $f = \dot{\theta}$.
7. Damage localization usually causes low stress failure and appears to be responsible for disaster.

Acknowledgment

This work is granted by the National Natural Science Foundation of China and Chinese Academy of Sciences. Discussions with Prof. N.G. Liang are greatly appreciated.

REFERENCES

Bai, Y.L., Ke, F.J. and Xia, M.F. (1991) Formulation of statistical evolution of microcracks in solids, Acta Mechanica Sinica Volume **7**, 59-66.

Bai, Y.L., Han, W.S. and Bai, J. (1997) A statistical evolution equation of microdamage and its application, ASTM STP **1315**

Chaboche, J. L. (1988) Continuum damage mechanics, Trans. J. Appl. Mech. **55**, 59-64, 65-72.

Chan, K.S. and Page, R.A. (1993), Creep damage development in structural ceramics, J. Am.Ceramic Soc., **76**, 803-826.

Curran, D.R., Seaman, L. and Shockey, D.A. (1987) Dynamic failure of solids, Physics Reports **147**, 253-388.

Davison, L. and Stevens, A.L (1972) Continuum measures of spall damage, J. Appl. Phys. **43**, 988-994.

Hayhurst, D.R., Dimmer, P.R. and Morrison, C.J. (1984) Development of continuum damage in the creep rupture of notched bars, Phil. Trans. R. Soc. London **A311**, 103-129.

Kachanov, L.M. (1986) Introduction to continuum damage mechanics, Martinus Nijhoff, The Netherlands

Lemaitre, J. and Plumtree, A. (1979) Application of damage concepts to predict creep-fatigue failures, ASME Trans. J. Engng. Matl. & Tech. **101**, 284-292.

Pantelides, S.T. (1992) What is materials physics, any way? , Physics Today, Sept., 67-69.

Taylor, G.I. and Tankin, R.S. (1958) Gas dynamics of detonation, in Fundamentals of gas dynamics, (Section G), ed., H.W. Emmons, Princeton University Press,

Xia, M.F., Han, W.S., Ke, F.J. and Bai, Y.L (1995) Statistical meso-scopic damage mechanics and damage evolution induced catastrophe, Advances in Mechanics **25**, 1- 40.

ENERGY ESTIMATES FOR PIECEWISE SMOOTH RATE TYPE THERMO-VISCOELASTIC MODELS WITH VAN DER WAALS TYPE EQUILIBRIUM SURFACE

M. MIHĂILESCU-SULICIU AND I. SULICIU
Institute of Mathematics of Romanian Academy,
P.O.BOX 1-764, RO-70700 Bucharest, Romania
E-Mail: isuliciu@stoilow.imar.ro

Abstract. This paper discusses a rate–type thermo–viscoelastic model with a piecewise linear van der Waals type equilibrium surface. The thermodynamic restrictions on the constitutive equations are obtained in the regions where they are smooth as well as the jump conditions they must satisfy across the curves of discontinuities. An energy identity/inequality is derived which allows us to obtain energy estimates of the solutions when phase transitions take place or defects may generate and propagate. A numerical scheme and its stability in energy is briefly discussed.

1. Introduction

The purpose of this work is to model formation and evolution of the defects in bodies exhibiting softening properties on a strain interval. Here we extend the results discussed in Suliciu (1995) for isothermal processes to the dynamic thermomechanical ones. The prototype we consider is a rate–type thermo-viscoelastic model with a piecewise linear equilibrium surface of van der Waals type.

We assume that such an equilibrium surface can be experimentally determined as a piecewise linear stress–strain-temperature relation Van Humbeeck and Delaey (1982), Otsuka and Shimizu (1982) and also Fu *et al.* (1993). We also suppose that the rate of work and the rate of heat are known differential forms. We consider that the specific heat at constant strain and the latent heat at constant temperature as defined for instance in Truesdell and Bharatha (1977) can be determined by calorimetric measurements at any state, at least in principle.

R. Wang (ed.), IUTAM Symposium on Rheology of Bodies with Defects, 67–86.

Based on the above assumptions we formulate the two laws of thermodynamics in a very classical spirit (see Suliciu (1997a) and the references given there in). This formulation allows us to determine in an explicit form the internal energy and the entropy as functions of state variables even if the usual theorems on smooth differential forms can not be directly applied to get restrictions on the constitutive functions. Thus the fact that the normal to the equilibrium stress–strain–temperature relation may have jump discontinuities across certain lines in the strain–temperature plane implies that the heat capacity and the latent heat must have jump discontinuities too as a requirement of the two laws of thermodynamics. However the internal energy and the entropy result continuous and piecewise smooth functions of the state variables.

Next we construct the free energy for any process and the relative free energy for isothermal processes defined for all equilibrium states but with respect to a fixed equilibrium state. It was shown in Suliciu (1992b) that this relative free energy function is positive for all isothermal states whenever the strain of that fixed state lies outside the Maxwell's interval at the temperature of that given fixed state.

We establish an energy identity/inequality valid for all continuous and piecewise smooth solutions of the governing system of the partial differential equations. This identity/inequality is a consequence of the two laws of thermodynamics and it is expressed in terms of the availability at a given reference state. We prove that the availability is a positive valued function whenever the relative free energy is positive.

This property allows us to obtain energy estimates and an approach to equilibrium for this piecewise smooth case as it was done in Suliciu (1997a) for the smooth constitutive equations. The equilibrium stress–strain–temperature relation may be taken as an elastic constitutive equation to obtain a complete system governing the one dimensional thermomechanical motion of our body. As it has negative values of the derivative with respect to strain on some regions some initial boundary value problems may be ill posed in the sense of Hadamard thus our thermo–viscoelastic problem may be thought also as a viscoelastic regularization of the elastic ill posed problem. The energy identity/inequality allows us to justify such a regularization.

One can also extend the results obtained for rate–type thermo–viscoelasticity to rate–type thermo–viscoplasticity where the piecewise smoothness is a hypothesis very often made. As the yield limit moves here with the temperature, one can get an equilibrium surface of the type considered in Suliciu (1995) if for instance one takes a constant temperature large enough such that $\sigma_E(\varepsilon_2(\theta), \theta) < 0$. Finally we show how the energy inequality can be used here as in Mihăilescu-Suliciu and Suliciu (1995) to obtain stability conditions on the numerical schemes for computation of the solutions of these combined hyperbolic–parabolic initial–boundary value problems and study thus the generation and propagation of defects in softening materials.

2. The Constitutive Equations and the Two Laws of Thermodynamics

The thermodynamical structure we consider here is similar to the one presented in Serrin (1986) and Šilhavý (1986) and discussed in details in Suliciu (1997a). As the aim of these studies is to determine the thermodynamic potentials, as internal energy and entropy, such that the two laws of thermodynamics be satisfied, the smoothness assumption on the constitutive functions allows us to apply well known mathematical results. However many models from plasticity and pseudo-elasticity do not satisfy such smoothness requirements at least concerning the form of the equilibrium response. Our aim here is to obtain energy estimates applicable to the non smooth constitutive equations describing the thermomechanical behavior of shape memory alloys.

2.1. STRESS–STRAIN–TEMPERATURE RELATION, RATE OF WORK AND RATE OF HEAT

We adopt the following rate type thermo-viscoelastic stress-strain-temperature relation

$$\dot{\sigma} = E\dot{\varepsilon} - F\dot{\theta} + G(\varepsilon,\theta,\sigma),\ G(\varepsilon,\theta,\sigma) = -k\left(\sigma - \sigma_E(\varepsilon,\theta)\right),\ \sigma_E(\varepsilon,\theta) = E_o[\varepsilon - \alpha(\theta-\theta_r)] - (E_o + M)\begin{cases} 0 & \text{if } \varepsilon \le \varepsilon_1(\theta), \\ \varepsilon - \varepsilon_1(\theta) & \text{if } \varepsilon_1(\theta) < \varepsilon < \varepsilon_2(\theta), \\ \varepsilon_2(\theta) - \varepsilon_1(\theta) & \text{if } \varepsilon_2(\theta) \le \varepsilon. \end{cases} \tag{1}$$

Here

$$\begin{gathered} E > 0,\quad E_o = \text{const} > 0,\quad F > 0,\quad M = \text{const} > 0,\quad \alpha = \text{const} > 0 \\ \varepsilon_i(\theta) = \alpha_o\theta + \beta_i \quad \alpha_o = \text{const} > 0,\quad \beta_i = \text{const},\quad i = 1,2 \quad \beta_1 < \beta_2, \end{gathered} \tag{2}$$

and these material parameters have the following meaning: E and E_o are the dynamic and quasistatic Young's moduli respectively; F and αE_o are the dynamic and quasistatic linear expansion coefficients while α_o and β_i, $i = 1,2$ are the austenitic$\leftrightarrow$martensitic transformation coefficients. Finally, θ is the absolute temperature and θ_r is a reference absolute temperature. Of course σ is the stress and ε is the strain. We assume that E and F are functions of the state variable $s = (\varepsilon,\theta,\sigma)$. Their smoothness properties will be discussed in the next sections. As one observes $G(\varepsilon,\theta,\sigma)$ is not smooth since $\sigma_E(\varepsilon,\theta)$ is not smooth.

We take the form $(1)_1$ of the rate–type constitutive equation for simplicity reasons. The linear dependence on θ of $\varepsilon_i(\theta)$ $i = 1,2$ is experimentally established for many shape memory alloys Otsuka and Shimizu (1982), Van Humbeeck and Delaey (1982) and also Fu *et al.* (1993) and the above simplified form of the equilibrium stress–strain–temperature relation $\sigma = \sigma_E(\varepsilon, \theta)$ agrees with the experimentally determined one (see for instance Fu *et al.* (1993)). The case when the straight lines $\varepsilon = \varepsilon_i(\theta)$, $i = 1,2$ are not parallel will be treated elsewhere. For these materials the contribution of $\alpha E_o(\theta - \theta_r)$ in $\sigma_E(\varepsilon, \theta)$ is often neglected. The form (1) of the constitutive equation extends to the non isothermal processes the constitutive equation proposed in Suliciu (1989) and discussed in some details in (1992a), (1992b), Făciu and Suliciu (1994) and also in Făciu (1996).

We follow Suliciu (1997a) (see also the references given there in) and assume that the work and heat rates are defined by

$$\dot{W} = -\sigma\dot{\varepsilon}, \quad \dot{Q} = C\dot{\theta} + \nu\dot{\varepsilon}. \tag{3}$$

Here the heat capacity at constant strain C and the latent heat ν depend on the state variable $\boldsymbol{s}= (\varepsilon, \theta, \sigma)$.

2.2. THE TWO LAWS OF THERMODYNAMICS

We take now the form of the two laws of thermodynamics as it was done in Suliciu (1997a) (see also Serrin (1986) and Šilhavý (1986)):
The first law: There is a function $U : \mathcal{D} \mapsto \mathbb{R}$ called the internal energy such that the difference between the rate of heat and the rate of work during any process is exactly the differential of the internal energy function $U(\varepsilon, \theta, \sigma)$, i.e.,

$$\dot{Q} - \dot{W} = C\dot{\theta} + (\sigma + \nu)\dot{\varepsilon} = \varrho\dot{U}(\varepsilon, \theta, \sigma). \tag{4}$$

The second law: There is a function $\eta : \mathcal{D} \mapsto \mathbb{R}$ called the entropy such that the differential form of heat divided by the absolute temperature θ can not exceed the differential of the entropy function $\eta(\varepsilon, \theta, \sigma)$, i.e.,

$$\dot{Q} = C\dot{\theta} + \nu\dot{\varepsilon} \leq \varrho\theta\dot{\eta}(\varepsilon, \theta, \sigma). \tag{5}$$

during any process.

In the domains where the constitutive functions E, F, ν, C are smooth and G is continuous one can prove that: *the first law of thermodynamics is satisfied if and only if there is a function $U(\varepsilon, \theta, \sigma)$ verifying*

$$\frac{\partial U}{\partial \varepsilon} + E\frac{\partial U}{\partial \sigma} = \frac{\sigma + \nu}{\varrho}, \quad \frac{\partial U}{\partial \theta} - F\frac{\partial U}{\partial \sigma} = \frac{C}{\varrho}, \quad \frac{\partial U}{\partial \sigma}G = 0 \tag{6}$$

and *the second law of thermodynamics is satisfied if and only if there is a function $\eta(\varepsilon, \theta, \sigma)$ verifying*

$$\frac{\partial \eta}{\partial \varepsilon} + E\frac{\partial \eta}{\partial \sigma} = \frac{\nu}{\varrho\theta}, \quad \frac{\partial \eta}{\partial \theta} - F\frac{\partial \eta}{\partial \sigma} = \frac{C}{\varrho\theta}, \quad 0 \leq \frac{\partial \eta}{\partial \sigma}G\,. \tag{7}$$

As the function $G(\varepsilon, \theta, \sigma)$ is assumed to vanish only on $\sigma = \sigma_E(\varepsilon, \theta)$ the first law of thermodynamics (6) implies that

$$\frac{\partial U}{\partial \sigma} = 0, \quad \frac{\partial C}{\partial \sigma} = 0, \quad \frac{\partial \nu}{\partial \sigma} = -1. \tag{8}$$

These properties will be used in what follows.

3. On the Instantaneous Elastic Response in Stress, in Internal Energy and in Entropy and on the Equilibrium Response

3.1. ON THE INSTANTANEOUS ELASTIC RESPONSE

A more detailed discussion on the response of a rate–type material in fast processes is given in Suliciu (1997a), see also (1997b) where the case when the instantaneous response in stress may be path dependent is discussed too. We consider an arbitrary but fixed state $\boldsymbol{s}_o = (\varepsilon_o, \theta_o, \sigma_o)$ in the constitutive domain $\mathcal{D}$ and assume that in a neighborhood of that state the constitutive functions E, F, ν, C are smooth functions. We say that we have instantaneous elastic response in stress and respectively in internal energy and in entropy if there are the functions

$$\sigma = \sigma_I(\varepsilon, \theta; s_o), \quad U = U_I(\varepsilon, \theta; s_o, U_o), \quad \eta = \eta_I(\varepsilon, \theta; s_o, \eta_o), \tag{9}$$

such that

$$\begin{aligned}
&\sigma_I(\varepsilon_o, \theta_o; s_o) = \sigma_o, \quad U_I(\varepsilon_o, \theta_o; s_o, U_o) = U_o, \quad \eta_I(\varepsilon_o, \theta_o; s_o, \eta_o) = \eta_o, \\
&\dot{\sigma}_I = E(\varepsilon, \theta, \sigma_I)\dot{\varepsilon} - F(\varepsilon, \theta, \sigma_I)\dot{\theta} \\
&\varrho \dot{U}_I = C(\varepsilon, \theta, \sigma_I)\dot{\theta} + (\nu(\varepsilon, \theta, \sigma_I) + \sigma_I)\dot{\varepsilon}, \\
&\varrho \dot{\eta}_I = \frac{C(\varepsilon, \theta, \sigma_I)}{\theta}\dot{\theta} + \frac{\nu(\varepsilon, \theta, \sigma_I)}{\theta}\dot{\varepsilon}.
\end{aligned} \tag{10}$$

One way to imagine an instantaneous process is as follows. First we observe that the constitutive equation $(1)_1$ written in an incremental way $d\sigma = Ed\varepsilon - Fd\theta + Gdt$ determines the increment $d\sigma$ of the stress when the increments $d\varepsilon$ of strain and $d\theta$ of temperature corresponding to increment dt of time are given. One can consider that the increment of strain and of temperature take place in a vanishing increment of time (i.e. when $dt \mapsto 0$). Physically, such a process may take place exactly across a shock wave for instance or approximately whenever Gdt is negligible with respect to $Ed\varepsilon$ for instance (for a more detailed discussion on this matter see Suliciu (1997a) and (1997b)).

As a consequence of (9) and (10) we have

$$\begin{aligned} \frac{\partial \sigma_I}{\partial \varepsilon} =& E(\varepsilon,\theta,\sigma_I), \quad \frac{\partial \sigma_I}{\partial \theta} = -F(\varepsilon,\theta,\sigma_I). \\ \varrho\frac{\partial U_I}{\partial \varepsilon} =& \nu(\varepsilon,\theta,\sigma_I)+\sigma_I, \quad \varrho\frac{\partial U_I}{\partial \theta} = C(\varepsilon,\theta,\sigma_I), \\ \varrho\theta\frac{\partial \eta_I}{\partial \varepsilon} =& \nu(\varepsilon,\theta,\sigma_I), \quad \varrho\theta\frac{\partial \eta_I}{\partial \theta} = C(\varepsilon,\theta,\sigma_I). \end{aligned} \tag{11}$$

and thus the functions E, F, C and ν must be related by (see Suliciu (1997a))

$$\frac{\partial E}{\partial \theta} - F\frac{\partial E}{\partial \sigma} = -\frac{\partial F}{\partial \varepsilon} - E\frac{\partial F}{\partial \sigma}\,, \tag{12}$$

and

$$\nu = \theta F, \quad \theta\left(\frac{\partial F}{\partial \theta} - F\frac{\partial F}{\partial \sigma}\right) = \frac{\partial C}{\partial \varepsilon} + E\frac{\partial C}{\partial \sigma}. \tag{13}$$

In other words as a consequence of the existence of the instantaneous elastic response the two dynamic moduli E and F must be related by (12), the latent heat ν is related to the dynamic linear thermal expansion coefficient F by the so called Clapeyron formula $(13)_1$ and the material functions E, F, C must also satisfy $(13)_2$.

3.2. ON THE EQUILIBRIUM RESPONSE

The states

$$(\varepsilon,\theta,\sigma=\sigma_E(\varepsilon,\theta)) \quad \text{such that} \quad G(\varepsilon,\theta,\sigma_E(\varepsilon,\theta))=0, \tag{14}$$

are called equilibrium states since if one holds ε and θ constant in time and takes the initial stress $\sigma=\sigma_E(\varepsilon,\theta)$ then the stress remains also constant in time.

Now, we define

$$\dot{W}_E = -\sigma_E\dot{\varepsilon}, \quad \dot{Q}_E = C_E\dot{\theta} + \nu_E\dot{\varepsilon}, \tag{15}$$

where

$$C_E(\varepsilon,\theta) = C(\varepsilon,\theta,\sigma_E(\varepsilon,\theta)), \quad \nu_E(\varepsilon,\theta) = \nu(\varepsilon,\theta,\sigma_E(\varepsilon,\theta)). \tag{16}$$

We ask ourselves which are the restrictions on σ_E, C_E, ν_E such that there are the functions

$$U = U_E(\varepsilon,\theta), \qquad \eta = \eta_E(\varepsilon,\theta) \tag{17}$$

verifying (the two laws of thermodynamic at equilibrium states)

$$\begin{aligned}\varrho\dot{U}_E(\varepsilon,\theta) &= \dot{Q}_E - \dot{W}_E = C_E\dot{\theta} + (\nu_E + \sigma_E)\dot{\varepsilon},\\ \varrho\theta\dot{\eta}_E(\varepsilon,\theta) &= \dot{Q}_E = C_E\dot{\theta} + \nu_E\dot{\varepsilon}.\end{aligned} \tag{18}$$

When U_E and η_E exist and they are of C^2 class, we have

$$\frac{\partial U_E}{\partial\varepsilon} = \frac{\nu_E+\sigma_E}{\varrho}, \quad \frac{\partial U_E}{\partial\theta} = \frac{C_E}{\varrho}, \quad \frac{\partial\eta_E}{\partial\varepsilon} = \frac{\nu_E}{\varrho\theta}, \quad \frac{\partial\eta_E}{\partial\theta} = \frac{C_E}{\varrho\theta}. \tag{19}$$

and

$$\frac{\partial\nu_E}{\partial\theta} + \frac{\partial\sigma_E}{\partial\theta} = \frac{\partial C_E}{\partial\varepsilon}, \quad -\frac{\nu_E}{\theta^2} + \frac{1}{\theta}\frac{\partial\nu_E}{\partial\theta} = \frac{1}{\theta}\frac{\partial C_E}{\partial\varepsilon}, \tag{20}$$

respectively. As a consequence of (20) we have

$$\nu_E = -\theta\frac{\partial\sigma_E}{\partial\theta}. \tag{21}$$

which may be called the equilibrium Clapeyron's formula. Thus the relations (20) are equivalent to (21) and

$$\frac{\partial C_E}{\partial\varepsilon} = -\theta\frac{\partial^2\sigma_E}{\partial\theta^2}. \tag{22}$$

at the points where $\sigma_E(\varepsilon,\theta)$ is of C^2 class with respect to θ which in our case is true for any (ε,θ) with $\varepsilon \neq \varepsilon_i(\theta),\ i = 1,2$.

From (21), (22) and $(1)_3$ we have the global form of the functions $\nu_E(\varepsilon,\theta)$ and $C_E(\varepsilon,\theta)$

$$\nu_E(\varepsilon,\theta) = \alpha E_o\theta - \alpha_o(E_o + M)\theta\begin{cases}0 & \text{if } \varepsilon < \varepsilon_1(\theta),\\ 1 & \text{if } \varepsilon_1(\theta) < \varepsilon < \varepsilon_2(\theta),\\ 0 & \text{if } \varepsilon_2(\theta) < \varepsilon.\end{cases} \tag{23}$$

and

$$C_E(\varepsilon,\theta) = \begin{cases}C_1(\theta) & \text{if } \varepsilon < \varepsilon_1(\theta),\\ C_2(\theta) & \text{if } \varepsilon_1(\theta) < \varepsilon < \varepsilon_2(\theta),\\ C_3(\theta) & \text{if } \varepsilon_2(\theta) < \varepsilon.\end{cases} \tag{24}$$

Here $C_E(\varepsilon,\theta)$ may depend on θ but is piecewise constant with respect to ε.

The two laws of thermodynamics impose further restrictions on the form of the function $C_E(\varepsilon,\theta)$. Indeed as the differential forms $C_E d\theta + (\nu_E + \sigma_E)d\varepsilon$ and $(C_E/\theta)d\theta + (\nu_E/\theta)d\varepsilon$ in (18) are exact their integration along any closed curve must vanish. These facts together with the continuity of $\sigma_E(\varepsilon,\theta)$ lead to

$$[\![C_E]\!] + \alpha_o[\![\nu_E]\!] = 0 \quad \text{across } \varepsilon = \varepsilon_i(\theta), \quad i = 1,2, \tag{25}$$

where $[\![f]\!]$ denotes the jump of f across $\varepsilon = \varepsilon_i(\theta)$. Now, by using (23)–(24) in (25) we arrive at the desired restrictions

$$C_2(\theta) = C_1(\theta) + \alpha_o^2(E_o + M)\theta, \quad C_3(\theta) = C_1(\theta). \tag{26}$$

Thus, once $C_1(\theta)$ is known (say!) then $C_2(\theta)$ and $C_3(\theta)$ are determined by (26) as a consequence of the laws of thermodynamics and if $C_1(\theta) > 0$ so are $C_2(\theta)$ and $C_3(\theta)$. If one assumes that $C_E = C_o$ =const> 0 in the region 1 ($\varepsilon < \varepsilon_1(\theta)$) then C_E is constant in the region 3 ($\varepsilon > \varepsilon_2(\theta)$) with the same value of the constant but C_E has a linear variation with temperature in the region 2 ($\varepsilon_1(\theta) < \varepsilon < \varepsilon_2(\theta)$) where the phase transition takes place. Therefore

$$C_E(\varepsilon, \theta) = C_o + \alpha_o^2(E_o + M)\theta \begin{cases} 0 & \text{if } \varepsilon < \varepsilon_1(\theta), \\ 1 & \text{if } \varepsilon_1(\theta) < \varepsilon < \varepsilon_2(\theta), \\ 0 & \text{if } \varepsilon_2(\theta) < \varepsilon. \end{cases} \tag{27}$$

We use (23), $(1)_3$ and (27) in (18) to find the continuous and piecewise smooth expressions of the equilibrium potentials $U_E(\varepsilon, \theta)$, $\eta_E(\varepsilon, \theta)$, i.e., to write

$$\begin{aligned} \varrho U_E(\varepsilon, \theta) &= \varrho U_E(\varepsilon_o, \theta_o) + \frac{E_o\,(\varepsilon^2 - \varepsilon_o^2)}{2} + E_o\alpha\theta_r(\varepsilon - \varepsilon_o) + C_o(\theta - \theta_o) \\ &- (E_o + M) \begin{cases} 0, & \text{if } \varepsilon \leq \varepsilon_1(\theta), \\ [(\varepsilon - \beta_1)^2 - \alpha_o^2\theta^2]/2, & \text{if } \varepsilon_1(\theta) < \varepsilon < \varepsilon_2(\theta), \\ (\beta_2 - \beta_1)\,[\varepsilon - (\beta_2 + \beta_1)/2] & \text{if } \varepsilon_2(\theta) \leq \varepsilon; \end{cases} \\ \varrho\eta_E(\varepsilon, \theta) &= \varrho\eta_E(\varepsilon_o, \theta_o) + \alpha E_o(\varepsilon - \varepsilon_o) + C_o \ln(\theta/\theta_o) - \\ &(E_o + M)\alpha_o \begin{cases} 0, & \text{if } \varepsilon \leq \varepsilon_1(\theta), \\ \varepsilon - \varepsilon_1(\theta), & \text{if } \varepsilon_1(\theta) < \varepsilon < \varepsilon_2(\theta), \\ \beta_2 - \beta_1, & \text{if } \varepsilon_2(\theta) \leq \varepsilon. \end{cases} \end{aligned} \tag{28}$$

where the initial state $(\varepsilon_o, \theta_o)$ was chosen in the region 1, i.e., such that $\varepsilon_o < \varepsilon_1(\theta_o)$. For other choices of the initial state the equilibrium potentials are constructed in a similar way. We observe that across $\varepsilon = \varepsilon_i(\theta)$, $i = 1, 2$ both potentials $U_E(\varepsilon, \theta)$, $\eta_E(\varepsilon, \theta)$ are continuous while their partial derivatives have jump discontinuities.

4. The Potentials of Thermodynamics at Arbitrary States

First of all we collect the restrictions obtained so far on the constitutive functions $(C, \nu, F, E)(\varepsilon, \theta, \sigma)$ and on the form of the internal energy U. Due to (8), (12)–(13) and (21)–(22) we can write for all states

$$U(\varepsilon,\theta,\sigma)=U_E(\varepsilon,\theta),\quad C(\varepsilon,\theta,\sigma)=C_E(\varepsilon,\theta),$$
$$\nu(\varepsilon,\theta,\sigma)=-\theta\frac{\partial\sigma_E(\varepsilon,\theta)}{\partial\theta}+\sigma_E(\varepsilon,\theta)-\sigma=\theta F(\varepsilon,\theta,\sigma), \tag{29}$$
$$\frac{\partial E}{\partial\theta}+\left(\frac{\partial\sigma_E}{\partial\theta}+\frac{\sigma-\sigma_E}{\theta}\right)\frac{\partial E}{\partial\sigma}-\frac{E}{\theta}=-\frac{1}{\theta}\frac{\partial\sigma_E}{\partial\varepsilon}+\frac{\partial^2\sigma_E}{\partial\varepsilon\partial\theta}.$$

Thus the internal energy at any state $(\varepsilon,\theta,\sigma)$ is the same as the internal energy at equilibrium states $(\varepsilon,\theta,\sigma_E(\varepsilon,\theta))$ and it is given by $(28)_1$. The constitutive functions ν and F are determined by $(29)_{2,3}$ and they have jump discontinuities across the straight lines $\varepsilon=\varepsilon_i(\theta)$, $i=1,2$. All these conclusions are the consequence of this strong form of the first law (for a weak form of the first law see Serrin (1986)) and the existence of the instantaneous elastic response.

For our purpose here we choose as in Suliciu (1997a) a simple solution of the equation $(29)_4$ which will be positive even when $\partial\sigma_E(\varepsilon,\theta)/\partial\varepsilon$ is negative if the absolute temperature stays away from zero, i.e., we choose

$$E(\varepsilon,\theta,\sigma)=\frac{\partial\sigma_E(\varepsilon,\theta)}{\partial\varepsilon}+\gamma\theta>0\ \text{ for }\ \theta>\theta_m=\frac{M}{\gamma},\quad \theta_m\ll\theta_r \tag{30}$$

if γ is properly chosen (see next section for a justification on the way γ is selected). This function E has also jump discontinuities across the lines $\varepsilon=\varepsilon_i(\theta)$, $i=1,2$, but we have

$$[\![F]\!]+\alpha_o[\![E]\!]=0,\quad \text{across }\ \varepsilon=\varepsilon_i(\theta),\ i=1,2, \tag{31}$$

automatically satisfied without additional restrictions. Therefore $(\sigma_I,U_I,\eta_I)(\varepsilon,\theta,\sigma)$ can be determined as continuous solutions of (11).

In order to construct the entropy function (since we already have $U(\varepsilon,\theta,\sigma)=U_E(\varepsilon,\theta)$) we follow now a procedure similar to the one in Suliciu (1997a). We need first the instantaneous elastic response functions in stress with respect to the equilibrium state $s_1=(\varepsilon_1,\theta_1,\sigma_E(\varepsilon_1,\theta_1))$; by taking into account the expression (30) for E and $(29)_4$ for F from equation $(10)_4$ we easily find that

$$\sigma_I(\varepsilon,\theta;s_1)=\sigma_E(\varepsilon,\theta)+\gamma\theta(\varepsilon-\varepsilon_1) \tag{32}$$

is the instantaneous elastic response in stress verifying $(11)_{1,2}$. The instantaneous elastic response in entropy with respect to s_1 and $\eta_1=\eta_E(\varepsilon_1,\theta_1)$ is determined from

$$\varrho\frac{\partial\eta_I}{\partial\varepsilon}=\frac{\nu(\varepsilon,\theta,\sigma_I)}{\theta}=-\frac{\partial\sigma_E(\varepsilon,\theta)}{\partial\theta}-\gamma(\varepsilon-\varepsilon_1),\quad \varrho\frac{\partial\eta_I}{\partial\theta}=\frac{C_E}{\theta} \tag{33}$$

and it has the form

$$\varrho\eta_I(\varepsilon,\theta;\boldsymbol{s}_1,\eta_1) = \varrho\eta_E(\varepsilon,\theta) - \frac{\gamma}{2}(\varepsilon-\varepsilon_1)^2\,. \tag{34}$$

The entropy in our case is obtained from (34) eliminating ε_1 with the help of (32), i.e, it is given by

$$\varrho\eta(\varepsilon,\theta,\sigma) = \varrho\eta_E(\varepsilon,\theta) - \frac{[\sigma-\sigma_E(\varepsilon,\theta)]^2}{2\gamma\theta^2}\,. \tag{35}$$

We shall need in what follows the free energy potential defined as

$$\psi = U - \theta\eta, \tag{36}$$

which by (7) verifies the relations

$$\frac{\partial\psi}{\partial\varepsilon} + E\frac{\partial\psi}{\partial\sigma} = \frac{\sigma}{\varrho},\quad \frac{\partial\psi}{\partial\theta} - F\frac{\partial\psi}{\partial\sigma} = -\eta,\quad \frac{\partial\psi}{\partial\sigma}G \le 0. \tag{37}$$

Now by using first (35) and relations (28) subsequently, we find

$$\begin{gathered}\varrho\psi(\varepsilon,\theta,\sigma) = \varrho\psi_E(\varepsilon,\theta) + \frac{[\sigma-\sigma_E(\varepsilon,\theta)]^2}{2\gamma\theta} \ge \varrho\psi_E(\varepsilon,\theta) = \\ \varrho\psi_E(\varepsilon_o,\theta_o) + (\theta_o-\theta)\varrho\eta_E(\varepsilon_o,\theta_o) + \frac{E_o(\varepsilon^2-\varepsilon_o^2)}{2} - \\ \alpha E_o(\varepsilon-\varepsilon_o)(\theta-\theta_r) + C_o\theta\left[\frac{\theta-\theta_o}{\theta} - \ln\left(\frac{\theta}{\theta_o}\right)\right] - \\ (E_o+M)\begin{cases}0 & \text{if } \varepsilon\le\varepsilon_1(\theta),\\ [\varepsilon-\varepsilon_1(\theta)]^2/2 & \text{if } \varepsilon_1(\theta)<\varepsilon<\varepsilon_2(\theta),\\ (\beta_2-\beta_1)\{\varepsilon-[\varepsilon_1(\theta)+\varepsilon_2(\theta)]/2\} & \text{if } \varepsilon_2(\theta)\le\varepsilon.\end{cases}\end{gathered} \tag{38}$$

Another potential called the equilibrium relative free energy with respect to a fixed state $\boldsymbol{s}_o = (\varepsilon_o,\theta_o,\sigma_o=\sigma_E(\varepsilon_o,\theta_o))$ was introduced in Suliciu (1989) and discussed in some details in Suliciu (1992b) and will also be needed here. It is defined as

$$\begin{gathered}\varrho\psi_R(\varepsilon;\varepsilon_o,\theta_o) = \varrho\psi_E(\varepsilon,\theta_o) - \varrho\psi_E(\varepsilon_o,\theta_o) - \sigma_o(\varepsilon-\varepsilon_o) = \frac{E_o(\varepsilon-\varepsilon_o)^2}{2} - \\ (E_o+M)\begin{cases}0 & \varepsilon\le\varepsilon_1(\theta_o),\\ [\varepsilon-\varepsilon_1(\theta_o)]^2 & \varepsilon_1(\theta_o)<\varepsilon<\varepsilon_2(\theta_o),\\ (\beta_2-\beta_1)\{\varepsilon-[\varepsilon_1(\theta_o)+\varepsilon_2(\theta_o)]/2\} & \varepsilon_2(\theta_o)\le\varepsilon.\end{cases}\end{gathered} \tag{39}$$

where the fact that $\varepsilon_o < \varepsilon_1(\theta_o)$ was taken into account. One can observe that the relative free energy $\psi_R(\varepsilon;\varepsilon_o,\theta_o)$ is a smooth function of ε.

5. Energy Identity/Inequality, Energy Estimates

Under the constitutive assumptions stated in this paper the governing system of PDE's for our rate–type viscoelastic material is

$$\begin{gathered}\varrho\frac{\partial v}{\partial t}-\frac{\partial\sigma}{\partial X}=\varrho f,\ \ \frac{\partial\varepsilon}{\partial t}-\frac{\partial v}{\partial X}=0,\\ C_E\frac{\partial\theta}{\partial t}+\theta F\frac{\partial\varepsilon}{\partial t}+\frac{\partial H}{\partial X}=\varrho r,\ \ H=-K\frac{\partial\theta}{\partial X},\\ \frac{\partial\sigma}{\partial t}-E\frac{\partial\varepsilon}{\partial t}+F\frac{\partial\theta}{\partial t}=G(\varepsilon,\theta,\sigma),\end{gathered}\tag{40}$$

where K =const.> 0 and C_E, E, F are given by (29)–(30); X is the initial spatial coordinate of the bar, between 0 and its length L, t is the time, f is the body force and r is the heat suply.

Now, we choose a fixed reference equilibrium state $\boldsymbol{s}_o=(\varepsilon_o,\theta_o,\sigma_o=\sigma_E(\varepsilon_o,\theta_o))$ and define the availability relative to that equilibrium state $\boldsymbol{s}_o=(\varepsilon_o,\theta_o,\sigma_o=\sigma_E(\varepsilon_o,\theta_o))$ by

$$\begin{aligned}\varphi(\varepsilon,\theta,\sigma;\boldsymbol{s}_o)=U_E(\varepsilon,\theta)-U_E(\varepsilon_o,\theta_o)-\theta_o[\eta(\varepsilon,\theta,\sigma)-\eta(\varepsilon_o,\theta_o,\sigma_o)]-\\ \frac{\sigma_o(\varepsilon-\varepsilon_o)}{\varrho}.\end{aligned}\tag{41}$$

One can prove as in Suliciu (1997a) that all the continuous solutions of the system (40) verify the following energy identity/inequality.

$$\begin{aligned}\frac{\partial}{\partial t}\left(\varrho\frac{v^2}{2}+\varrho\varphi\right)-\frac{\partial}{\partial X}\left((\sigma-\sigma_o)v-\frac{\theta-\theta_o}{\theta}H\right)-\varrho f v-\\ \varrho\frac{\theta-\theta_o}{\theta}r=\frac{\theta_o}{\theta^2}H\frac{\partial\theta}{\partial X}+\varrho\frac{\theta_o}{\theta}\frac{\partial\psi}{\partial\sigma}G\le 0.\end{aligned}\tag{42}$$

The key point in exploiting the identity/inequality (42) is that the availability $\varphi(\varepsilon,\theta,\sigma;\boldsymbol{s}_o)$ is positive for all states $(\varepsilon,\theta,\sigma)$ if the reference state $\boldsymbol{s}_o$ is chosen to verify certain restrictions which will be discussed below. First we observe from (35) that the availability $\varphi(\varepsilon,\theta,\sigma;\boldsymbol{s}_o)$ exceeds the equilibrium availability, i.e., we have

$$\begin{aligned}\varrho\varphi(\varepsilon,\theta,\sigma;\boldsymbol{s}_o)\ge\varrho\varphi(\varepsilon,\theta,\sigma_E(\varepsilon,\theta);\boldsymbol{s}_o)=\varrho\varphi_E(\varepsilon,\theta;\boldsymbol{s}_o)=\\ \varrho[U_E(\varepsilon,\theta)-U_E(\varepsilon_o,\theta_o)]-\varrho\theta_o[\eta_E(\varrho,\theta)-\eta_E(\varrho_o,\theta_o)]-\sigma_o(\varepsilon-\varepsilon_o)\end{aligned}\tag{43}$$

Now, as

$$\varrho\frac{\partial\varphi_E(\varepsilon,\theta;\boldsymbol{s}_o)}{\partial\theta}=\frac{\theta-\theta_o}{\theta}C_E\ \text{ with }\ C_E(\varepsilon,\theta)\ge 0\ \text{ for any }\ (\varepsilon,\theta)$$

there results

$$\varrho\varphi_E(\varepsilon,\theta;\boldsymbol{s}_o) \geq \varrho\varphi_E(\varepsilon,\theta_o;\boldsymbol{s}_o) = \varrho\psi_R(\varepsilon;\varepsilon_o\theta_o).$$

The relative free energy $\psi_R(\varepsilon;\varepsilon,\theta_o)$ defined in (39) was studied in the isothermal case and, according to Suliciu (1992b), it is positive for any ε if and only if ε_o is outside the Maxwell's interval at $\theta = \theta_o$

We introduce the Maxwell's line $\sigma = \sigma_\mu(\theta)$, $\sigma_E(\varepsilon_2(\theta),\theta) \leq \sigma_\mu(\theta) \leq \sigma_E(\varepsilon_1(\theta),\theta)$ which determines together with $\sigma = \sigma_E(\varepsilon,\theta)$ two triangles of equal areas. Then the Maxwell's interval $[\varepsilon_a(\theta),\varepsilon_b(\theta)]$ is determined such that

$$\sigma_E(\varepsilon_a(\theta),\theta) = \sigma_E(\varepsilon_\mu(\theta),\theta) = \sigma_E(\varepsilon_b(\theta),\theta) = \sigma_\mu(\theta), \quad \varepsilon_a(\theta) \leq \varepsilon_\mu(\theta) \leq \varepsilon_b(\theta)$$

and we have

$$\varepsilon_a(\theta) = \varepsilon_1(\theta) - \frac{M}{2E_o}\left(\varepsilon_2(\theta) - \varepsilon_1(\theta)\right), \quad \varepsilon_b(\theta) = \varepsilon_2(\theta) + \frac{M}{2E_o}\left(\varepsilon_2(\theta) - \varepsilon_1(\theta)\right),$$
$$\varepsilon_\mu(\theta) = \frac{1}{2}\left(\varepsilon_2(\theta) + \varepsilon_1(\theta)\right), \quad \sigma_\mu(\theta) = E_o\left[\varepsilon_1(\theta) - \alpha\theta\right] - \frac{M}{2}\left[\varepsilon_2(\theta) - \varepsilon_1(\theta)\right].$$

Therefore the availability $\varphi(\varepsilon,\theta,\sigma;\boldsymbol{s}_o)$ is positive for all states $(\varepsilon,\theta,\sigma)$ if and only if the reference state $\boldsymbol{s}_o$ is chosen such that ε_o is outside the Maxwell's interval at $\theta = \theta_o$. Indeed the same condition is necessary and sufficient for $\psi_R(\varepsilon;\varepsilon_o,\theta_o) = \varphi(\varepsilon,\theta_o,\sigma_E(\varepsilon_o,\theta_o);\boldsymbol{s}_o)$ to be positive and this explains in fact the choice we made for (ε_o,θ_o) when constructing the equilibrium potentials $U_E(\varepsilon,\theta)$ and $\eta_E(\varepsilon,\theta)$ in § 3. In particular, when $\varepsilon_o < \varepsilon_1(\theta_o)$ the expression of $\varphi(\varepsilon,\theta;\boldsymbol{s}_o)$ follows from (43) and (28)

$$\varrho\varphi_E(\varepsilon,\theta;\boldsymbol{s}_o) = C_o\theta_o\left[\frac{\theta-\theta_o}{\theta_o} - \ln\left(1+\frac{\theta-\theta_o}{\theta_o}\right)\right] + \frac{E_o(\varepsilon-\varepsilon_o)^2}{2} -$$
$$(E_o+M)\begin{cases} 0 & \varepsilon \leq \varepsilon_1(\theta), \\ \{[\varepsilon-\varepsilon_1(\theta_o)]^2 - \alpha_o^2(\theta-\theta_o)^2\}/2, & \varepsilon_1(\theta) < \varepsilon < \varepsilon_2(\theta), \\ (\beta_2-\beta_1)\{\varepsilon - [\varepsilon_1(\theta_o)+\varepsilon_1(\theta_o)]/2\}, & \varepsilon_2(\theta) \leq \varepsilon, \end{cases} \tag{44}$$

and it is positive if $\varepsilon_o \leq \varepsilon_a(\theta_o) \leq \varepsilon_1(\theta_o)$.

Before going farther we remark that the temperature dependent part of the availability $\varphi_E(v,\theta;v_o,\theta_o)$, i.e. its first term in (44) has the following property:

$$f(\theta) = C_o\left[\frac{\theta-\theta_o}{\theta_o} - \ln\left(1+\frac{\theta-\theta_o}{\theta_o}\right)\right] - \frac{\gamma}{2}\left(\frac{\theta-\theta_o}{\theta}\right)^2 \geq 0 \text{ for } \theta \geq \theta_m > 0 \tag{45}$$

where γ is chosen such that

$$0 < \frac{\gamma}{C_o} < 1 \quad \text{and} \quad 0 < \theta_m < \sqrt{\frac{\gamma}{C_o}}\,\theta_o\,. \tag{46}$$

Indeed, as

$$f'(\theta) = \frac{C_o}{\theta^3}(\theta - \theta_o)\left(\theta^2 - \frac{\gamma}{C_o}\theta_o^2\right) \tag{47}$$

it follows there exists one and only one root $\theta = \theta_m \in (0, (\gamma/C_o)^{1/2}\theta_o)$ of $f(\theta)$ such that (45) holds (of course $f(\theta)$ vanishes also at $\theta = \theta_o$ but it is non-negative for all $\theta \geq \theta_m$).

Once we have the energy identity/inequality (42) and the positiveness of the availability (41) the energy estimates and the approach to equilibrium results of the type discussed in Suliciu (1997a) see also Mihăilescu-Suliciu and Suliciu (1992), Suliciu (1992b), Mihăilescu-Suliciu and Suliciu (1995), can be obtained.

We illustrate that here on an example. For that we add to the system (40) of partial differential equations describing the motion of our thermomechanical body the following initial conditions

$$(v, \varepsilon, \theta, \sigma)(X, 0) = (v_i, \varepsilon_i, \theta_i, \sigma_i)(X), \quad \text{for} \quad X \in (0, L) \tag{48}$$

for a body of length L. We choose the boundary conditions such that

$$\left(v\sigma - \frac{\theta - \theta_r}{\theta} H\right)(0, t) = \left(v\sigma - \frac{\theta - \theta_r}{\theta} H\right)(L, t) = 0 \quad \text{for all} \quad t \geq 0, \tag{49}$$

i.e., such that there is no exchange of energy with the surroundings of the body through its ends $X = 0$ and $X = L$. Such conditions are of course realized if $v = 0$ or $\sigma = 0$ and $\theta = \theta_o$ or $H = 0$ at $X = 0$ and $X = L$. These conditions are not necessary in fact for obtaining the estimates presented below. One can take non vanishing boundary data and make changes of dependent variables such that the new introduced variables verify conditions of the form (48), as it was done for instance by Suliciu (1984) for the isothermal case.

The total energy $e(t)$ and its density $e^*(X, t)$ are defined by

$$e(t) = \int_0^L \varrho e^*(x, t)\, dx \quad \text{and} \quad e^* = e^*(\varepsilon, \theta, \sigma) = \left(\frac{v^2}{2} + \varphi(\varepsilon, \theta, \sigma)\right). \tag{50}$$

Now, we can state and prove the following result:
Whenever the availability $\varphi(\varepsilon, \theta, \sigma)$ *is positive* (which is so if the reference state $(\varepsilon_o, \theta_o, \sigma_E(\varepsilon_o, \theta_o))$ is properly chosen) *and* $\theta \geq \theta_m$ *where* θ_m *is defined above, the continuous solution of the initial–boundary value problem (40)+(48)+(49) is bounded in energy by the energy of the input data,* i.e., *the following energetic bound holds*

$$\sqrt{e(t)} \leq \sqrt{e(0)} + b(t) \tag{51}$$

where

$$b(t)=\frac{1}{2}\int_{0}^{t}\left[\int_{0}^{L}\left(2\varrho f^{2}(X,s)+\varrho\gamma r^{2}(X,s)\right)dX\right]^{1/2}ds. \tag{52}$$

Proof of (51). By applying Schwarz inequality and taking into account the positiveness of $\varphi(\varepsilon,\theta,\sigma)$, the inequality (45) and (52), we can write

$$\int_{0}^{L}\left(\varrho f v+\varrho\left(\frac{\theta-\theta_{r}}{\theta}\right)r\right)dX\leq\left[\int_{0}^{L}\left(\varrho\frac{v^{2}}{2}+\frac{\varrho}{\gamma}\left(\frac{\theta-\theta_{r}}{\theta}\right)^{2}\right)dX\right]^{1/2}\times$$
$$\left[\int_{0}^{L}\left(2\varrho f^{2}+\varrho\gamma r^{2}\right)dX\right]^{1/2}\leq 2\sqrt{e(t)}\,\dot{b}(t).$$

By integration of (44) with respect to X on the interval $(0,L)$ and taking into account (49) and the above inequality we arrive at

$$\dot{e}(t)\leq 2\sqrt{e(t)}\,\dot{b}(t). \tag{53}$$

Now, integrating the above inequality on a time interval $(0,t)$ we obtain the desired result.

Based on the same energy identity/inequality (42) and by taking into account $(37)_3$ one can obtain energetic bounds on dissipative effects or equivalently on the deviation from the equilibrium (see Suliciu (1984) for the isothermal case), i.e. we can write

$$\int_{0}^{t}\int_{0}^{L}\left[\frac{\theta_{r}K}{\theta^{2}}\left(\frac{\partial\theta}{\partial X}\right)^{2}-\frac{\varrho\theta_{r}}{\theta}\frac{\partial\psi}{\partial\sigma}G\right]dX\,dt\leq\left[e(0)+b(t)\right]^{2}. \tag{54}$$

Among other things this inequality shows the close connection between rate dependent and rate independent models. We do not discuss this problem in detail here (see the above quoted papers for discussions).

Before closing this section we observe that we can take $\theta_m\mapsto 0$ by taking $\gamma\mapsto 0$ but then the inequality (45) can not be applied to get the estimate (53). It is one reason why we must consider processes for which the absolute temperature stays bounded from below.

6. A numerical integration scheme and energy control of its stability

For the sake of simplicity we discuss here a prototype model which is linear. It is close to our piecewise smooth model discussed in the previous sections whenever the strain ε lies outside the phase transition interval $(\varepsilon_1(\theta),\varepsilon_2(\theta))$.

We do that in order to get energetic stability restrictions on the time and space integration steps. The restrictions thus obtained must also apply in the case when ε moves through the phase transition interval.

The main assumptions we lay down are that the temperature does not deviate too much from the reference temperature θ_r, the stress does not deviate too much from the equilibrium stress and disregard the phase transition. More precisely we assume that.

$$|\theta - \theta_r| \ll \theta_r, \quad |\sigma - \sigma_E(\varepsilon, \theta)| \ll E_o \alpha \theta_r \quad \text{and} \quad \beta_1 = \beta_2. \tag{55}$$

We also introduce the notations

$$E = E_o + \gamma\theta_r, \quad R = \sigma - E\varepsilon + \alpha E_o(\theta - \theta_r), \quad \lambda = \sqrt{\frac{E}{\varrho}}. \tag{56}$$

Based on the above assumptions and notations $(1)_3$ and $(29)_2$ can be written as

$$\sigma_E(\varepsilon, \theta) = E_o[\varepsilon - \alpha(\theta - \theta_r)], \quad F = \alpha E_o. \tag{57}$$

Now, the system (40), when the body force f and the heat supply r are neglected, takes the following linear form (see for a comparison Suliciu (1997a))

$$\begin{aligned} &\varrho\frac{\partial v}{\partial t} - E\frac{\partial \varepsilon}{\partial X} + \alpha E_o \frac{\partial \theta}{\partial X} - \frac{\partial R}{\partial X} = 0, \quad \frac{\partial \varepsilon}{\partial t} - \frac{\partial v}{\partial X} = 0, \\ &C_o\frac{\partial \theta}{\partial t} + \alpha\theta_r E_o \frac{\partial \varepsilon}{\partial t} - K\frac{\partial^2 \theta}{\partial X^2} = 0, \quad \frac{\partial R}{\partial t} = G(\varepsilon, \theta, R). \end{aligned} \tag{58}$$

In order to put the system (58) in a form closer to the system of linear thermoelasticity as discussed in Mihăilescu-Suliciu and Suliciu (1995) we make the following change of dependent variables

$$v + \lambda\varepsilon = 2P, \quad v - \lambda\varepsilon = 2Q, \quad \theta - \theta_r = \varrho\sqrt{\frac{2\varrho\theta_r}{C_o}}\,T \tag{59}$$

and notations

$$a = \alpha E_o\sqrt{\frac{\theta_r}{2\varrho C_o}} \quad b = \frac{K}{C_o}. \tag{60}$$

The system (58) takes the following form

$$\begin{aligned} &\frac{\partial P}{\partial t} - \lambda\frac{\partial P}{\partial X} + a\frac{\partial T}{\partial X} - \frac{1}{2\varrho}\frac{\partial R}{\partial X} = 0, \quad (dX = -\lambda dt), \\ &\frac{\partial Q}{\partial t} + \lambda\frac{\partial Q}{\partial X} + a\frac{\partial T}{\partial X} - \frac{1}{2\varrho}\frac{\partial R}{\partial X} = 0, \quad (dX = \lambda dt), \\ &\frac{\partial R}{\partial t} = G(P, Q, R, T) = -k\left[\frac{\gamma\theta_r}{\lambda}(P - Q) + R\right], \quad (dX = 0), \\ &\frac{\partial T}{\partial t} + \frac{a}{\lambda}\left(\frac{\partial P}{\partial t} - \frac{\partial Q}{\partial t}\right) - b\frac{\partial^2 T}{\partial X^2} = 0, \end{aligned} \tag{61}$$

or the equivalent form

$$\begin{gathered}\frac{\partial}{\partial t}\left(P+\frac{1}{2\varrho\lambda}R\right)-\lambda\frac{\partial}{\partial x}\left(P+\frac{1}{2\varrho\lambda}R\right)+a\frac{\partial T}{\partial X}=\frac{G}{2\varrho\lambda},\\ \frac{\partial}{\partial t}\left(-Q+\frac{1}{2\varrho\lambda}R\right)-\lambda\frac{\partial}{\partial x}\left(-Q+\frac{1}{2\varrho\lambda}R\right)-a\frac{\partial T}{\partial X}=\frac{G}{2\varrho\lambda},\\ \frac{1}{2\varrho\lambda}\frac{\partial R}{\partial t}=\frac{1}{2\varrho\lambda}G,\\ \frac{\partial T}{\partial t}+\frac{a}{\lambda}\frac{\partial}{\partial x}(P-Q)-b\frac{\partial^2 T}{\partial X^2}=0,\quad\text{or}\\ \frac{\partial T}{\partial t}+a\frac{\partial}{\partial x}\left[\left(P+\frac{1}{2\varrho\lambda}R\right)-\left(-Q+\frac{1}{2\varrho\lambda}R\right)\right]-b\frac{\partial^2 T}{\partial x^2}=0\end{gathered}\tag{62}$$

which is more appropriate for the numerical scheme we intend to discuss.

The first three equations in (62) have a form closed to the characteristic form of Mihăilescu-Suliciu and Suliciu (1992) (see also (1985)) for the case when $\alpha = 0$. In fact with the change $p = \sigma + \varrho\lambda v$, $q = \sigma - \varrho\lambda v$ and taking into account that $P = (p - R)/(2\varrho\lambda)$, $Q = -(q - R)/(2\varrho\lambda)$ the above system takes exactly the form discussed in the above quoted papers. When the viscous effects are neglected (i.e. when $G = 0$) then we have exactly the system of linear thermoelasticity as discussed in Mihăilescu-Suliciu and Suliciu (1995). The form (61) of the governing system of equations allow us to take into account in the numerical scheme in an easier way the boundary conditions.

The numerical scheme we present here extends the scheme of Mihăilescu-Suliciu and Suliciu (1995) to the linear thermo-viscoelasticity modeled by (61). With τ the time integration step and $h = L/N$ the space integration step, where L is the length of the bar and N the number of divisions along the bar and the notations

$$f^i = f(ih,(j+1)\tau),\quad f_i = f(ih, j\tau)\ ,i=0,1,\cdots,N,\ j=0,1,\cdots,\qquad \mu=\frac{b\tau}{h^2},\quad \kappa=\frac{a\tau}{h},\quad \nu=\frac{\lambda\tau}{h}\tag{63}$$

this scheme at an inner point reads

$$P^i = (1-\nu)P_i + \nu\left(P_{i+1} + \frac{1}{2\lambda\varrho}(R_{i+1}-R_i) - \frac{a}{\lambda}(T_{i+1}-T_i)\right) + \frac{\tau}{2\varrho\lambda}(G_{i+1}-G_i),\tag{64)a$$

$$Q^i = (1-\nu)Q_i + \nu\left(Q_{i-1} + \frac{1}{2\lambda\varrho}(R_i-R_{i-1}) - \frac{a}{\lambda}(T_i-T_{i-1})\right) - \frac{\tau}{2\varrho\lambda}(G_i-G_{i-1}),\tag{64)b$$

$$
\begin{aligned}
R^i &= R_i + \tau G_i, \\
T^i = T_i + \mu(T_{i+1} - 2T_i + T_{i-1}) &+ \frac{a\tau}{2\varrho\lambda^2}(G_{i+1} - 2G_i + G_{i-1}) - \\
\kappa\left(P_{i+1} - P_i - Q_i + Q_{i-1} - \frac{a}{\lambda}(T_{i+1} - 2T_i + T_{i-1})\right) &+ \\
&\frac{\kappa}{2\varrho\lambda}(R_{i+1} - 2R_i + R_{i-1}).
\end{aligned}
\tag{64c}
$$

In the case when the viscous effect may be neglected (i.e., when $G = 0$) it was shown in Mihăilescu-Suliciu and Suliciu (1995) that the energy inequality (51) requires the integration step τ to satisfy

$$0 < \tau \leq \tau_1(h) = \frac{h^2}{2(b + ah)} \quad, 0 < \tau \leq \tau_2(h) = \frac{\lambda h^2}{2ab + (\lambda^2 + 2a^2)h} \tag{65}$$

in order to get numerical stability. On the other hand if the thermo-mechanical coupling coefficient is negligible (i.e. when $\alpha = 0$) then we must have Mihăilescu-Suliciu and Suliciu (1985) (see also (1992))

$$\nu \leq 1, \quad \tau \leq \frac{2}{k}. \tag{66}$$

Finally, if neither G nor α are negligible then both conditions (65) and (66) are necessary conditions for numerical stability of our scheme. They remain necessary conditions for numerical stability in the case when the phase transition is present as we may discuss elsewhere.

Strictly speaking the conditions (66) valid for the uncoupled case ($\alpha = 0$) or the isothermal case ($\theta = \theta_r$, i.e. $T = 0$) are proved in Mihăilescu-Suliciu and Suliciu (1985) when the numerical scheme is the method of characteristics. In this case $\nu = 1$ and thus $\tau = \bar{\tau} = h/\lambda$. Therefore if we assume that the quantities p_i, q_i (which are defined below the formulas (61)) and R_i are known at the time level $t = j\tau$ then they are determined at $t = j\tau + \bar{\tau}$ by the following formulas

$$\bar{p}^i = p_{i+1} + \bar{\tau}G_{i+1}, \quad \bar{q}^i = q_{i-1} + \bar{\tau}G_{i-1}, \quad \bar{R}^i = R_i + \bar{\tau}G_i. \tag{67}$$

We want to prove now that the stability condition (66) must hold for any $\nu \neq 1$. First we observe that the quantities p, q, R resulting at time level $t = (j+1)\tau$ from formulas for arbitrary $\nu > 0$ can be written as

$$p^i = (1-\nu)p_i + \nu\bar{p}^i, \quad q^i = (1-\nu)q_i + \nu\bar{q}^i, \quad R^i = (1-\nu)R_i + \nu\bar{R}^i. \tag{68}$$

Next, we notice that the density of the total energy in this linear case has the quadratic expression

$$\varrho e^* = \varrho\frac{v^2}{2} + \frac{\sigma^2}{2E} + \frac{E_o R^2}{2E(E - E_o)} = \frac{p^2 + q^2}{4E} + \frac{E_o R^2}{2E(E - E_o)} \tag{69}$$

and the numerical total energy at time level $t = j\tau$ is given by

$$e_j = \sum_i e^*(p_i, q_i, R_i). \tag{70}$$

We can state now precisely a result from Mihăilescu-Suliciu and Suliciu (1985). For $\nu = 1$ the numerical energy of a Cauchy problem verifies the inequality

$$\sum_i e^*(\bar{p}^i, \bar{q}^i, \bar{R}^i) \leq \sum_i e^*(p_i, q_i, R_i) \ \text{ if and only if } \ \bar{\tau} \leq \frac{2}{k}\,. \tag{71}$$

When $\nu \in (0,1)$ from the convexity of $e^*(p,q,R)$ we can write

$$\sum_i e^*(p^i, q^i, R^i) - \sum_i e^*(p_i, q_i, R_i) \leq$$
$$\nu\left(\sum_i e^*(\bar{p}^i, \bar{q}^i, \bar{R}^i) - \sum_i e^*(p_i, q_i, R_i)\right)$$

which implies that for the scheme with $\nu \in (0,1)$ the total energy of the Cauchy problem decreases if $\tau < \bar{\tau} \leq 2/k$.

For $\nu > 1$ we may choose the initial data (p_i, q_i, R_i) such that $G_i = 0$ (one example is $(v, \varepsilon, \sigma)(X, 0) = (AX/L, 0, 0)$, for $X \in (0, L)$) then

$$\bar{p}^i = p_{i+1}, \quad \bar{q}^i = q_{i-1}, \quad \bar{R}^i = R_i \ \text{ and}$$
$$p^i = (1-\nu)p_i + \nu p_{i+1}, \quad q^i = (1-\nu)q_i + q_{i-1}, \quad R^i = R_i$$

Thus

$$\sum_i e^*(p^i, q^i, R^i) - \sum_i e^*(p_i, q_i, R_i) = -$$
$$\frac{\nu(1-\nu)}{4E} \sum_i \left[(p_{i+1} - p_i)^2 + (q_i - q_{i-1})^2\right] > 0$$

if there is at least one index i such that $p_{i+1} - p_i \neq 0$ or $q_{i+1} - q_i \neq 0$ and therefore there is no stability in energy as the energy increases in time.

7. Some remarks on a Sokolovskii type thermo–viscoelastic perfect plastic model

A Sokolovskii rate–type elastic perfect plastic model with a temperature dependent yield function as considered by Bell (1973) § 4.21, §4.23

$$\sigma_Y(\theta) = \begin{cases} \sigma_Y^o(1 - \theta/\theta_m) & \theta < \theta_m \\ 0 & \theta \geq \theta_m, \end{cases} \tag{72}$$

where θ_m is the melting point absolute temperature and σ_Y^o is the yield stress at zero temperature, was recently discussed by Suliciu (1997a). There the relation

$$\sigma_Y^o = \alpha E_o \theta_m \tag{73}$$

was imposed as a technical requirement in order to obtain the latent heat $\nu(\varepsilon, \theta, \sigma)$ as a continuous function of its arguments. Based on the results obtained in the preceding sections we show that such a restriction can be removed and still construct the thermodynamics potentials and get "good" energy estimates.

The governing system of equations is (40) but the function $G(\varepsilon, \theta, \sigma)$ is taken slightly more general than in Suliciu (1997a) (see also Cristescu and Suliciu (1982)) and it is

$$G(\varepsilon, \theta, \sigma) = -k \begin{cases} 0 & \text{if } |\sigma| < \sigma_Y(\theta), \\ \operatorname{sgn}(\sigma)|\sigma - \sigma_Y(\theta)| & \text{if } \sigma \geq \sigma_Y(\theta), \end{cases} \tag{74}$$

where again k is a positive Maxwell's type viscosity coefficient.

One can attach to the above viscoplastic model a viscoelastic one (for $\theta < \theta_m$ as for $\theta > \theta_m$ is not well defined) by taking a positive constant E_o (an equilibrium Young's modulus) and define

$$\sigma_E(\varepsilon, \theta) = \begin{cases} -\sigma_Y(\theta) & \text{if } \varepsilon < \varepsilon_Y^-(\theta), \\ E_o[\varepsilon - \alpha(\theta - \theta_r)] & \text{if } \varepsilon_Y^-(\theta) \leq \varepsilon \leq \varepsilon_Y^+(\theta), \\ \sigma_Y(\theta) & \text{if } \varepsilon_Y^+(\theta) < \varepsilon \end{cases} \tag{75}$$

where

$$\varepsilon_Y^-(\theta) = -\frac{\sigma_Y(\theta)}{E_o} + \alpha(\theta - \theta_r), \quad \varepsilon_Y^+(\theta) = \frac{\sigma_Y(\theta)}{E_o} + \alpha(\theta - \theta_r). \tag{76}$$

Then a relaxation function $G_v(\varepsilon, \theta, \sigma) = -k[\sigma - \sigma_E(\varepsilon, \theta)]$ will lead to a viscoelastic model with a piecewise smooth equilibrium curve as considered in the previous sections. Any availability function constructed for the viscoelastic model is an availability function for the viscoplastic model and thus we may obtain energy estimates for the viscoplastic model too. The lines $\varepsilon = \varepsilon_Y^{\pm}(\theta)$ are curves of discontinuity for latent heat ν and for specific heat C_E. When the condition (73) is satisfied the functions ν and C_E are continuous across the line $\varepsilon = \varepsilon_Y^+(\theta)$.

For the time being we conclude our considerations here.

References

Bell, J. F. (1973) *The experimental foundations of solid mechanics*, in Handbuch der Physik, vol. VIa/1, Springer–Verlag, Berlin

Cristescu, N. and Suliciu, I. (1982) *Viscoplasticity*, Nijhoff, The Hague

Făciu, C. (1996) Initiation and growth of strain bands in rate-type viscoelastic materials, Euro. J. Mech. A/Solids **15**, Part I, 968–988, Part II, 989–1011.

Făciu, C. and Suliciu, I. (1994) A Maxwellian model for pseudoelastic materials, Scripta Metall., **31**, pp. 1399–1404.

Fu, S., Huo, Y. and Müller, I. (1993) Thermodynamics of pseudoelasticity, Acta Mech., **99**, pp.1-19.

Van Humbeeck, J. and Delaey, L. (1982) The strain rate, orientation and temperature dependence of the strain–induced beta to martensite transformation in Cu-Zn-Al single crystals, in H.I. Aaronson *et al* (eds.),*Proceedings of an International Conference on Solid → Solid Phase Transformation.*, The Metallurgical Society of AIME, pp.1317–1321.

Mihăilescu-Suliciu, M. and Suliciu, I. (1985) On the method of characteristics in rate–type viscoelasticity, ZAMM Z. angew. Math. Mech. **65**, pp.479–486.

Mihăilescu-Suliciu, M. and Suliciu, I. (1992) On the method of characteristics in rate–type viscoelasticity with non–monotone equilibrium curve, ZAMM Z. angew. Math. Mech. **72**, pp.667–674.

Mihăilescu-Suliciu, M. and Suliciu, I. (1995) Energy estimates and energy control of numerical stability in coupled dynamic thermoelasticity, Mech. Res. Comm., **22** pp.467–472.

Otsuka,K. and Shimizu, K. (1982) Stress-induced martensitic transformations martensite-to-martensite transformations in H.I. Aaronson *et al* (eds.),*Proceedings of an International Conference on Solid → Solid Phase Transformation.*, The Metallurgical Society of AIME, pp.1267–1286.

Serrin, J. (1986) An outline of Thermodynamical Structure in Serrin, J., (ed.), *New Perspective in Thermodynamics*, Springer-Verlag, Berlin, pp.3–32.

Šilhavý, M. (1986) Foundation of Continuum Thermodynamics, in Serrin, J., (ed.), *New Perspective in Thermodynamics*, Springer-Verlag, Berlin, pp.33–48.

Suliciu I. (1984) Some Energetic Properties of Smooth Solutions in Rate–Type Viscoelasticity, Int. J. Nonlinear Mech. **19**, 525–544

Suliciu, I. (1989) On the description of the dynamics of phase transitions by means of rate-type constitutive equations, in A. S. Khan and M. Tokuda (eds.) *Proceedings of Plasticity '89*, Pergamon Press pp.417–420.

Suliciu, I. (1992a) Some stability–instability problems in phase transitions modelled by piece-wise linear elastic or viscoelastic constitutive equations, Int. J. Engng. Sci. **30**, pp.483–494.

Suliciu, I. (1992b) The relative free energy and energy estimates in in modelling phase transitions by rate–type constitutive equations, Int. J. Engng. Sci. **30**, pp.495–505.

Suliciu, I. (1995) Some numerical results for a viscoplastic softening problem, Arch. Mech. **47**, pp.391–403.

Suliciu, I. (1997a) Energy estimates in rate–type thermo–viscoplasticity, Int. J. Plasticity, to appear

Suliciu, I. (1997b) On the thermodynamics of rate–type fluids and and phase transitions. I–Rate–type fluids. II–Phase transitions, to be published

Truesdell, C. and Bharatha, S. (1977) *The Concepts and Logic of Classical Thermodynamics*, Springer–Verlag, Berlin

SOME REMARKS ON THERMODYNAMIC THEORY OF VISCOUS ELASTOPLASTIC MEDIA

ZHEN-BANG KUANG
Shanghai Jiaotong University, Shanghai, 200240, P.R. China

Abstract

In this paper some aspects of the thermodynamic theory with internal variables are discussed. A method to determine the constitutive equation of viscous elastoplastic media with or without damage is proposed. This method is more flexible than traditional one and more conformable to microscopic deformation mechanism.

1. Introduction

Let ρ, e, T, S, $\mathbf{q}$ and h be the mass density, the internal energy per unit mass, the thermodynamic temperature, the entropy per unit mass, the heat influx per unit area and the heat supply per unit mass, then the first and second laws of thermodynamics for a continuum are (Kuang, 1989)

$$\rho \dot{e} = \sigma : \dot{\varepsilon} - \nabla \cdot \mathbf{q} + \rho h \tag{1}$$

and

$$\rho \dot{S} \geq \rho h / T - \nabla \cdot (\mathbf{q} / T) \tag{2}$$

where σ and ε are stress and strain tensors respectively.
Introducing the Helmholtz free energy f per unit mass by

$$f = e - ST \tag{3}$$

We can write the Clausius-Duhem inequality as

$$\rho T \sigma^* = \sigma : \dot{\varepsilon} - \rho\left(\dot{f} + S\dot{T}\right) + T\mathbf{q} \cdot \nabla(1/T) \geq 0 \tag{4}$$

where σ^* is the irreversible entropy production rate per unit mass. Introducing internal variables η in current thermodynamics we assume that

$$f = f(\varepsilon.T.\eta) \tag{5}$$

R. Wang (ed.), IUTAM Symposium on Rheology of Bodies with Defects, 87–98.

In the plastic and viscoplastic analysis it is usually assumed that(Chaboche 1993, Nemes et. al. 1990, Lubarda & Krajcinovic 1995)

$$\left.\begin{array}{l}\dot{\boldsymbol{\varepsilon}}=\dot{\boldsymbol{\varepsilon}}^{e}+\dot{\boldsymbol{\varepsilon}}^{i}, \boldsymbol{\sigma}=\boldsymbol{\sigma}^{e}=\boldsymbol{\sigma}^{i} \\ \dot{W}=\dot{W}^{e}+\dot{W}^{i}, \dot{W}^{e}=\boldsymbol{\sigma}:\dot{\boldsymbol{\varepsilon}}^{e}, \dot{W}^{i}=\boldsymbol{\sigma}:\dot{\boldsymbol{\varepsilon}}^{i}\end{array}\right\} \tag{6}$$

where $\dot{\boldsymbol{\varepsilon}}^e$ and $\dot{\boldsymbol{\varepsilon}}^i$ are elastic and inelastic strain rates respectively; $\dot{W}, \dot{W}^e$ and $\dot{W}^i$ are the total, reversible and irreversible work rates respectively. In the viscoelastic analysis it is often assumed that(Valanis 1971, Lemaitre & Chaboche 1985)

$$\left.\begin{array}{l}\boldsymbol{\sigma}=\boldsymbol{\sigma}^{e}+\boldsymbol{\sigma}^{i}, \dot{\boldsymbol{\varepsilon}}=\dot{\boldsymbol{\varepsilon}}^{e}=\dot{\boldsymbol{\varepsilon}}^{i} \\ \dot{W}=\dot{W}^{e}+\dot{W}^{i}, \dot{W}^{e}=\boldsymbol{\sigma}^{e}:\dot{\boldsymbol{\varepsilon}}, \dot{W}^{i}=\boldsymbol{\sigma}^{i}:\dot{\boldsymbol{\varepsilon}}\end{array}\right\} \tag{7}$$

where $\boldsymbol{\sigma}^e$ and $\boldsymbol{\sigma}^i$ are elastic and inelastic stresses respectively. In more general case some authors assumed (Maugin et. al. 1992)

$$\left.\begin{array}{l}\boldsymbol{\sigma}=\boldsymbol{\sigma}^{e}+\boldsymbol{\sigma}^{i}, \dot{\boldsymbol{\varepsilon}}=\dot{\boldsymbol{\varepsilon}}^{e}+\dot{\boldsymbol{\varepsilon}}^{i} \\ \dot{W}=\dot{W}^{e}+\dot{W}^{i}, \quad \dot{W}^{e}=\boldsymbol{\sigma}^{e}:\dot{\boldsymbol{\varepsilon}}^{e}, \quad \dot{W}^{i}=\boldsymbol{\sigma}^{i}:\dot{\boldsymbol{\varepsilon}}^{i}+\boldsymbol{\sigma}^{e}:\dot{\boldsymbol{\varepsilon}}^{i}+\boldsymbol{\sigma}^{i}:\dot{\boldsymbol{\varepsilon}}^{e}\end{array}\right\} \tag{8}$$

From the above discussion it is clear that in current literatures there is no consistent way to decompose the work rate into reversible and irreversible parts, but this is an essential problem in the thermodynamic theory with internal variables. In the present paper we shall discuss this problem in more detail.

2. The Basic Postulates of Work Rate, Strain and Stress

In the constitutive equation of dielectric media the electric permittivity is effected by two internal mechanisms: polarization and the leakage of electricity. In a condenser with dielectric the polarization is equivalent to adding a resistor in series, and the leakage of electricity is equivalent to adding a resistor in parallel with the condenser. It is clear that in a dielectric there exist two different power dissipation mechanisms. The Voigt and Maxwell models of viscoelasticity are consistent with dashpots connected in parallel and in series, respectively, to a spring. More complicated models in viscoelasticity have been introduced. In the plastic-creep deformation at high temperature we can consider that the diffusion of atomic defects and the motion of dislocations are linked in the same way. Summarizing the above discussions we propose the following assumptions for a deformation version of thermodynamic theory:

(1) In all cases the work rate can be divided into reversible and irreversible parts.

$$\dot{W}=\dot{W}^{e}+\dot{W}^{i} \tag{9}$$

(2) The total stress tensor is composed of several internal stress tensors, i.e. the total stress is the sum of several partial stresses:

$$\boldsymbol{\sigma} = \sum_{j=1}^{n} \boldsymbol{\sigma}_j \tag{10}$$

(3) Corresponding to each $\boldsymbol{\sigma}_j$, there is also a partial strain rate $\dot{\boldsymbol{\varepsilon}}_j$. All the partial strain rates $\dot{\boldsymbol{\varepsilon}}_j$ are equal to each other and equal to $\dot{\boldsymbol{\varepsilon}}$, but the different partial strain rates $\dot{\boldsymbol{\varepsilon}}_j$ may be composed of several parts, such as $\dot{\boldsymbol{\varepsilon}}_j^e, \dot{\boldsymbol{\varepsilon}}_j^p, \dot{\boldsymbol{\varepsilon}}_j^c$ ($\dot{\boldsymbol{\varepsilon}}_j^c$ is the creep strain rate) etc.:

$$\dot{\boldsymbol{\varepsilon}} = \dot{\boldsymbol{\varepsilon}}_1 = \dot{\boldsymbol{\varepsilon}}_2 = \cdots = \dot{\boldsymbol{\varepsilon}}_n, \dot{\boldsymbol{\varepsilon}}_j = \dot{\boldsymbol{\varepsilon}}_j^e + \dot{\boldsymbol{\varepsilon}}_j^p + \dot{\boldsymbol{\varepsilon}}_j^c + \cdots \tag{11}$$

But the stresses $\boldsymbol{\sigma}$ and $\boldsymbol{\sigma}_j$ are not divided into reversible and irreversible parts.

(4) All kinds of irreversible parts of strain rates (such as $\dot{\boldsymbol{\varepsilon}}^p, \dot{\boldsymbol{\varepsilon}}^c$) produce irreversible parts of the work rate. All the elastic parts $\dot{\boldsymbol{\varepsilon}}^e$ of strain rates produce reversible parts of the work rate.

(5) For some internal variables there may exist a bounding surface $f_j(\boldsymbol{\sigma}_j, T, \boldsymbol{\eta}) = 0$, such as the yield surface for the plastic strain, the damage surface for damage variables, the creep-hardening bound for creep strain (Murakami & Ohno 1982). The bounding surface is an idealized narrow region where the physical variables significantly change.

(6) The decomposition version is illustrated in figure 1, and

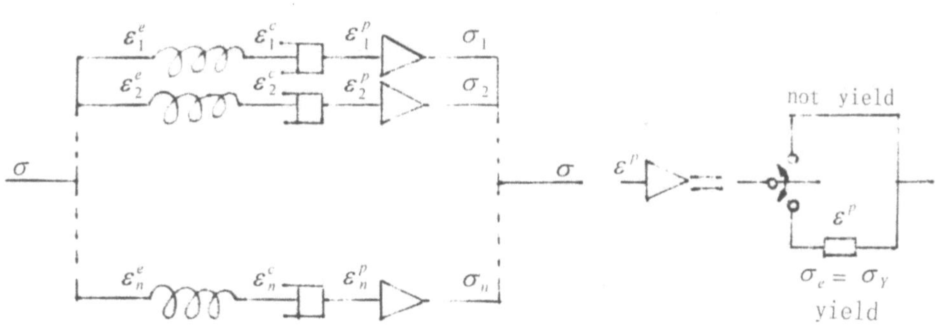

Fig. 1 the decomposition sketch of stress and strain

$$\left.\begin{aligned} &\dot{W} = \dot{W}^e + \dot{W}^i, \ \dot{W}^e = \sum_{j=1}^{n} \boldsymbol{\sigma}_j : \dot{\boldsymbol{\varepsilon}}_j^e, \ \dot{W}^i = \sum_{j=1}^{n} \boldsymbol{\sigma}_j : (\dot{\boldsymbol{\varepsilon}}_j^p + \dot{\boldsymbol{\varepsilon}}_j^c + \cdots) \\ &\dot{\boldsymbol{\varepsilon}} = \dot{\boldsymbol{\varepsilon}}_j = \dot{\boldsymbol{\varepsilon}}_j^e + \dot{\boldsymbol{\varepsilon}}_j^i (j = 1, 2, \cdots, n) \end{aligned}\right\} \tag{12}$$

3. The Proposed Constitutive Equations.

It is noted that $\boldsymbol{\varepsilon}_j^i$ are dependent on internal variables $\boldsymbol{\eta}$, but the $\boldsymbol{\varepsilon}_j^e$ are not. Equation(5) is modified to

$$f = f(\boldsymbol{\varepsilon}^e, T, \boldsymbol{\eta}) \tag{13}$$

The model shown in Fig. 1 was very useful in one-dimensional viscoelasticity. It is important to extend this model to three dimensional continuum mechanics and connect it with internal microscopic deformation mechanisms; this is the aim of this paper. It is emphasized that in (13) the internal variables $\boldsymbol{\eta}$ may be divided into two types: ① $\boldsymbol{\eta}$ are direct variables which describe internal deformation mechanisms, such as the dislocation density and pattern, twin density and distribution etc. In this case $\boldsymbol{\varepsilon}^i$ are dependent only on $\boldsymbol{\eta}$ but not included in $\boldsymbol{\eta}$. ② $\boldsymbol{\eta}$ are indirect variables which describe internal deformation mechanisms, such as the flow stress, hardening parameters, etc. In this case some special combinations of $\boldsymbol{\varepsilon}^i$ may be included in $\boldsymbol{\eta}$, and $\boldsymbol{\varepsilon}^i$ may be dependent on $\boldsymbol{\eta}$ and $\boldsymbol{\sigma}$ or other variables.

In Ghosh(1980) the deformation is divided into grain deformation and grain boundary deformation; his constitutive structure is similar to ours. In his model the grain boundary element has a constant stress term and a viscous term and is in series with the grain element. Holzapfel and Simo (1996) assumed

$$f = f_\infty(\boldsymbol{\varepsilon}, T) + \sum_{j=1}^{n} \left\{ f_j(\boldsymbol{\varepsilon}, T) - 2\nabla_\varepsilon f_j(\boldsymbol{\varepsilon}, T) + l_j |\boldsymbol{\eta}_j|^2 \right\} \tag{14}$$

where ∇_ε is a divergence operator on $\boldsymbol{\varepsilon}$, and the l_j are constants. Their model also is essentially similar to ours.

Substituting equations (12) and (13) into (4) we get

$$\rho T\sigma^* = \sum_{j=1}^{n} \left(\sigma_j : \dot{\varepsilon}_j^e + \sigma_j : \dot{\varepsilon}_j^i \right) - \rho S \dot{T} - \rho \sum_{j=1}^{n} \left(\frac{\partial f}{\partial \varepsilon_j^e} : \dot{\varepsilon}_j^e \right) - \rho \frac{\partial f}{\partial T} \dot{T} - \rho \frac{\partial f}{\partial \eta} \bullet \dot{\eta} - \frac{\partial T}{\partial x_k} \frac{q_k}{T} \geq 0 \tag{15}$$

Because (15) always holds, we obtain

$$\boldsymbol{\sigma}_j = \rho\left(\partial f / \partial \boldsymbol{\varepsilon}_j^e\right), \quad S = -\partial f / \partial T \tag{16}$$

and

$$\rho T \sigma^* = \sum_{j=1}^{n} \boldsymbol{\sigma}_j : \dot{\boldsymbol{\varepsilon}}_j^i + \boldsymbol{A} \cdot \dot{\boldsymbol{\eta}} - \nabla T \cdot \mathbf{q} / T \geq 0 \tag{17}$$

$$\boldsymbol{A} = -\rho\left(\partial f / \partial \boldsymbol{\eta}\right) \tag{18}$$

In (17) $\boldsymbol{\sigma}_j$, $\mathbf{A}$ and ∇T are considered as irreversible forces corresponding to irreversible fluxes $\dot{\boldsymbol{\varepsilon}}_j^i$, $\dot{\boldsymbol{\eta}}$ and $\mathbf{q}$ respectively. In the following, two possible versions are discussed.

(1) First version In this version internal variables are assumed as indirect

variables describing deformation mechanisms. As in usual thermodynamic theory, we assume that the irreversible fluxes are functions of irreversible forces, i.e.

$$\left.\begin{aligned}\dot{\varepsilon}^i_j = \dot{\varepsilon}^i_j(\sigma_k, A),\ \ \dot{\eta} = \dot{\eta}\ (\sigma_k, A),\ \ \mathbf{q} = \lambda(\nabla T)\\ \sigma_k = \sigma_k(\dot{\varepsilon}, \dot{\eta}),\ \ A = A(\dot{\varepsilon}^i_j, \dot{\eta}),\ \ \nabla T = (1/\lambda)\cdot\mathbf{q}\end{aligned}\right\} \tag{19}$$

Because f does not depend on $\partial T / \partial x_k$ (Coleman 1964), we assume in (19) that $\dot{\varepsilon}^i_j$ and $\dot{\eta}$ also do not depend on $\partial T / \partial x_k$. Since f is dependent on T, so $\sigma_j, \dot{\varepsilon}^i_j, \eta$ and λ are all dependent on T. Equations (19) should be determined by experimental data and experience. For most engineering materials it is possible to introduce a general flow potential $F(\sigma_k, A)$ such that

$$\dot{\varepsilon}^i_j = \partial F / \partial \sigma_j, \dot{\eta} = \partial F / \partial A \tag{20}$$

Introducing general flow·potential we can establish the constitutive equation in a simplified manner. Equations (16), (18) and (19) or (16), (18) and (20) give the complete constitutive equation system. Usually the equations for $\dot{\eta}$ are called the evolution equations of internal variables.

(2) Second version In the first version the Helmholtz free energy f is used, actually the specific Gibbs character function $g(\sigma_j, T, \eta)$ (or the complementary energy) can also be used. Because

$$\rho g = \rho f - \sigma_j : \varepsilon_j,\ \ \rho \dot{g} = \rho \dot{f} - \sigma_j : \dot{\varepsilon}_j - \dot{\sigma}_j : \varepsilon_j \tag{21}$$

The Clausius-Dinhem inequality is

$$\begin{aligned}\rho T \sigma^* &= -\sum_{j=1}^{n} \varepsilon_j : \dot{\sigma}_j - \rho(\dot{g} - S\dot{T}) - \mathbf{q}\nabla T / T\\ &= -\sum_{j=1}^{n} (\varepsilon_j + \rho \frac{\partial g}{\partial \sigma_j}) : \dot{\sigma}_j - \rho(S + \frac{\partial g}{\partial T})\dot{T} - \rho \frac{\partial g}{\partial \eta}\dot{\eta} - \mathbf{q}\nabla T / T \geq 0\end{aligned} \tag{22}$$

When η remain constant we get the usual thermoelastic constitutive structure:

$$\varepsilon_j = -\rho \partial g / \partial \sigma_j,\ \ S = -\partial g / \partial T \tag{23}$$

and (22) becomes

$$\rho T \sigma^* = \boldsymbol{B} \cdot \dot{\eta} - \mathbf{q} \cdot \nabla T / T \tag{24}$$

$$\boldsymbol{B} = -\rho \partial g / \partial \eta \tag{25}$$

where $\boldsymbol{B}$ are the irreversible forces corresponding to η. From (24) the evolution equation of η may be written as

$$\dot{\eta} = \dot{\eta}(\boldsymbol{B}) \tag{26}$$

If $g(\sigma_j, T, \eta)$ is known, then equations (23), (25) and (26) form a complete constitutive equation system. However, because it is not easy in practice, an alternative method will be used (Rice 1971). Equation(23), can be rewritten in the form

$$\dot{\varepsilon}_j = \dot{\varepsilon}_j^e + \dot{\varepsilon}_j^i = -\rho\left(\frac{\partial^2 g}{\partial\sigma_j\partial\sigma_k}:\dot{\sigma}_k + \frac{\partial^2 g}{\partial\sigma_j\partial T}\dot{T} + \frac{\partial^2 g}{\partial\sigma_j\partial\eta}\cdot\dot{\eta}\right) \tag{27}$$

If η are direct variables describing deformation mechanisms, then the irreversible strain rates $\dot{\varepsilon}_j^i$ can occur only if the η are changed. Then (27) gives

$$\dot{\varepsilon}_j^e = -\rho\left(\partial^2 g/\partial\sigma_j\partial\sigma_k\right):\dot{\sigma}_k - \rho\left(\partial^2 g/\partial\sigma_j\partial T\right)\dot{T} \tag{28}$$

$$\dot{\varepsilon}_j^i = -\rho\left(\partial^2 g/\partial\sigma_j\partial\eta\right)\dot{\eta} \tag{29}$$

and

$$\dot{S}^e = -\left(\partial^2 g/\partial\sigma_j\partial T\right):\dot{\sigma}_j - \left(\partial^2 g/\partial T^2\right)\dot{T} \tag{30}$$

$$\dot{S}^i = -\left(\partial^2 g/\partial T\partial\eta\right)\cdot\dot{\eta} = (1/\rho)(\partial B/\partial T)\dot{\eta} \tag{31}$$

If there exists a "flow potential" H such that

$$H(\sigma_j, T, \eta) = \int_0^{B(\sigma_j,T,\eta)} \dot{\eta}(B, T, \eta)dB \tag{32}$$

then

$$\dot{\varepsilon}_j^i = \partial H/\partial\sigma_j \tag{33}$$

Geometrically $\dot{\varepsilon}_j^i$ is normal to the surface of constant flow potential in the partial stress space.

4. Isothermal Viscoelasticity with Damage

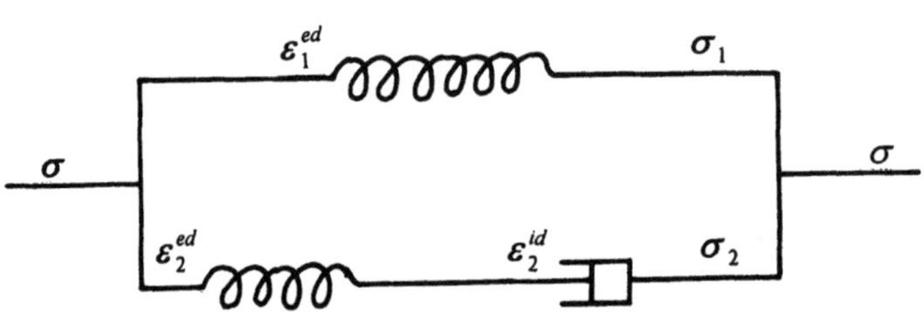

Fig. 2 A model of viscoelastic media with damage

As a first example, an isothermal viscoelastic media with damage is discussed (Fig. 2). For simplicity, the damage variable is considered as a scalar, and denoted by D=(V-Vs)/V, where V and Vs are the total and solid volumes of an element respectively. In Fig. 2 the superscripts "ed", "id", …etc. will be used to denote the strain in the elastic and inelastic elements respectively arising from the damage effect. We note that the damage theory is a developing theory and the definition of 'damage' is not fully clear. In fatigue damage, Cheng et. al. (1996) define damage as the ductility exhaustion of material. In that model the dislocation structures also belong to the damage. In this paper the phenomenon of damage is represented by microcracks and/or microcavities, and in this sense it involves a rheological process quite different from deformation. Microcracks and microcavities themselves do not directly affect the strength of a material (matrix), but they affect the local stiffness of the material. The strength of a material is influenced indirectly through the local deformation around these defects. Plastic flow can occur without damage and similarly, damage can occur without noticeable macroplastic flow. So the damage variable is different from other internal variables.

① *Version 1* Let

$$\rho f = \frac{1}{2}\boldsymbol{\varepsilon}_1^{ed}:(1-D)\boldsymbol{E}_1:\boldsymbol{\varepsilon}_1^{ed} + \frac{1}{2}\boldsymbol{\varepsilon}_2^{ed}:(1-D)\boldsymbol{E}_2:\boldsymbol{\varepsilon}_2^{ed} \tag{34}$$

where $\boldsymbol{E}_1$ and $\boldsymbol{E}_2$ are the elastic moduli of parts 1 and 2 respectively. Equation (34) represents the elastic strain energy stored in a damaged material, and it shows that the effective stiffness is influenced by damage. According to the above explanation and the method given in version 1, we have.

$$\left.\begin{aligned} &\boldsymbol{\sigma}_1 = \rho\partial f / \partial\boldsymbol{\varepsilon}_1^{ed} = (1-D)\boldsymbol{E}_1:\boldsymbol{\varepsilon}_1^{ed}, \quad \boldsymbol{\sigma}_2 = \rho\partial f / \partial\boldsymbol{\varepsilon}_2^{ed} = (1-D)\boldsymbol{E}_2:\boldsymbol{\varepsilon}_2^{ed} \\ &A = -\rho\partial f / \partial D = \frac{1}{2}\boldsymbol{\varepsilon}_1^{ed}:\boldsymbol{E}_1:\boldsymbol{\varepsilon}_1^{ed} + \frac{1}{2}\boldsymbol{\varepsilon}_2^{ed}:\boldsymbol{E}_2:\boldsymbol{\varepsilon}_2^{ed} \end{aligned}\right\} \tag{35}$$

If we assume that $\boldsymbol{\varepsilon}_j^i$ is determined only by the variables of its own part, but D is a function of variables in all parts, then

$$\left.\begin{aligned} &\rho T\sigma^* = \boldsymbol{\sigma}_2:\dot{\boldsymbol{\varepsilon}}_2^{id} + A\dot{D} \geq 0 \\ &\dot{\boldsymbol{\varepsilon}}_2^{id} = \dot{\boldsymbol{\varepsilon}}_2^{id}(\boldsymbol{\sigma}_2, A), \quad \dot{D} = \dot{D}(\boldsymbol{\sigma}_1, \boldsymbol{\sigma}_2, A) \end{aligned}\right\} \tag{36}$$

The simplest case is

$$\dot{\boldsymbol{\varepsilon}}_2^{id} = \frac{1}{(1-D)\mu_2}:\boldsymbol{\sigma}_2 \qquad \dot{D} = 2^n C A^n = \frac{C}{(1-D)^{2n}}(\boldsymbol{\sigma}_1:\frac{1}{\boldsymbol{E}_1}:\boldsymbol{\sigma}_1 + \boldsymbol{\sigma}_2:\frac{1}{\boldsymbol{E}_2}:\boldsymbol{\sigma}_2)^n \tag{37}$$

where μ_2 is a visco-sity coefficient, and C, n are constants.

Fig. 2 shows that

$$\boldsymbol{\sigma} = \boldsymbol{\sigma}_1 + \boldsymbol{\sigma}_2, \quad \dot{\boldsymbol{\varepsilon}}_1^{ed} = \dot{\boldsymbol{\varepsilon}}_2^{ed} + \dot{\boldsymbol{\varepsilon}}_2^{id} = \dot{\boldsymbol{\varepsilon}} \tag{38}$$

Using (35), (37) and (38) we get the following coupled differential equations to determine $\boldsymbol{\sigma}$, $\boldsymbol{\varepsilon}$ and D:

$$\left.\begin{aligned}&\sigma = (1-D)E_1{:}\varepsilon + \sigma_2\\ &\dot{\sigma}_2 + (\frac{\dot{D}I}{1-D} + E_2{:}\frac{1}{\mu_2}){:}\sigma_2 = (1-D)E_2{:}\dot{\varepsilon}\\ &\dot{D} = \frac{C}{(1-D)^{2n}}\left[(\sigma-\sigma_2){:}\frac{1}{E_1}{:}(\sigma-\sigma_2) + \sigma_2{:}\frac{1}{E_2}{:}\sigma_2\right]^n\end{aligned}\right\} \tag{39}$$

or

$$\left.\begin{aligned}&\dot{\sigma} + \left(\frac{\dot{D}I}{1-D} + E_2{:}\frac{1}{\mu_2}\right){:}\sigma = (1-D)(E_1+E_2){:}\dot{\varepsilon} + \left[(1-D)E_2{:}\frac{1}{\mu_2}\right]{:}E_1{:}\varepsilon\\ &\dot{D} = \frac{C}{(1-D)^{2(n-1)}}\left\{\varepsilon{:}E_1{:}\varepsilon + \left[\frac{\sigma}{1-D} - E_1{:}\varepsilon\right]{:}\frac{1}{E_2}{:}\left[\frac{\sigma}{1-D} - E_1{:}\varepsilon\right]\right\}^n\end{aligned}\right\} \tag{40}$$

In one-dimensional case, (40) reduces to

$$\left.\begin{aligned}&\dot{\sigma} + \left(\frac{\dot{D}}{1-D} + \frac{E_2}{\mu_2}\right)\sigma = (1-D)(E_1+E_2)\dot{\varepsilon} + (1-D)\frac{E_2}{\mu_2}E_1\varepsilon\\ &\dot{D} = \frac{C}{(1-D)^{2(n-1)}}\left\{E_1\varepsilon^2 + \frac{1}{E_2}\left(\frac{\sigma}{1-D} - E_1\varepsilon\right)^2\right\}^n\end{aligned}\right\} \tag{41}$$

In the above equations the evolution form of D is only an example, the more precise form should be given by a further research.

② *Version 2* In version 2 we let

$$-\rho g = \frac{1}{2(1-D)}\sigma_1{:}\frac{1}{E_1}{:}\sigma_1 + \frac{1}{2(1-D)}\sigma_2{:}\frac{1}{E_2}{:}\sigma_2 + \rho\tilde{g}(\sigma_2,\eta) \tag{42}$$

In equation (42) we have assumed that the damage is associated with the stiffness and the inelastic strain is also associated with internal variables η as indicated at the beginning of this section.

According to (23) and (25) we have

$$\left.\begin{aligned}&\varepsilon_1^{ed} = \frac{1}{1-D}\sigma_1{:}\frac{1}{E_1},\ \varepsilon_2^{ed} = \frac{1}{1-D}\sigma_2{:}\frac{1}{E_2},\ \varepsilon_2^{id} = \frac{\partial B}{\partial\sigma_2}\dot{\eta},\ B = -\rho\frac{\partial\tilde{g}}{\partial\eta}\\ &A = -\rho\frac{\partial g}{\partial D} = \frac{1}{2(1-D)^2}\left\{\sigma_1{:}\frac{1}{E_1}{:}\sigma_1 + \sigma_2{:}\frac{1}{E_2}{:}\sigma_2\right\}\end{aligned}\right\} \tag{43}$$

According to the theory of viscoelasticity the flow potential H is assumed as

$$H = \int_0^B \dot{\eta}(B,\eta)dB = \sigma_2{:}\frac{1}{2(1-D)\mu_2}{:}\sigma_2 \tag{44}$$

where μ_2 is the viscosity coefficient tensor. In this case (33) gives

$$\dot{\varepsilon}_2^{id} = \frac{1}{1-D}\sigma_2 : \frac{1}{\mu_2} \tag{45}$$

Using (43), (45), (38) and (37), we again obtain equation (39). Version 1 and 2 can give the same constitutive equation under the appropriate assumptions.

③ *Comments on damaged material* Equations (5) and (43) show that ε^{ed} can be divided into elastic strain ε^e and damage strain ε^d, *i.e.*

$$\left.\begin{aligned} \mathbf{e}^{ed} &= \frac{1}{1-D}\mathbf{s} : \frac{1}{\mathbf{E}} = \mathbf{s} : \frac{1}{\mathbf{E}} + \frac{D}{1-D}\mathbf{s} : \frac{1}{\mathbf{E}} = \mathbf{e}^e + \mathbf{e}^d \\ \dot{\mathbf{e}}^{ed} &= \dot{\mathbf{s}} : \frac{1}{\mathbf{E}} + \left[\frac{D}{1-D}\dot{\mathbf{s}} : \frac{1}{\mathbf{E}} + \frac{\dot{D}}{(1-D)^2}\mathbf{s} : \frac{1}{\mathbf{E}}\right] = \dot{\mathbf{e}}^e + \dot{\mathbf{e}}^d \end{aligned}\right\} \tag{46}$$

The damage strain ε^d is an inelastic strain which is produced through changing the stiffness from $\mathbf{E}$ to $(1-D)\mathbf{E}$, but its deformation mechanism is still the same as in the elastic media. In (46) an ε^{ed} is simply the sum of ε^e and ε^d, so differentiating ρf with ε^{ed} or ε^e in (34), we will get the same σ. This illustration for ε^{id} is also correct.

5. Coupled Creep-Elastoplastic Analysis

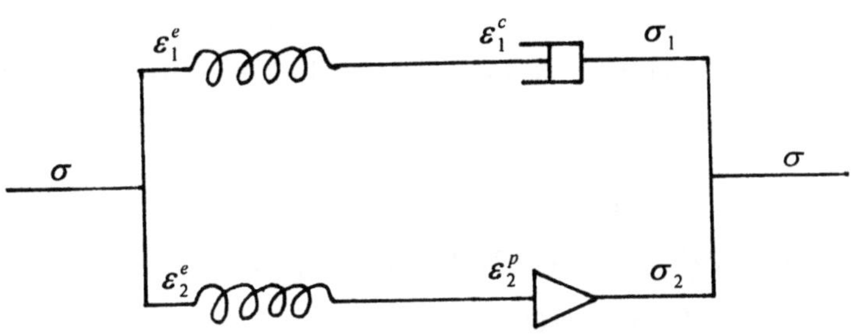

Fig. 3 A model of coupled creep-elastoplastic analysis

As a second example, an isothermal coupled creep-elastoplastic analysis is considered (Fig. 3). In this analysis Norton's law for creep and Mises-flow theory with combined hardening rules for elastoplasticity are assumed, so the flow potential F will take the form given in equation (47). Note that the time-independent plastic behavior is a particular case of viscoplasticity when the rate of flow (inelastic strian) approaches infinite. In version 1, the Helmholtz free energy f and a general flow potential F should be assumed. According to the above explanation we select

$$\left.\begin{aligned} \rho f &= (1/2)\boldsymbol{\varepsilon}_1^e:\boldsymbol{E}_1:\boldsymbol{\varepsilon}_1^e + (1/2)\boldsymbol{\varepsilon}_2^e:\boldsymbol{E}_2:\boldsymbol{\varepsilon}_2^e + \rho\tilde{f}(p_2,\boldsymbol{\alpha}_2) \\ F &= \frac{a}{N+1}\left(\frac{\sigma_{1e}}{a}\right)^{N+1} + F_{2p} \\ F_{2p} &= J(\boldsymbol{\sigma}_2 - \boldsymbol{Z}_2) - R_2 - \sigma_Y \end{aligned}\right\} \tag{47}$$

where

$$J(\boldsymbol{\sigma}) = \sqrt{\frac{3}{2}\boldsymbol{\sigma}':\boldsymbol{\sigma}'}, \quad \sigma_e = J(\boldsymbol{\sigma}), \quad \boldsymbol{\sigma}' = \boldsymbol{\sigma} - \frac{1}{3}\sigma_{kk} \tag{48}$$

a and N are constants, σ_e is equivalent stress, σ_Y is an initial yield stress , $\mathbf{Z}_2$ is a back stress, R_2 is an isotropic hardening parameter , p_2 and $\boldsymbol{\alpha}_2$ are internal variables. Variables of parts 1 and 2 are distinguished by subscripts 1 and 2 respectively. According to version 1 we have

$$\boldsymbol{\sigma}_1 = \boldsymbol{E}_1:\boldsymbol{\varepsilon}_1^e, \quad \boldsymbol{\sigma}_2 = \boldsymbol{E}_2:\boldsymbol{\varepsilon}_2^e, \quad \boldsymbol{Z}_2 = \rho\partial f/\partial\boldsymbol{\alpha}_2, \quad R_2 = \rho\partial f/\partial p_2 \tag{49}$$

$$\dot{\boldsymbol{\varepsilon}}_1^c = \frac{\partial F}{\partial\boldsymbol{\sigma}_1} = \frac{3}{2}\left(\frac{\sigma_{1e}}{a}\right)^N \frac{\boldsymbol{\sigma}_1'}{\sigma_{1e}} \tag{50}$$

$$\dot{\boldsymbol{\varepsilon}}_2^p = \dot{m}\frac{\partial F}{\partial\boldsymbol{\sigma}_2} = \frac{3}{2}\dot{m}\frac{\boldsymbol{\sigma}_2' - \boldsymbol{Z}_2'}{J(\boldsymbol{\sigma}_2 - \boldsymbol{Z}_2)} \tag{51}$$

$$\dot{p}_2 = -\dot{m}\frac{\partial F}{\partial R_2} = \dot{m} = \sqrt{\frac{2}{3}\dot{\boldsymbol{\varepsilon}}_2^p:\dot{\boldsymbol{\varepsilon}}_2^p}, \quad \dot{\boldsymbol{\alpha}}_2 = -\dot{m}\frac{\partial F}{\partial\boldsymbol{Z}_2} = \dot{\boldsymbol{\varepsilon}}_2^p \tag{52}$$

where the plasticity multiplier $\dot{m}$ is determined by the consistency condition:

$$F_{2p} = 0, \quad \dot{F}_{2p} = \frac{\partial F_{2p}}{\partial\boldsymbol{\sigma}_2}:\dot{\boldsymbol{\sigma}}_2 + \frac{\partial F_{2p}}{\partial\boldsymbol{Z}_2}:\dot{\boldsymbol{Z}}_2 + \frac{\partial F_{2p}}{\partial R_2}\dot{R}_2 = 0 \tag{53}$$

From (49) we assume $\boldsymbol{Z}_2$ and R_2 in a simple form

$$\left.\begin{aligned} \dot{\boldsymbol{Z}}_2 &= \rho\partial\tilde{f}/\partial\boldsymbol{\alpha}_2 = (2/3)C(p_2)\dot{\boldsymbol{\varepsilon}}_2^p - \gamma(p_2)\boldsymbol{Z}_2\dot{p}_2 \\ \dot{R}_2 &= \rho\partial\tilde{f}/\partial p_2 = b(Q - R_2)\dot{p}_2 \end{aligned}\right\} \tag{54}$$

Combining (53) and (54) we get

$$\dot{m} = \frac{1}{h}\frac{\frac{3}{2}(\boldsymbol{\sigma}_2' - \boldsymbol{Z}_2'):\dot{\boldsymbol{\sigma}}_2}{J(\boldsymbol{\sigma}_2 - \boldsymbol{Z}_2)}, \quad h = C - \frac{\frac{3}{2}\gamma\boldsymbol{Z}_2:(\boldsymbol{\sigma}_2' - \boldsymbol{Z}_2')}{J(\boldsymbol{\sigma}_2 - \boldsymbol{Z}_2)} + b(Q - R_2) \tag{55}$$

where C and γ are selected functions of p_2, b and Q are constants. Equation (51) is homogeneous in time of order one , so $\boldsymbol{\varepsilon}_2^p$ is an implicit function of time. Further study on plastic deformation is omitted here , the reader can find it in many textbooks

(Kuang 1989, Lemaitre & Chaboche 1985)
The model shown in Fig .3 gives

$$\sigma = \sigma_1 + \sigma_2, \quad \dot{\varepsilon}_1^e + \dot{\varepsilon}_1^c = \dot{\varepsilon}_2^e + \dot{\varepsilon}_2^p = \dot{\varepsilon} \tag{56}$$

So the constitutive equation of coupled creep -elastoplastic media for loading case is

$$\left.\begin{aligned} &\sigma = \sigma_1 + \sigma_2 \\ &\frac{1}{E_1} : \dot{\sigma}_1 + \frac{3}{2}\left(\frac{\sigma_{1e}}{a}\right)^N \frac{\sigma_1}{\sigma_{1e}} = \dot{\varepsilon} \\ &\frac{1}{E_2} : \dot{\sigma}_2 + \frac{3\left(\sigma_2' - Z_2'\right) : \dot{\sigma}_2}{2h\left(\sigma_2' - Z_2'\right) : \left(\sigma_2' - Z_2'\right)}\left(\sigma_2' - Z_2'\right) = \dot{\varepsilon} \end{aligned}\right\} \tag{57}$$

In a simple tension loading case, the above equation is reduced to

$$\sigma = \sigma_1 + \sigma_2, \quad \frac{\dot{\sigma}_1}{E_1} + \left(\frac{\sigma_1}{a}\right)^N = \dot{\varepsilon}, \quad \frac{\dot{\sigma}_2}{E_2} + \frac{\dot{\sigma}_2}{h} = \dot{\varepsilon} \tag{58}$$

or

$$\left.\begin{aligned} &\sigma = a\dot{\varepsilon}^{\frac{1}{N}} + \int_0^{\varepsilon} \frac{E_2 h d\varepsilon}{E_2 + h} - a\left(\frac{\dot{\sigma}}{E_1} - \frac{E_2 h \dot{\varepsilon}}{E_1\left(E_2 + h\right)}\right)^{\frac{1}{N}} \\ &\sigma_2 = \int_0^{\varepsilon} \frac{E_2 h}{E_2 + h} d\varepsilon, \quad h = C - \gamma Z_2 + b(Q - R_2) \end{aligned}\right\} \tag{59}$$

Acknowledgement This work was supported by the National Natural Science Foundation of China

References

Chaboche, J.-L. (1993)Cyclic viscoplastic constitutive equations, part 1: A thermodynamically consistent formulation *J. Appl. Mech.* **60**, 813-821

Cheng, G-X, Zuo, J-Z, Luo, Z-W and Kuang, Z-b. (1996), Continuum damage model of low-cycle fatigue and fatigue damage analysis of welded joint, *Engng. Fract. Mech.* **55**, 155-161

Ghosh, A. K, (1980) A physically-based constitutive model for metal deformation, *Acta Metallurgica* **28**, 1443-1465.

Germain, P. Nguyen, Q.S and Suquet P. (1983) Continuum thermodynamics. *J. Appl. Mech.* **50**, 1010-1020

Hansen, N.R. and schreyer, H.L. (1994) A thermodynamically consistent framework for theories of elastoplasticity coupled with damage , *Int. J. Solids Structures* **31**, 359-389.

Holzapfel, G.A. and Simo. J.C. (1996) A new viscoelastic constitutive model for continuous media at finite thermomechanical changes, *Int. J. Solids structures* **33**, 301903034.

Kuang, Z-B. (1989) The foundation of continuum mechanics , Xian Jiaotong University Publishing House, Xian. (in Chinese)

Lemaitre, J. and Chaboche J.-L. (1985) Mecanique des materiaux solids, Dunod. paris.

Lubrada, V.A. and krajcinovic (1995) Some fundamental issues in rate theory of damage-elastoplasticity, *Int. J. Plasticity* **11**, 763-797.

Maugin, G.A. et. al. (1992) Nonlinear electromechanical couplings, John wiley & Sons, Chichester.

Nemes, J.A., Eftis, J. and Randles, P.W. (1990) Viscoplastic constitutive modeling of high strain-rate deformation, material damage and spall fracture, *J. Appl. Mech.* **57**, 282-291.

Rice, J.R, (1971) Inelastic constitutive relations for solids:an internal-variable theory and its application to metal plasticity, *J. Mech. Phys. Solids* **19** , 433-455.

STOCHASTIC RESPONSE OF DEGRADING ELASTIC SYSTEMS

K. SOBCZYK, J. TRĘBICKI

Center of Mechanics, Institute of Fundamental Technological Research, Polish Academy of Sciences, Świętokrzyska 21, 00-049 Warsaw, Poland

Abstract

The dynamic response of mechanical/structural systems and degradation processes due to dynamics has mostly been analysed without the appropriate mutual coupling. In this paper a joint (coupled) description of the system dynamics and the associated degradation process is presented. After general formulation of the problem we consider a randomly vibrating linear elastic system with stiffness degradation due to fatigue. The vibration - degradation problem is represented by the vibration equation with a special (nonlinear) stiffness degradation term. The solution of this equation (the approximate probability density) is obtained by the modified maximum entropy method and the results of associated numerical calculations are illustrated graphically.

1. Introduction

In the last decades the dynamics of elastic and inelastic bodies and systems has been extensively studied, including the stochastic dynamics, which accounts additionally for uncertainties and random irregularities in external and parametric excitations. Various models and methods of analysis have been elaborated which provide the effective tools for characterization of the stress distribution, and in the consequence for the reliability estimation (cf. [1], [2], [3]). Also, various deterioration processes (which take place in the material due to time-varying stress conditions) have been modelled and studied. Especially, modelling and analysis of fatigue process (both, in deterministic and stochastic setting) has been of a great research effort (cf. [4]). Fatigue accumulation causes a progressive degrading of the stiffness of a material and it affects directly the system performance. However, the studies mentioned above (particularly, those concerned with stochastic dynamics and fatigue) have been conducted without the

R. Wang (ed.), IUTAM Symposium on Rheology of Bodies with Defects, 99–108.

appropriate mutual coupling. It is no doubt that a more satisfactory approach should consists in joint (coupled) description of the system dynamics and the associated degradation process.

The objective of this paper, is the modelling and analysis of a coupled stochastic dynamics - deterioration process. In particular, we consider randomly vibrating linear system with stiffness degradation due to fatigue. Defining the fatigue degradation process in terms of an instantaneous crack size and its critical value the coupled problem is represented in the form of a vibration equation with a special (nonlinear) stiffness degradation term. Solution of this equation (the approximate probability density of the response) is obtained by the modified maximum entropy method.

2. General formulation of the response-degradation problem.

For a wide class of nonlinear vibratory systems with random excitation (both, external or parametric) the coupled response - degradation problem can be formulated in the following form:

$$\ddot{Y}(t) + F[\dot{Y}(t), Y(t), D(t), X(t,\gamma)] = 0 \tag{1}$$

$$\mathscr{F}[\dot{D}(t), D(t), Y(t), \dot{Y}(t)] = 0 \tag{2}$$

$$Y(t_0) = Y_0\ , \quad \dot{Y}(t_0) = Y_{1,0}\ , \quad D(t_0) = D_0 \tag{3}$$

where:

$Y(t)$ - unknown response process,

$D(t)$ - degradation process,

$F[.]$ - given function of indicated variables satisfying the appropriate conditions for the existence and uniqueness of the solution,

$X(t,\gamma)$ - given stochastic process characterizing the excitation; $\gamma \in \Gamma$, and Γ is the space of elementary events in the basic scheme $(\Gamma, \mathscr{B}, P)$ of probability theory,

$\mathscr{F}[.]$ - symbolizes the relationship between degradation and response process; its specific mathematical form depends on the particular situation.

An important special class of the response-degradation problems is obtained if relationship (2) takes the form of differential equation, that is, equations (1), (2) are

$$\ddot{Y}(t) + F[\dot{Y}(t), Y(t), D(t), X(t,\gamma)] = 0 \tag{4}$$

$$\dot{D}(t) = G[D(t), Y(t), \dot{Y}(t)] \tag{5}$$

where G is the appropriate function specifying the evolution of degradation; its mathematical form is inferred from the elaboration of empirical data, or - it is derived from the analysis of the physics of the process. In equation (5) dependence on $Y(t)$ is regarded here in more relaxed sense then it is usual. Degradation rate $\dot{D}(t)$ may depend on the actual values of $Y(t)$, but it can also depend on some functionals of $Y(t)$, for example - on the integral of $Y(\tau)$, $\tau \in [t_0, t]$. In fatigue degradation problem with $D(t)$ interpreted as a "normalized" crack size, the most common evolution equation is the Paris equation which includes not $Y(t)$ itself, but the stress range, i.e. quantity related to $Y_{max} - Y_{min}$. In particular case, when the right-hand side of equation (5) depends solely on the response process $Y(t)$ and, possibly on $\dot{Y}(t)$ then this equation gives the following integral expression for the variability of degradation in time

$$D(t) = D_0 + \int_{t_0}^{t} G[Y(\tau), \dot{Y}(\tau)] \, \mathrm{d}\tau \tag{6}$$

Of course, if equation (4) does not depend on $D(t)$ or equation (5) does not depend on $Y(t)$, $\dot{Y}(t)$ then the response - degradation problem becomes uncoupled.

Another special class of problems characterized generally by equations (1), (2) is obtained if functional relationship (2) does not include explicitly $D(t)$ and $\dot{D}(t)$ depends on some statistical characteristics of the response process $Y(t)$ or the joint process $[Y(t), \dot{Y}(t)]$. A good example is a vibrating systems in which a degradation process depends on the time which the response $Y(t)$ spends above some critical level y^*.

3. Vibrating System with Stiffness Degradation due to fatigue.

Let us consider linear vibrations of a structural mechanical component subjected to Gaussian white excitation. Assume that during the vibration process a macroscopic fatigue crack develops in the system which affects the stiffness (or natural frequency) of the vibrating system. Let us denote by $k(D)$ the stiffness dependence on the degradation measure D. The governing equation has the form

$$M\ddot{y}(t) + c\dot{y}(t) + k(D)y(t) = \xi(t,\gamma) \tag{7}$$

where $\xi(t,\gamma)$ is a random process (white noise); γ denotes an element of space of elementary events Γ on which probability is defined. It is clear that $y(t)$ denotes the displacement (response process) and M, c are the mass and damping coefficients, respectively. Dividing, both sides of equation (7) by M and then introducing new variables: $Y = y / \sigma_y$, $\tau = \omega_0 t$ where σ_y denotes a standard deviation of the stationary displacement of equation (7) without degradation (i.e. when $k(D) = \omega_0^2$) we obtain a dimensionless form of equation (7)

$$\ddot{Y}(t) + 2\zeta\dot{Y}(t) + q(D)Y(t) = \xi_1(t,\gamma) \tag{8}$$

where $k_0 = M\omega_0^2$ denotes the initial (non-degraded) stiffness and $k(D) = \omega_0^2 q(D)$; function $q(D)$ is monotonically decreasing function of the degradation parameter D; $\xi_1(t,\gamma) = \xi(\tau/\omega_0,\gamma)/k_0\sigma$ is the white noise with correlation function $K_{\xi_1}(\tau_1,\tau_2) = 4\zeta\delta(\tau_2 - \tau_1)$ where $\delta(.)$ means the Dirac delta function. It should be noticed, that for system (8) without degradation has in the stationary state $\sigma_y = \sigma_{\dot{y}} = 1$, where $\sigma_{\dot{y}}$ is the standard deviation of the velocity.

A kinetic equation - empirically motivated - for fatigue crack size $A(t)$ can be taken in various forms (cf. [4]). The most common is the Paris equation

$$\frac{\mathrm{d}A}{\mathrm{d}N} = C(\Delta K)^m \quad , \quad \Delta K > 0 \tag{9}$$

where ΔK is the stress intensity range, C and m are empirical constants. The stress intensity factor K characterizes the magnitude of the local stress around the crack tip, and depends on variables characterizing the loading, crack size and shape, and geometry of the specimen. In general, it has the form

$$K = B(A)S\sqrt{\pi A} \tag{10}$$

where A is the crack length, S represents the far-field stress resulting from the applied load, $B(A)$ is a factor which accounts for the shape of the specimen and the crack geometry. The Paris equation when combined with the expression (10) leads to the following equation for the fatigue crack growth rate

$$\frac{\mathrm{d}A}{\mathrm{d}N} = CB^m(A)(\sqrt{\pi A})^m(\Delta S)^m \tag{11}$$

where ΔS is the stress range (associated with the response process $Y(t)$), and N denotes the number of fatigue cycles.

Equations (8) and (11) constitute the coupled system of equations for the response-degradation process. When equation (8) is represented in the form of two first order equations (for processes $Y_1(t) = Y(t)$, $Y_2(t) = \dot{Y}(t)$):

$$\begin{aligned} \dot{Y}_1(\tau) &= Y_2(\tau) \\ \dot{Y}_2(\tau) &= -2\zeta Y_2(\tau) - q(D)Y_1(\tau) + \xi_1(\tau,\gamma) \end{aligned} \tag{12}$$

then the response - degradation process $[Y_1(t), Y_2(t), D(t)]$ is governed by the system of three first-order equations (12) and (11) where the crack size A is additionally (by appropriate function) related to the degradation measure D.
This system of stochastic equations can be analyzed by use of various methods. The method proposed in this paper consists in reducing the nonlinear system of three equations to the system of two first order equations. To this end, let us introduce the function $\psi(A)$ as

$$\psi(A) = \int_{A_0}^{A} \frac{\mathrm{d}x}{B^m(x)(\sqrt{\pi x})^m} \tag{13}$$

where A_0 is the initial crack size. Let us denote by ψ^* the value of $\psi(A)$ for the critical crack length $A = A^*$, and define the degradation measure D as

$$D = \frac{\psi(A)}{\psi^*} \quad , \quad \psi^* = \psi(A) \tag{14}$$

which takes its values between zero and one. Hence, we have

$$\mathrm{d}D = \frac{1}{\psi^*}\,\mathrm{d}\,\psi(A) = \frac{1}{\psi^*}\frac{\mathrm{d}A}{B^m(A)(\sqrt{\pi A})^m} = \frac{1}{\psi^*}C(\Delta S)^m\,\mathrm{d}N \tag{15}$$

Therefore, the kinetic equation for the degradation due to fatigue crack growth takes the form

$$\frac{\mathrm{d}D}{\mathrm{d}N} = \frac{1}{\psi^*}C(\Delta S_Y)^m \tag{16}$$

where ΔS_Y is the stress range generated by the response process $Y(t)$.

Since the vibratory system in question is subjected to random excitation $\xi_1(\tau,\gamma)$ then the response $Y(t)$ is a random process as well. In this case equation (16) has to be modified (cf. [4]). Here, under assumption that the response process is narrow band, the random stresses range will be characterized by double amplitude of the envelope of the dimensionless response of the system (12), i.e.

$$\Delta S_Y = 2\sqrt{Y_1^2(t) + Y_2^2(t)} \tag{17}$$

and change from cycles to time is based on the relation

$$\mathrm{d}N = \mu^+\,\mathrm{d}\tau \tag{18}$$

where μ^+ is the average number of up-crosings (cf. [4]) of a fixed (critical) level by process $Y(t)$. Therefore, equation (16) is represented as

$$\frac{\mathrm{d}D}{\mathrm{d}\tau} = \mu^+ C_1\left(\sqrt{Y_1^2(t)+Y_2^2(t)}\right)^m \quad , \quad D(\tau_0)=D_0 \quad , \quad C_1 = \frac{C2^m}{\psi^*} \tag{19}$$

The equations (12) and (19) constitute explicitly coupled response-degrading problem for linearly vibrating system. It turns out that equation (19) does not depend explicitly on D, so it belongs to the class of the degradation relations of the general form (6). The exact analysis of the coupled system (12), (19) is involved. To make the further treatment of the problem possible we will take advantage of the fact that the envelope amplitude varies slowly in time. This agrees with the observation that the degradation process is slow in comparison with the response itself. Therefore, making the "linear approximation" $D(\tau)=\dot{D}(\tau)\,\tau$ where $\dot{D}(\tau)$ is given by (19) we can regard D occurring in dynamic equations (12) as explicitly expressed by $Y(\tau), \dot{Y}(\tau)$ and τ. Hence, equations (12) can be written in the form of the following Itô stochastic equations

$$\begin{aligned} \mathrm{d}Y_1(\tau) &= Y_2(\tau)\mathrm{d}\tau \\ \mathrm{d}Y_2(\tau) &= -[2\zeta Y_2(\tau) - g(Y_1,Y_2,\tau)]\mathrm{d}\tau + 2\sqrt{\zeta}\,\mathrm{d}W(\tau) \end{aligned} \tag{20}$$

where $W(\tau)$ is the standard Wiener process, and $g(Y_1,Y_2,\tau)$ is the nonlinear term accounting for the dependence of stiffness on the degradation, i.e.

$$g(y_1,y_2,\tau) = y_1\,q(D) = y_1\,q(y_1,y_2,\tau) \tag{21}$$

4. Solution of the problem

In order to obtain quantitative results, one has to assume the specific function characterizing the dependence of the stiffness on the degradation due to fatigue. Such a dependence can be inferred from the empirical data. In the literature (cf. [1]) a polynomial $a_k D^k$, (k=1,2,...,10) of degree ten has been proposed. Empirical data can also be approximated by the equation

$$q(D) = \alpha_1 + \alpha_2 \exp(-\alpha_3 D^{\alpha_4}) \tag{22}$$

where α_1, α_2, α_3, α_4 are positive constants, such that $\alpha_1+\alpha_2=1$ (to have $q(D_0=0)=1$).

In order to obtain a probabilistic characterization of the response with the degradation of stiffness (22) the moment equations for the stochastic system (20) are generated. If we denote $m_{ij}(\tau)=\left\langle Y_1^{\,i}(\tau)Y_2^{\,j}(\tau)\right\rangle$ where $\langle . \rangle$ is the symbol of the probabilistic mean value, then we have

$$\begin{aligned}\frac{\mathrm{d}}{\mathrm{d}\,\tau}\left\langle Y_1^i Y_2^j\right\rangle &= \left\langle iY_1^{i-1}Y_2^{j+1} - j\left[2\zeta Y_2 + g(Y_1,Y_2,\tau)\right]Y_1^i Y_2^{j-1} + 2\zeta j(j-1)Y_1^i Y_2^{j-2}\right\rangle \\ &\equiv \left\langle G_{ij}(Y_1,Y_2;\tau)\right\rangle, \qquad \left\langle Y_1^i(\tau_0)Y_2^j(\tau_0)\right\rangle = m_{ij}(\tau_0)\end{aligned} \tag{23}$$

i,j = 1,2,... . The information on the behaviour of the system is taken in the form of five first equations from the above hierarchy of equations, i.e. equations for the first order and second order moments $(m_{10}, m_{20}, m_{01}, m_{02}, m_{11})$

$$\begin{aligned}
&i=1,\quad j=0:\quad \frac{\mathrm{d}}{\mathrm{d}t}\langle Y_1\rangle = \langle Y_2\rangle \\
&i=2,\quad j=0:\quad \frac{\mathrm{d}}{\mathrm{d}t}\left\langle Y_1^2\right\rangle = 2\langle Y_2 Y_1\rangle \\
&i=0,\quad j=1:\quad \frac{\mathrm{d}}{\mathrm{d}t}\langle Y_2\rangle = -\langle 2\zeta Y_2 + g(Y_1)\rangle \\
&i=0,\quad j=2:\quad \frac{\mathrm{d}}{\mathrm{d}t}\left\langle Y_2^2\right\rangle = -2\left\langle 2\zeta Y_2^2 + Y_2 g(Y_1,Y_2;\tau)\right\rangle + 2\zeta \\
&i=1,\quad j=1:\quad \frac{\mathrm{d}}{\mathrm{d}t}\langle Y_1 Y_2\rangle = \left\langle Y_2^2 - 2\zeta Y_1 Y_2 - Y_1 g(Y_1,Y_2;\tau)\right\rangle
\end{aligned} \tag{24}$$

The approximate probability density $p(y_1,y_2;\tau)$ is determined via the modified maximum entropy method (cf. [5], [6]). This density has the form

$$\begin{aligned}p(y_1,y_2;\tau) &= \frac{1}{\tilde{C}}\exp\{-[\lambda_{10}y_1 + \lambda_{20}\, y_1\, g(y_1,y_2;\tau) + \lambda_{01}y_2 + \lambda_{02}y_2^2 + \lambda_{11}y_1y_2\,]\} \\ &= \frac{1}{\tilde{C}}\,\tilde{p}(y_1,y_2;\tau)\end{aligned} \tag{25}$$

where λ_{ij} are unknown Lagrange coefficients and

$$\tilde{C} = \int_{-\infty}^{+\infty} \tilde{p}(y_1,y_2;\tau)\,\mathrm{d}\,y_1\,\mathrm{d}\,y_2 \tag{26}$$

is the normalizing constant.

Let us discretize system (24) using (for instance) the Euler scheme (to make further equations more clear) with the step $\Delta\tau$. As the result, the system (24) can be rewritten as

$$m_{ij}(\tau_{k+1}) = m_{ij}(\tau_k) + \Delta\tau \int_{-\infty}^{+\infty} G_{ij}(y_1,y_2;\tau_k)\,p(y_1,y_2;\tau_k)\,\mathrm{d}y_1\,\mathrm{d}y_2 \qquad i,j=0,1,2 \quad 0<i+j\le 2 \tag{27}$$

with $m_{ij}(\tau_0)$ assumed to be given initial condition. In our consideration these initial moments were taken as the moments of stationary solution of the system (8) without degradation.

The Lagrange coefficients λ_{ij} are determined at each discrete time τ_k numerically from the following system of algebraic nonlinear equations

$$m_{ij}(\tau_k) = \int_{-\infty}^{+\infty} y_1^i y_2^j\, p(y_1,y_2;\tau_k)\,\mathrm{d}y_1\,\mathrm{d}y_2 \ , \qquad i,j=0,1,2 \quad 0<i+j\le 2 \tag{28}$$

Taking into account (25) and (26) this system can be written as

$$\int_{-\infty}^{+\infty} [y_1^i y_2^j - m_{ij}(\tau_k)]\,\tilde{p}(y_1,y_2;\tau_k)\,\mathrm{d}y_1\,\mathrm{d}y_2 = 0 \ , \qquad i,j=0,1,2 \quad 0<i+j\le 2 \tag{29}$$

and can be solved using (for example) the five dimensional Newton method.

In the calculations the following values of parameters were used: $\alpha_1=\alpha_2=0.5$, $\alpha_4=2.0$, $\zeta=0.125$, m=3.0, $C_1=1.3298\cdot10^{-5}$. It is assumed that the degradation starts when dimensionless system reaches the stationary state. In the absence of degradation $(q(D)=1)$ the response of the system (which is linear) is Gaussian. Degradation introduces nonlinear and time-dependent stiffness and, therefore leads - in general - to non-Gaussian behaviour of the system.

The results of numerical calculations are illustrated on figures 1-3. Figure 1 vizualizes function $q(D)$ versus D for selected values of parameters α_3: for curve (1) $\alpha_3=0.1$, for curve (2) $\alpha_3=0.5$, for curve (3) $\alpha_3=1$. Generally, the form of degradation function (22) is very flexible and many kinds of possible types of degradation (from linear to strongly nonliniear) can be obtained. In practice, the values of parameters α_1, α_2, α_3, α_4 should be estimated from experimental data.

Figure 2 shows the variance σ_y^2 of the displacement in the system with degradation for the same different values of parameters α_3 as in Fig. 1. As was mentioned above, in the dimensionless system without degradation we have $\sigma_y^2=1$. From this figure we see

influence of degradation of the stiffness on the response of the system. The increasing of degradation causes the increase of the variance of displacement of the system.
Figure 3 illustrates the non-Gaussian probability density of the distribution of the displacement of the system with degradation for different number of response cycles.

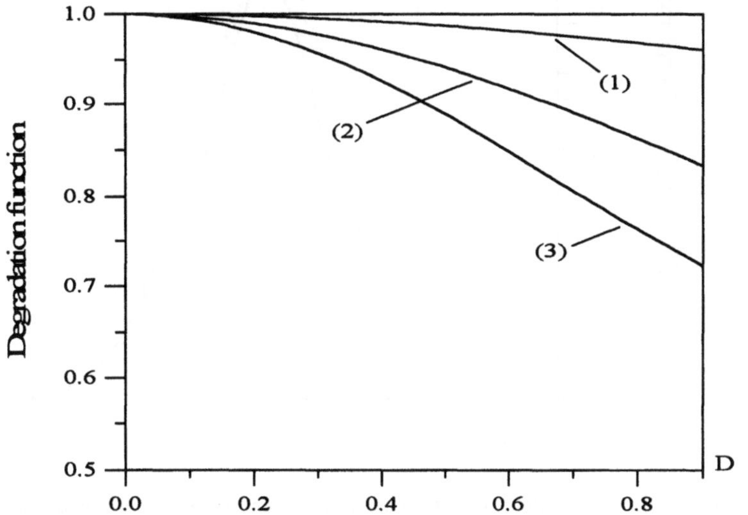

Figure 1. Function of degradation $q(D)$ given in (22) for different values of parameter α_3: (1) - $\alpha_3 = 0.1$; (2) - $\alpha_3 = 0.5$; (3) - $\alpha_3 = 1$. Values of other parameters in the text.

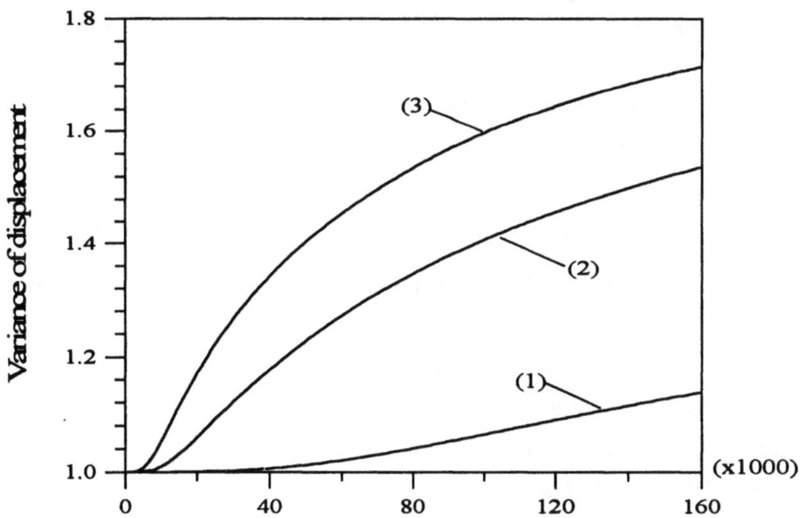

Figure 2. The variance of the displacement in system with degradation for functions of degradation corresponding to the Figure 1.

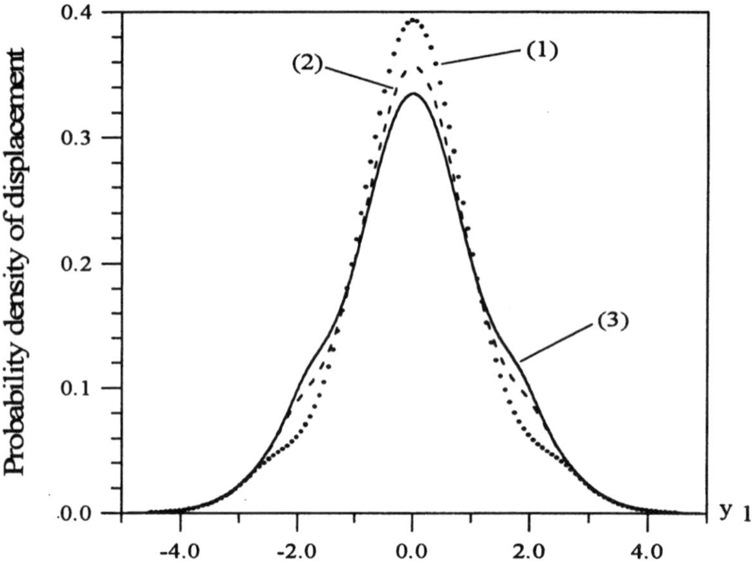

Figure 3. Non-Gaussian probability density of the displacement in the system with degradation for different numbers of cycles N of the response: (1) – $N = 5\times10^3$; (2) – $N = 10\times10^3$; (3) – $N = 15\times10^3$

References:

1. Soong T.T, Grigoriu M., (1993), *Random Vibrations of Structural and Mechanical Systems*, Prentice Hall, New Jersey.

2. Lin Y.K., Cai G.Q., (1995), *Advanced Probabilistic Dynamics*, McGraw Hill, New York.

3. Sobczyk K., (1991), *Stochastic Differential Equations with Applications to Physics and Engineering*, Kluwer, Dordrecht-Boston.

4. Sobczyk K., Spencer B.F., (1992), *Random Fatigue: From Data to Theory*, Academic Press, Boston.

5. Sobczyk K., Trębicki J., (1993), Maximum entropy principle and nonlinear stochastic oscillators, *Physica A*, 193, 448-468.

6. Trębicki J., Sobczyk K., (1996), Maximum entropy principle and non-stationary distributions of stochastic systems, *Probab. Engng Mechanics*, Vol.11, 169-178

7. Trębicki J., et. al. , Response of nonlinear degrading oscillator to random excitation, (submitted for publication in *J. Sound and Vibration*)

EXPERIMENTAL STUDIES ON THE EVOLUTION OF DEFECT TEMPERATURE FIELD DURING DEFORMATION OF ABS*

WENBO LUO
Institute of Rheological Mechanics, Xiangtan University, Hunan 411105, P.R.C.

Abstract The process of defect evolution is complicated which cuts across micro-, meso- and macroscale levels, and should be inherently characterized by its rheology and dissipation. Experiments of heat generation induced by any irreversible deformation in the tensile failure process of thermoplastics with prefabricated defects have been made in details. The initiation and evolution laws of the local temperature field near defects are observed and recorded with infrared photography, and then preliminarily analyzed. It is shown that the heat generated during defect evolution is significant in our experiments and comprises some 25 to 70 per cent of the of external work, so its contribution to the failure of materials is not negligible. Considering the micro- and mesoscopic characters of deformation, a preliminary and qualitative explanation to the cooling and heating phenomena observed in the experiments is also presented in this paper.

1. Introduction

The mechanical properties and failure laws of polymeric materials have been extensively studied not only because of their importance in industrial applications but also because of their theoretical complications. At present, it has been realized that the damage and fracture process of thermoplastics is dissipative and compatible with irreversible thermodynamics. It cuts across micro-, meso- and macroscale levels, that is, it starts from the rearrangement, slippage, reorientation, disentanglement and scission of the entangled chain segments among molecular chains at microlevel, passes through the mesolevel including the craze initiation, craze growth, fracture of craze matters, the initiation and propagation of microcracks, and ends in macroscopic gross failure induced by microcrack cascades and macrocrack propagations. Energy dissipation will occur during the local molecular reorientation, rheological deformations in the materials which lead to heat generation and form a temperature field. Such temperature effect during the deformation of materials resulted from thermomechanical coupling, was first considered by Duhamel[1] early in 1837. Thereafter, Farren and Taylor[2] measured the heat generated by pulling a bar into the plastic region under quasi-static conditions and found that 85 per cent of the mechanical energy was converted into heat. The advances

* Project 19632030 supported by National Natural Science Foundation of China.

R. Wang (ed.), IUTAM Symposium on Rheology of Bodies with Defects, 109–119.

in irreversible thermodynamics[3] gave a keen impetus to the developments in the thermomechanical coupling. Dillon reported experimental results on the heat generated during the torsional oscillations of round aluminium and copper tubes[4,5], and gave an analysis of the heat generated in a nonlinear thermoelastic material by the deviatoric components of strain[6]. Cernocky and Krempl[7] proposed a coupled thermovisco-plasticity theory based on the overstress constitutive equation and applied it to investigate special types of loading: pure torsion, uniaxial loading and cyclic loading. Ghoneim[8-10] presented coupled thermoviscoplasticity equations and applied them to solve the problem of dynamic loading of a thick-walled cylinder as well as compression or cyclic loading of an end-constrained cylinder. Beginning in 1985, Allen[11-13] published a series of papers dealing with thermomechanical coupling in viscoplastic uniaxial metallic bars. So far as thermoelastic coupling models are concerned, some initial work was done by Schapery[14], Christensen[15,16] and Crochet and Naghdi[17]. Considering the changing between surface and volume energy in deformed bodies, Sih gave a surface/volume energy density theory[18] to account for the mutually thermomechanical coupling effects. Such theory has already been applied to a host of problems and predictions agreed with experiments[19-21]. Although great advances have been made in the theory and applications of thermomechanical coupling, most of them, except Sih's theory, were aimed at metallic continua[4,5,7,10-13], with a few dealing with nonmetallic materials[22-24] or cracked bodies[19,25,26]. Recently, Kinra and Bishop[27] presented an approximate analysis to a Griffith crack subjected to time harmonic loading in model I, II and III. During the last two decades in this century, taking the dissipation and rheology of defect evolution process as main research chains, Yuan emphatically studied the thermomechanical coupling and thermomagnetic effects, and proposed a new theory — rheology of bodies with defects[28-31]. It is generally pointed out by experiments and simulation results due to the thermomechanical coupling models mentioned above that during the initial elastic response the deformation causes cooling in uniaxial tension and heating in uniaxial compression, while pure torsion shows no temperature change for isotropic materials, and monotonous deformation in the inelastic region causes only self-heating. Temperature fluctuation in a thermoplastic specimen with prefabricated defects under uniaxial tensile plane stress is presented and preliminarily analyzed herein in view of micro- and mesoscopic physical characters of the deformation of thermoplastics, avoiding overelaborate procedures for modeling and simulating. The heat generated during inelastic deformation is calculated from the local temperature field and compared with the external work done.

2. Thermomechanical coupling equation

The coupled thermomechanical equations of a continuous media are defined by the balance of momenta and conservation of energy:

$$\sigma_{ij,i} + \rho f_j = \rho \ddot{u}_j \ , \qquad \sigma_{ij} = \sigma_{ji} \tag{1}$$

$$\rho \dot{e} = \rho r - q_{i,i} + \sigma_{ij} \dot{\varepsilon}_{ij} \tag{2}$$

where σ_{ij} express the stress tensor components, f_j stand for the body force vector components, ρ is the mass density, u_j are the displacement vector components, e denotes the internal energy per unit mass, q_j are the heat flux vector components per unit area of the boundary surface, r is the heat supply rate per unit mass, and ε_{ij} are the strain tensor components. Dots on top of the letters denote derivations with respect to time.

For any thermomechanical process, the above equations are restricted by the second law of thermodynamics, i.e.,

$$\rho T\dot{s} - \rho r + q_{i,i} - \frac{q_i}{T} T_{,i} \geq 0 \tag{3}$$

where s is the entropy per unit mass and T is the absolute temperature.

Considering that the thermomechanical state at a point of the inelastic material is defined by the current values of the independent variables ε_{ij}, T and ε_{ij}^I, where ε_{ij}^I is the inelastic strain tensor, we can postulate the following general constitutive relations:

$$\sigma_{ij} = \sigma_{ij}(\varepsilon_{ij}, T, \varepsilon_{ij}^I), \tag{4}$$

$$e = e(\varepsilon_{ij}, T, \varepsilon_{ij}^I) \tag{5}$$

$$s = s(\varepsilon_{ij}, T, \varepsilon_{ij}^I) \tag{6}$$

$$q_i = q_i(\varepsilon_{ij}, T, \varepsilon_{ij}^I) \tag{7}$$

Let Helmhotz free energy be

$$\phi = \phi(\varepsilon_{ij}, T, \varepsilon_{ij}^I) = e - Ts \tag{8}$$

therefore,

$$\dot{\phi} = \frac{\partial \phi}{\partial \varepsilon_{ij}} \dot{\varepsilon}_{ij} + \frac{\partial \phi}{\partial T} \dot{T} + \frac{\partial \phi}{\partial \varepsilon_{ij}^I} \dot{\varepsilon}_{ij}^I \tag{9}$$

Substituting eq. (8) into eq. (2) and utilizing eqs (9) and (3), we can obtain

$$\sigma_{ij} = \rho \frac{\partial \phi}{\partial \varepsilon_{ij}} \tag{10}$$

$$s = -\frac{\partial \phi}{\partial T} \tag{11}$$

$$\rho T\dot{s} + \rho \frac{\partial \phi}{\partial \varepsilon_{ij}^I} \dot{\varepsilon}_{ij}^I + q_{i,i} - \rho r = 0 \tag{12}$$

and

$$-\rho \frac{\partial \phi}{\partial \varepsilon_{ij}^I} \dot{\varepsilon}_{ij}^I - \frac{q_i}{T} T_{,i} \geq 0 \tag{13}$$

where the first term in the left hand side of eq.(13) represents internal dissipation and the second term is the heat conduction dissipation.

According to eq.(6)

$$\dot{s} = \frac{\partial s}{\partial \varepsilon_{ij}} \dot{\varepsilon}_{ij} + \frac{\partial s}{\partial T} \dot{T} + \frac{\partial s}{\partial \varepsilon_{ij}^I} \dot{\varepsilon}_{ij}^I \tag{14}$$

Using eq.(10) and substituting (14) into (12) there results

$$\rho \frac{\partial \phi}{\partial \varepsilon_{ij}^I} \dot{\varepsilon}_{ij}^I - \rho T \frac{\partial^2 \phi}{\partial \varepsilon_{ij}^I \partial T} \dot{\varepsilon}_{ij}^I - \rho T \frac{\partial^2 \phi}{\partial T^2} \dot{T} - \rho T \frac{\partial^2 \phi}{\partial T \partial \varepsilon_{ij}} \dot{\varepsilon}_{ij} + q_{i,i} - \rho r = 0 \tag{15}$$

If we assume the Fourier's law

$$q_i = -k_{ij}T_{,j} \tag{16}$$

where the thermal conductivity tensor k_{ij} may depend on ε_{ij}, T and ε_{ij}^I, it becomes clear that the only variable required to define the thermomechanical problem is ϕ, which can be expanded in a second order Taylor series in its arguments as follows[13]:

$$\begin{aligned}\phi = \rho^{-1}(A + B\hat{\theta} + C_{ij}\varepsilon_{ij} + \tfrac{1}{2}D_{ijkl}\varepsilon_{ij}\varepsilon_{kl} + E_{ij}\varepsilon_{ij}^I + F_{ijkl}\varepsilon_{ij}\varepsilon_{kl}^I \\ + \tfrac{1}{2}G_{ijkl}\varepsilon_{ij}^I\varepsilon_{kl}^I + H_{ij}\varepsilon_{ij}\hat{\theta} + I_{ij}\varepsilon_{ij}^I\hat{\theta} + K\hat{\theta}^2)\end{aligned} \tag{17}$$

where $\hat{\theta} = T - T_0$, and T_0 is the reference temperature at which no strain is observed. Substituting (17) into (10), we get

$$\sigma_{ij} = C_{ij} + D_{ijkl}\varepsilon_{kl} + F_{ijkl}\varepsilon_{kl}^I + H_{ij}\hat{\theta} \tag{18}$$

For many inelastic materials, we may define[13]

$$\begin{aligned}C_{ij} &= -E_{ij} \equiv \sigma_{ij}^0 \\ F_{ijkl} &= -D_{ijkl} = -G_{ijkl} \\ I_{ij} &= -H_{ij} \equiv D_{ijkl}\beta_{kl}\end{aligned} \tag{19}$$

where σ_{ij}^0 is the initial stress at zero strain and temperature change, D_{ijkl} is the elastic modulus tensor, and β_{kl} is called the thermal expansion tensor.
Thus

$$\sigma_{ij} = \sigma_{ij}^0 + D_{ijkl}(\varepsilon_{kl} - \varepsilon_{kl}^I - \beta_{kl}\hat{\theta}) \tag{20}$$

Substituting (17) into (15) and using (16) and (19), we can obtain

$$(k_{ij}T_{,j})_{,i} = \rho c_\rho \dot{T} - \rho r + D_{ijkl}\beta_{kl}T(\dot{\varepsilon}_{ij} - \dot{\varepsilon}_{ij}^I) - \sigma_{ij}\dot{\varepsilon}_{ij}^I \tag{21}$$

where $c_\rho = -T\dfrac{\partial^2\phi}{\partial T^2}$ is the specific heat at constant strain per unit mass.The above equation is the coupled heat conduction equation for anisotropic inelastic media. For an isotropic elastic material, the above equation reduces to

$$kT_{,ii} = \rho c_\rho \dot{T} - \rho r + (3\lambda + 2\mu)\beta T\dot{\varepsilon}_{kk} \tag{22}$$

Using eqs (17) and (19), eq.(11) is led to

$$\rho T(s - s_0) = TD_{ijkl}\beta_{kl}(\varepsilon_{ij} - \varepsilon_{ij}^I) + \rho c_\rho\hat{\theta} \tag{23}$$

where s_0 represents the initial entropy per unit mass. For an isotropic elastic material, that equation reduces to

$$\rho T(s - s_0) = T(3\lambda + 2\mu)\beta\varepsilon_{kk} + \rho c_\rho\hat{\theta} \tag{24}$$

Eq.(24) shows that the entropy change in an isotropic elastic material results from the temperature change and the volumetric strain.

3. Experimental procedure

3.1. SPECIMEN AND MATERIAL

Consider a thin flat defective specimen made of Acrylonitrile-Butadiene-Styrene (ABS), a typical thermoplastic copolymer. The dimensions of the specimen and distribution of defects are shown in figure 1, where d=5mm, $\alpha = 70°$, the specimen thickness h=4mm. The type and number of defects may be different as specimen requires. Table 1 lists the data for three different types of specimen. The physical properties of material tested are listed in table 2.

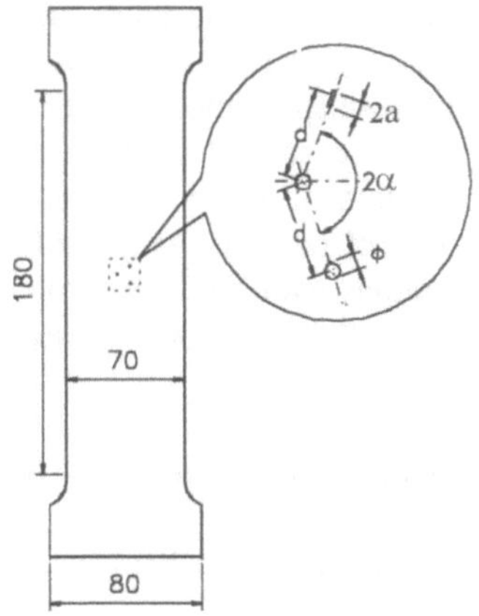

Figure 1. Shape,dimensions and defect distribution of specimen tested

TABLE 1. Defect type and number in different specimen

Specimen type	Defect type	Number of prefabricated defects
A	Mesoscopic void ($\Phi = 1.5\text{mm}$)	2
B	mesoscopic void ($\Phi = 1.5\text{mm}$)	3
C	mesoscopic crack (2a = 1.5mm)	3

TABLE 2. Physical properties of ABS

Density	Tensile strength	Specific heat capacity	Thermal conductivity	Thermal linear expansion
1040kgm^{-3}	$48 \pm 2.5\text{MPa}$	$1406\text{Jkg}^{-1}\text{K}^{-1}$	$0.2 \sim 0.33\text{Wm}^{-1}\text{K}^{-1}$	$9.0 \times 10^{-5}\text{K}^{-1}$

3.2. TEST METHOD

The load-measuring-record-analysis system consists of a LST-50S type electrohydraulic servosystem for fatigue test of materials (LST-50SEHS), a 6T61 type high sensitive infrared scanner (6T61HSIS), a digital/analog tape recorder (D/ATR), a four-channel tape recorder (FCTR), a dual channel spectrum analyzer (DCSA) and a microcomputer, their interconnection is diagrammatically sketched in figure 2. The specimen is loaded at constant displacement rate of $\dot{u} = 8.80 - 9.25\,\text{mm/min}$, the ambient temperature is 15.2℃.

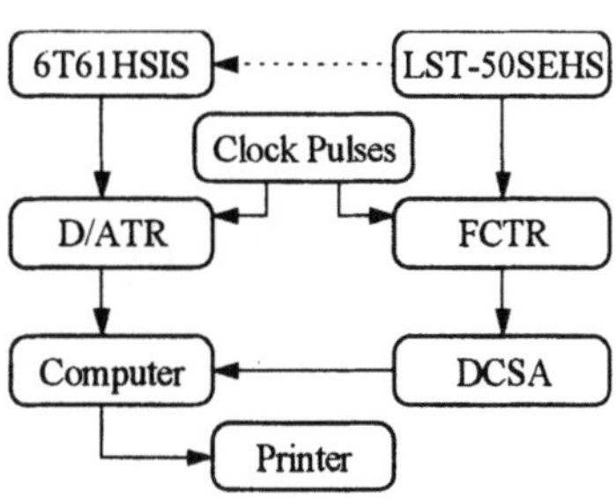

Figure 2. Diagrammatic sketch of test system

The electric signals of the loads and displacements during testing, together with the clock pulses, are recorded with tapes, then converted into digital values by DCSA, and finally the load-displacement curves can be obtained via computer. The temperature of specimen surface can be demarcated analogously with different colors, the analog signals of colors also can be converted into digital values by the analog-to-digital converter unit of D/ATR. As a result, all information of the surface temperature field and its variations with time are recorded in D/ATR, and from these information, some useful curves and data can be obtained as stated in the following sections.

4. Experimental results and analysis

4.1. LOCAL TEMPERATURE FIELD

The fluctuations of temperature at every spot near the defects in the specimen are observed and recorded in the experiments. Cooling of the specimen will occur in the initial stage of loading, this indicates there is a certain endothermic excitation embedded in the materials. Stress whitening surrounding the defects will be observed during the further load increasing stage, at this moment, the stress and strain response are already in the nonlinear range as the material is being deformed permanently. The local temperature returns to room condition and then rise steeply up to macroscopic rupture of the specimen. Such phenomena of cooling and heating for ten spots (A,B,C,...,J) near the defects in specimen B are clearly exhibited by the curves in figure 3 which indicate the maximum temperature drop is 1.3 ℃ and the maximum temperature rise is approximate 6.0℃. The close relation between the time history of local temperature field near the defects and the load-displacement (stress-strain) curve is figuratively shown in figure 4. During the elastic stage, the temperature drop is quite uniform over the entire surface of the specimen.

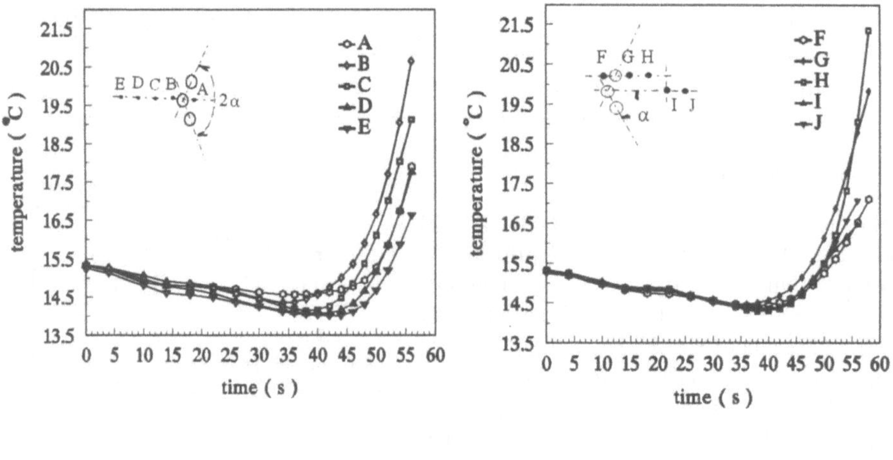

Figure 3. The temperature history of ten spots near defects in specimen B

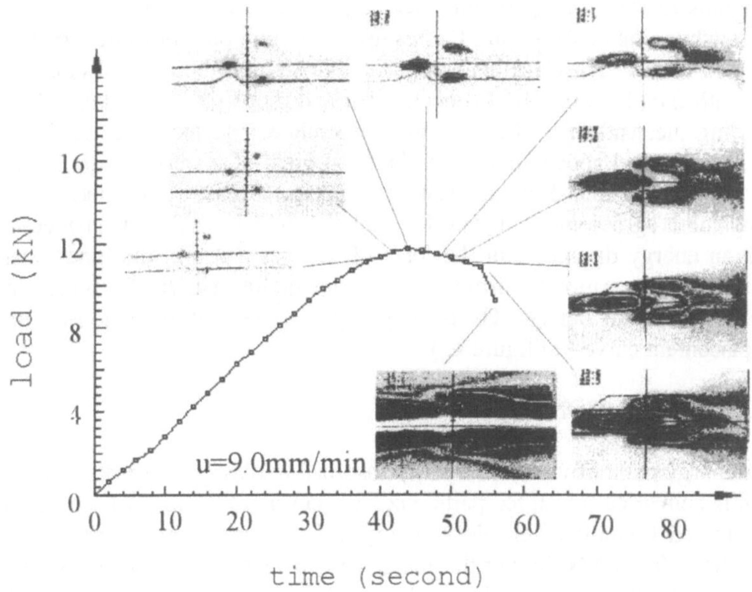

Figure 4. Variations of the local temperature field in specimen B with the load and deformation

Some investigators had drawn their attention to the peculiar phenomena of cooling and heating of materials during deformation in the past thirty years, they laid particular stress on metal materials[2-17]. Although a few experiments for non-metallic materials were reported[22-24], thorough investigations and studies are necessary. To understand the cause and mechanism of the phenomena, we would start with the micro- and mesoscopic physical characters of the deformation of materials.

As we know, in general , the molecular chains of polymers are asymmetric in geometry, their length-diameter ratio may vary at 2~4 orders of magnitude. The chain segments of these molecular chains will arrange in parallel to form an orientated structure. In fact, such orientation is the ordering process in which the molecules under external force field overcome the weak Van der Waals force, and it indicates the decrease of conformational entropy of material. The material is assumed to be isotropic initially, according to eq.(24) in which $\varepsilon_{kk} > 0$ for the uniaxial tension state in our experiments, during the initial stage of loading, the decrease of the conformational entropy of material, that is $s - s_0 < 0$, will give us

$$\hat{\theta} = T - T_0 < 0 \tag{25}$$

Eq.(25) agrees well with the cooling phenomenon observed and indicates the grouping of heat sinks.

As the applied load increases, the orientational motion will tend to reach the limit, and such state will firstly occur in the defective area with stress highly concentrated. At this moment, the local temperature of the material near the defects in specimen B will drop down to a valley at 36~38 sec, then the molecular chains subjected to high stresses will break, the stresses redistribute to the adjacent chains. Thus the chain rupture occurs concentratedly in a local area, and accumulates to form microvoids, when the quantity of microvoids reaches a critical value, a stable craze structure will come into being[33]. The stress whitening observed are the results of the grouping of quite small but highly concentrated crazes. Craze retains a considerable strength because of craze matters, it expands with meniscus instability mechanism and increases its thickness by interface cold-drawing mechanism in the direction perpendicular to the interface. The physical characters mentioned above show how the local yield of material at craze tip happens and the crazes are in cold-drawing plastic state, hence a plastic deformation zone comes into being and it agrees with Dugdale-Barenblatt's model in shape. Considerable plastic deformation energy dissipates in the form of heat, as a result, the inner heat sources provide the heat generation to form the local temperature field, they gathered in the area near the corresponding defects. The heating stage corresponds to the non-linear range of load-displacement curve (*cf* figure 4.).

4.2. HEAT GENERATION

From the energy point of view, the energy absorbed in plastic deformation during defect evolution is converted into three parts, viz., a small part of free elastic potential which can be released because of elastic recovery, a little larger part of restraint potential stored in the deformed body and the major part of plastic heat which forms the local temperature field. Without considering the heat transfer between deformed body and its environment, the generated heat, $\widetilde{Q}$,at time τ during the defect evolution can be calculated from the temperature field as follows

$$\widetilde{Q} = \int_V \rho\, c_p \hat{\theta}(\mathbf{x}, \tau)\, dV \tag{26}$$

in which, $\mathbf{x}$ denotes the position vector in the temperature field, and strictly speaking, c_p and ρ are also functions of temperature and the position vector, but since the temperature change is small, they are taken to be constants in the following analysis. It is difficult to get the exact expression of the temperature-rise function from the experimental data, considering the thin plate specimen tested, the temperature will be nearly the same along the thickness of the specimen, we may consider the isothermal surfaces to be perpendicular to the surfaces of the specimen, eq.(26) can be written as

$$\widetilde{Q} = \rho\, c_p h \sum_{i=1}^{n} (\mathrm{A}_i - \mathrm{A}_{i-1}) \hat{\theta}_i \tag{27}$$

where h denotes the thickness of specimen, A_i , A_{i-1} are the areas encircled by two adjacent isothermal lines respectively; the area located between two such isothermal lines is considered to be isothermal in which the temperature rise is $\hat{\theta}_i$.

If the areas, S_i, encircled by isothermal lines with regular temperature intervals (say $\hat{\theta}_0$) are measured, the generated heat revealed in the temperature field can be expressed by

$$\widetilde{Q} = \rho\, c_p h \hat{\theta}_0 \sum_{i=1}^{n} S_i \tag{28}$$

For $\hat{\theta}_0 = 0.25$ ℃ in our experiments, according to eq.(28), the variation of heat generated with the deformation displacement is calculated and compared with the external work for specimen A,B and C (*cf* figure 5.)

Figure 5(a) shows that, before the rupture of the specimen, the maximum heat generated is 24.3J(oule) while the external work is 56.35J at that moment for specimen A, in specimen B, 28.2J for heat generated and 63.74J for external work(figure 5(b)), in specimen C, 27.08J for heat generated and 38.7J for external work(figure 5(c)). It is shown that the heat generated in specimen with mesovoids comprises some 25~44 per cent of the external work, while that in specimen with mesocracks comprises some 45~70 per cent of the external work in our experiments. This indicates that a considerable part of external work is dissipated in irreversible deformation and converted into heat. The heat generated during defect evolution is significant and not negligible.

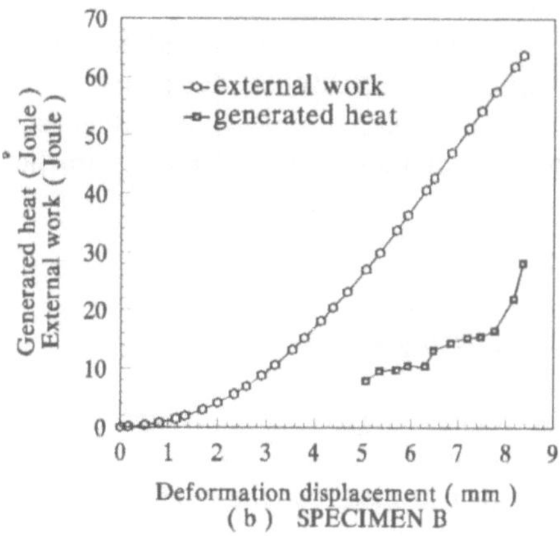

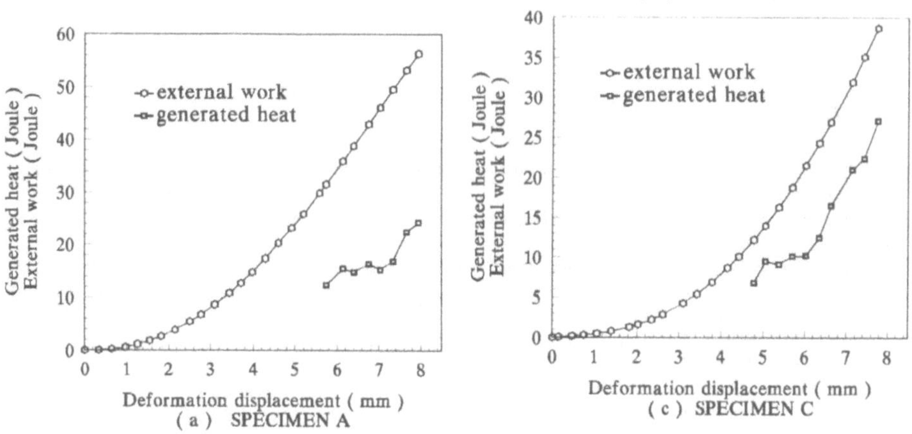

Figure 5. Variations of generated heat and external work with deformation displacement

5. Conclusions

The initiation and evolution of the local temperature field near the defects is observed in the paper. The cooling and heating phenomena and the heat generation in the process of defect evolution are experimentally and theoretically analyzed. Combining the micro- and mesoscopic physical characters of the deformation of materials, we consider it reasonable that the cooling of specimen under uniaxial tension occurs in the range of elastic deformation. It corresponds to the molecular re-orientations at microlevel, and indicates the grouping of inner heat sinks. The analysis of failure process which cuts across micro-, meso- and macrolevels shows that the plastic zone is formed, and the plastic deformation energy will normally dissipates by means of heat waves, the viscous drag to the propagation of the heat waves will result in the grouping of heat sources, therefore the non-uniform temperature field will be formed. Based on the experimental data, the heat generated is calculated which comprises some 25~70 per cent of the external work. So its contribution to the failure of materials is not negligible. However, it is obvious that the quantity of heat generated, the maximum temperature-rise and the temperature-decrease are dependent on the types of material, the load conditions, the type and number of defects, and even the shape and dimensions of the specimen, detail analyses of the influential factors in the defect evolution will be made in our later studies.

Acknowledgements

The author is grateful to Professor R. Wang (Peking University) and L.W. Yuan (Xiangtan University) for their useful discussions and suggestions. The financial support provided by the Natural Science Foundation of China is also sincerely acknowledged.

References

1 Duhamel, J.M.C.(1837) Second memoire sur les phenomenes thermo-mecaniques, *J de l'Ecole Polytechnique* **15**(25),1-57.
2 Farren, W.S. and Taylor G.I.(1925) The heat developed during plastic extension of metals, *Proc. Roy. Soc.* A**107**,422-451.
3 Prigogine I.(1962) *Introduction to nonequilibrium thermodynamics*, Wiley-Interscience, New York.
4 Dillon O.W.Jr.(1962) An experimental study of the heat generated during torsional oscillations, *J. Mech. Phys. Solids* **10**,235-244.
5 Dillon O.W.Jr. (1966) The heat generated during the torsional oscillations of copper tubes, *Int. J. Solids Structures* **2**,181-204.
6 Dillon O.W.Jr. (1962) A nonlinear thermoelastic theory, *J. Mech. Phys. Solids* **10**,123-131.
7 Cernocky E.P. and Krempl E.(1980) A theory of thermoviscoplasticity based on infinitesimal total strain, *Int. J. Solids Structures* **16**,723-741.
8 Ghoneim H.(1986) Thermoviscoplasticity by finite element: dynamic loading a thick-walled cylinder, *J. Thermal Stresses* **9**, 345-358.
9 Ghoneim H. and Matsuoka S.(1987) Thermoviscoplasticity by finite element: tensile and compression test, *Int. J. Solids Structures* **23**(8), 1133-1143.
10 Ghoneim H.(1990) Analysis and applications of a coupled thermoviscoplasticity theory, *J. Appl. Mech.* **57**,828-835.
11 Allen D.H.(1985) A prediction of heat generation in a thermoviscoplastic uniaxial bar, *Int. J Solids Structures* **21**(4),325-342.
12 Allen D.H.(1986) Predicted axial temperature gradient in a viscoplastic uniaxial bar due to thermomechanical coupling, *Int. J. Num. Methods Eng.* **23**(5),903-917.
13 Allen D.H.(1991) Thermomechanical coupling in inelastic solids, *Appl. Mech. Rev.* **44**(8),361-373.
14 Schapery R.A.(1969) On the characterization of nonlinear viscoelastic materials, *Poly. Eng. Sci.* **9**,295-310.
15 Christensen R.M.(1971) *Theory of Viscoelasticity, an Introduction*. Academic Press, New York.
16 Christensen R.M.(1967) Linear non-isothermal viscoelastic solids, *Acta Mechanica*, **3**(1),1-12.
17 Crochet M.I. and Naghdi P.M.(1969) A class of simple solids with fading m emory, *Int. J. Eng. Sci.*, **7**,1173-1198.

18 Sih G.C.(1985) Mechanics and physics of energy density theory, *Theoret. Appl. Fracture Mech*, **4**(3), 157-173.

19 Sih G.C. and Tzou D.Y.(1986) Heating predicted by cooling ahead of crack: macrodamage free zone, *Theoret. Appl. Fracture Mech.* **6**(2),103-111.

20 Sih G.C. and Chao C.K.(1989) Scaling of size/time/temperature part I : Progressive damage in uniaxial tensile specimen; part II : Progressive damage in uniaxial compressive specimen, *Theoret. Appl. Fracture Mech.* **12**(2),93-108; **12**(2),109-119.

21 Sih G.C., Lieu F.L. and Chao C.K.(1987) Thermal/mechanical damage of 6061-T6 aluminium tensile specimen, *Theoret. Appl. Fracture Mech.*7(2),67-78.

22 Tauchert T.R.(1967) The temperature generated during torsional oscillations of polyethylene rods, *Int. J. Eng. Sci.* **5**,353-365.

23 Tauchert T.R. and Afzal S.M.(1967) Heat generated during torsional oscillations of polymethylmethacrylate tubes, *J. Appl. Phys.* **38**,4568-4572.

24 Wu Z. and Glockner P.G.(1996) Thermo-mechanical coupling applied to plastics, *Int. J. Solids Structures* **33**(29),4431-4448.

25 Tzou D.Y. and Sih G.C.(1988) Thermal/mechanical interaction of subcritical crack growth in tensile specimen, *Theoret. Appl. Fracture Mech.* **10**(1),59-72.

26 Pippan R. and Stuwe H.P.(1982) The temperature field surrounding the fatigue crack tip, in K.L.Maurer and F.E.Matzer (eds.), *Proc. of the 4th E.C.F.* **2**, 457-459.

27 Kinra V.K. and Bishop J.E.(1996) Elastothermodynamic analysis of a griffith crack, *J. Mech. Phys. Solids* **44**(8),1305-1336.

28 Yuan Longwei, Swartz S.E. and Hu K.K.(1993) *Rheology of Defect Body*, Springer Verlag Press, NY.

29 Yuan Longwei.(1988) Theoretical and experimental research on nonlinear rheological fracture for polymer solids, *Proc. Xth Int. Cong. Rheology*, Monash University Printing Service, Sydney **2**,389-391.

30 Gao Yunxin and Yuan Longwei.(1991) Thermodynamical coupled relation of dissipative rheological solid, *Proc. China-Japan Int. Conf. Rheology*, Peking University Press, Beijing, 338-342.

31 Yuan Longwei, Li Zhida.(1989) On the rheological and dissipative phenomena in the process of crack propogation, *Natural Science Journal of Xiangtan University* **11**(1), 34-63.

32 Wang Ren.(1996) A review on creep failure of polymer and polymer composite, in T. Abe and T.Tsuta(eds.), Advances in Engineering Plasticity and its Applications, *Proc. of AEPA '96* (Hiroshima Japan), Pergamon Press, Amsterdam, 43-52.

RHEOLOGICAL-THERMAL FRACTURE BY LASER BEAM

Y. C. Zhou[1,2], Z. M. Zhu[1], Z. P. Duan[2] and Q. B. Yang[1]
[1]*Department of Physics, Xiangtan University, 411105, Hunan*
[2]*Laboratory for Laser and Dynamic Behaviors of Materials, Institute of Mechanics, CAS, Beijing, 100080, P. R. China*

ABSTRACT The bodies with defects subjected to laser beam thermal shock will successively undergo a variety of rheological processes. In this paper, the characteristics of rheological thermal fracture of materials with pre-existing crack are theoretically investigated by means of a simple model. Also, the behavior of rheological thermal fracture of metal matrix composites induced by laser beam thermal shock has been experimentally studied.

1. Introduction

Great attention has been paid to high power laser in the fields of materials and/or structure damage and laser processing[1-4]. Interaction of a high power laser with materials leads to two kinds of coupled damage, mechanical and thermal damage. Depending on the laser parameters and the mechanical properties of the target material, material damage could occur by spallation, melting and/or vaporization. However, in the previous studies on laser-induced material damage, people ignored more or less the fact that potential flaws may exist in components. The effect of pre-existing flaws and rapid heating environment must be considered in accurate damage tolerance analysis. If there are defects in such structures, the bodies subjected to laser beam thermal shock will successively undergo a variety of rheological processes. Such processes may be the absorption of light energy, temperature rise, thermal viscoplastic deformation, energy dissipation, intense thermal stress concentration around defects, the growth and coalescence of defects as well as the degradation of strength properties at elevated temperatures caused by the heating. The non-linear coupled effects of these rheological processes may result in catastrophic failure. The damage analysis for bodies with defects exposed to intense thermal shock is very important for determining the safety of

R. Wang (ed.), IUTAM Symposium on Rheology of Bodies with Defects, 121–132.

materials, components, especially the aerospace structural components .

In this paper, the characteristics of rheological thermal fracture of materials with pre-existing crack subjected to laser beam thermal shock are theoretically investigated by means of a simple model. Also, the behavior of rheological thermal failure of metal matrix composites induced by laser beam thermal shock is experimentally studied.

2. Characteristics of rheological thermal fracture

Many papers have been written on thermal shock problems. Gupta[5] determined the strength degradation and crack propagation in thermally shocked Al_2O_3. Schneider and Petzow[6] developed a new testing method to determine the thermal shock fracture toughness K_c^{TS} of Si_3N_4 up to 1000°C. The characteristics of thermal fracture induced by laser beam are different from those induced by general mechanical load. To understand the characteristics of rheological thermal fracture induced by laser beam we first analyze the transient thermal stress intensity factors (TSIFs) for a center crack of length 2c in an infinite plate subjected to laser beam thermal shock.

2.1 Theoretical model

Figure 1 shows the problem of interest. When the infinite plate is heated by laser beam the tangential thermal stresses $\sigma_{\theta\theta}(r,t)$ are illustrated in figure 2. From the results shown in Fig.2, we can see that tangential thermal stresses $\sigma_{\theta\theta}(r,t)$ take on negative values within the laser spot region and positive values around that region, respectively. Only tangential thermal stress $\sigma_{\theta\theta}(r,t)$ may contribute to the contact or opening of crack. From Fig.2, one can conclude that the compressive thermal stresses $\sigma_{\theta\theta}(r,t)$ within the heat affected zone cause the crack surface to come into contact with each other over a certain contact length $2b$ and the tensile stresses around the heat affected zone may tend to open the crack. Consequently, the problem may be treated as a pair of embedded cracks with a smooth closure condition at the center (Fig.1(c)).

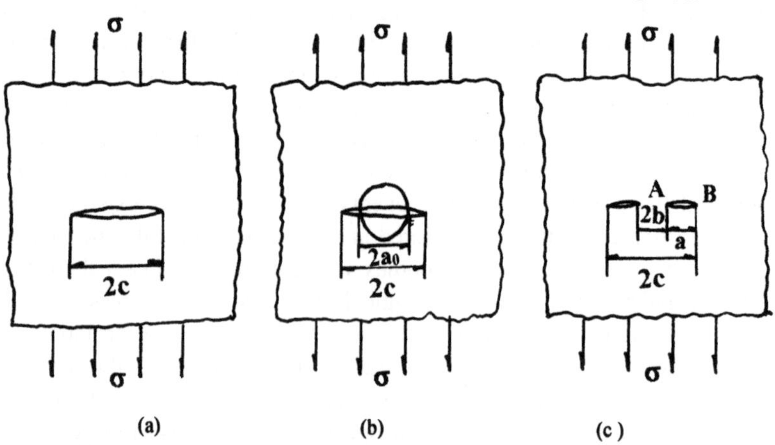

Figure 1. A center crack of length 2c in an infinite plate subjected to laser beam thermal shock: (a) center crack geometry; (b) center crack heated by laser beam; (c) the heated center crack with crack contact length 2b.

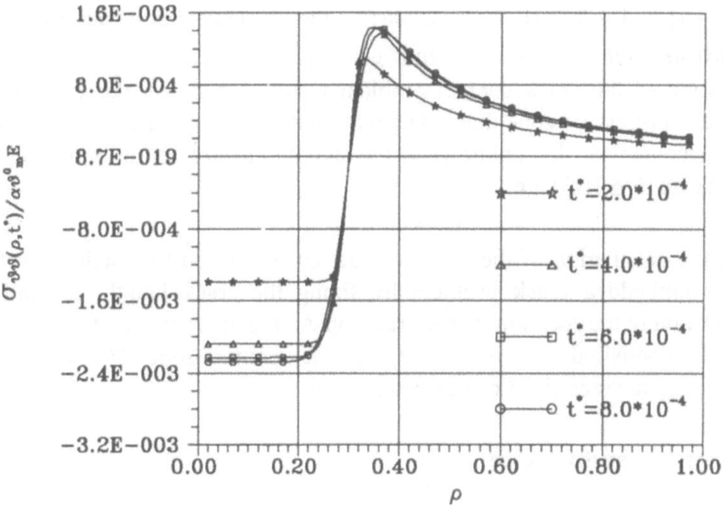

Figure 2. Transient tangential thermal stresses $\sigma_{\theta\theta}(r,t)$.

The weight function method proposed by Bueckner[7] and Rice[8] has proved to be a very useful and versatile method of calculating stress intensity factors, especially for cracks subjected to non-uniform stress fields. Once the weight function $m(x,a)$ for a particular cracked body in mode I is determined, the stress intensity factor can be calculated by integrating the weight function $m(x,a)$ and the stress distribution $\sigma(x)$ acting in the prospective crack plane. This results in the following expression for the stress intensity factor K,

$$K=\int_0^a \sigma(x)m(x,a)dx \tag{1}$$

where a is the crack length. After lengthy but straightforward derivation, we obtain the following transient thermal stress intensity factors K_{IA} and K_{IB} which are primary interest.

$$K_{IA}=-\frac{1}{\sqrt{2\pi a}F_{IAr}(a)}\frac{\partial}{\partial a}\int_0^a\left[4F_{IAr}(a)\sqrt{a(a-x)}+\frac{1}{\sqrt{a}}G_{IAr}(a)(a-x)^{3/2}\right]\sigma_{\theta\theta}^{A}(x,t)dx \tag{2}$$

$$K_{IB}=-\frac{1}{\sqrt{2\pi a}F_{IBr}(a)}\frac{\partial}{\partial a}\int_0^a\left[4F_{IBr}(a)\sqrt{a(a-x)}+\frac{1}{\sqrt{a}}G_{IBr}(a)(a-x)^{3/2}\right]\sigma_{\theta\theta}^{B}(x,t)dx \tag{3}$$

where $F_{IAr}(a)$ and $F_{IBr}(a)$ are the parameters of the crack opening displacement functions at the crack tips A and B, respectively. Also, $G_{IAr}(a)$ and $G_{IBr}(a)$ are functions of $F_{IAr}(a)$ and $F_{IBr}(a)$, respectively. Their detail expressions are found in [9].

To obtain $\sigma_{\theta\theta}^{A}(x,t)$ and $\sigma_{\theta\theta}^{B}(x,t)$, the abscissa x in the tangential hoop stresses $\sigma_{\theta\theta}(r,t)$ have to be transformed to $x = a + b - r$ and $x = r - b$, respectively.

The formulation of the crack contact problem would depend on the crack contact length 2b in the compressive zone that is an additional unknown variable. The physical condition that accounts for this unknown is the smooth closure condition of the crack surfaces at crack tip (Fig.1(c)), i.e.,

$$K_{IA} = 0 \tag{4}$$

Consequently, the calculation of the crack contact zone $2b$ can be obtained from the solution of the embedded crack problem by fixing the crack length $2c$ and then determining iteratively the location of the crack tip A (Fig.1(c)) at each time step such that equation (4) is satisfied. In this manner, the crack contact zone $2b$ and the stress intensity factor K_{IB} at crack tip B can be determined.

2.2 Thermal stress intensity factor

The normalized TSIFs $K_B^* = K_B / \alpha E \theta_m^0 \sqrt{c}$ at crack tip B are calculated and in this case the spatial shape $f(r)$ and temporal shape $g(t)$ are respectively assumed as

$$f(r) = \begin{cases} 1, & if \quad 0 \le r \le a_0, \\ 0, & if \quad a_0 < r < \infty \end{cases} \tag{5}$$

$$g(t) = e^{-\alpha_1 t}\left(1 - e^{-\beta_1 t}\right) \tag{6}$$

where a_0, α, E, θ_m^0 are the laser spot radius, thermal expansion coefficient, Young's modulus , reference temperature, α_1 and β_1 are determined experimentally,

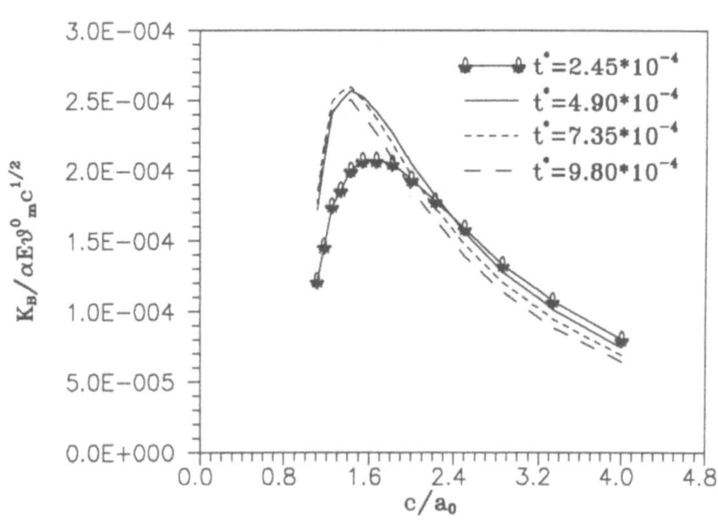

Figure 3. TSIFs versus the crack length.

The TSIFs originally increase as a function of time and reach maximum value at some time, then decrease. The crack-length-dependent stress intensity factor is shown in figure 3, where t* is non-dimensional time. The maximum in the stress intensity over the crack length curves is typical for thermal shock loading and separates the region of stable and unstable crack growth. This behavior is different from that under mechanical loading, where external forces are applied and the SIF increases with the square root of the crack length. The crack length c_m that corresponds to the maximum TSIF K_m decreases with time. This result is different in quenching experiments where c_m increase with time[10]. If for crack lengths $c \langle c_m$ the critical TSIF is exceeded, the cracks propagate in an unstable manner, because the TSIF increases until $c = c_m$. The effect of unstable crack propagation was very pronounced for initial notch lengths. This demonstrates that such cracks continued to grow also in the region where the static $K(c)$ was below K_c. Hence the static crack arrest criterion $K(c) \leq K_c$ does not hold in this case, and the crack arrest should be discussed by using the dynamic theory of elasticity.

3. Rheological thermal fracture of metal matrix composites

Particular interest in the aerospace industry has been directed towards ceramic-reinforced metal matrix composites (MMCs)[11-12]. Thermal shock may give rise to intense thermal stresses in the components around cracks and other kinds of defects and especially in the components of ceramic-reinforced metal matrix composites as a result of the thermoelastic mismatch in the metal and ceramic. The problem of the rheological thermal fracture of metal matrix composites is very complicated, the present investigation experimentally examines the fracture behavior of SiC particulate reinforced 6061 aluminum alloy under laser thermal shock.

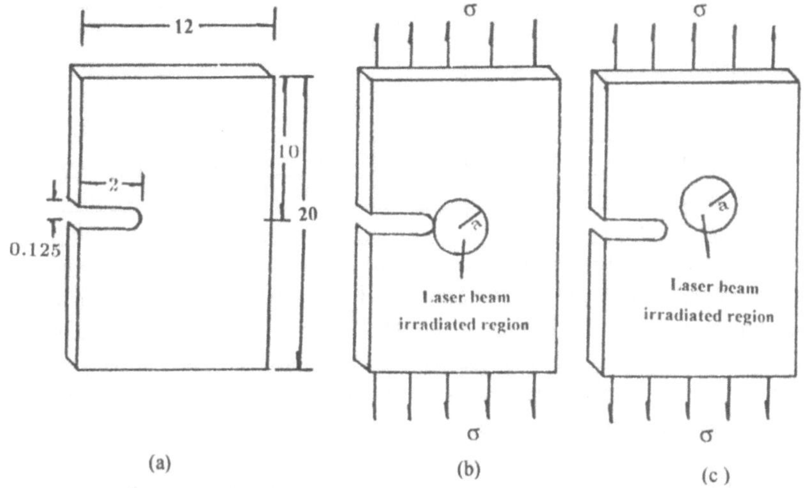

Figure 4. Specimen configuration: (a) the dimensions of the single edge notched specimen; (b) the notched-tip region subjected to laser beam heating; (c) the laser irradiated region away from the notched-tip region

3.1 Experimental procedure

Thermal shock is generated by an incident laser beam that impinges normally to a single edge notched specimen. The energy of the laser beam ranges from 1J to 5J with an intensity of the order of 1.0×10^4 to $9.0\times 10^4 W/cm^2$.It is a single pulse Nd:glass laser with a wavelength of 1.06 μm and a FWHM (full width at half of the maximum) of 1.0 ms and 250 μs. SiC particulate/6061 Al composite is chosen as a model MMC system for this study. The composites with 15wt% SiC were fabricated by a melt casting route, and as-cast ingots of the composite were subsequently extruded. The samples are cut from the hot-pressed billets with dimensions of $20\times 12\times 0.25mm$. The test samples were the single-edge notched specimens placed in the static tensile machine. So the thermal failure is induced by both laser thermal shock and far-field mechanical load. Figure 4 shows the specimen configuration and dimensions. The rear surface of the notched specimen which is polished is examined using both optical and scanning electron microscopy (SEM) and the front surface of the specimen is irradiated by the laser beam. The notched-tip region is subjected to laser beam heating. The mechanical load is tensile stress of $36.7\,MPa$ and the laser spot size is $5.0mm$ in diameter.

3.2 Feature of thermal fracture of metal matrix composites

Table I Macro-Phenomena and Threshold Intensity

No.	FWHM	Energy $E_J(J)$	Laser spot diameter (mm)	Energy density (J/cm^2)	Intensity (kW/cm^2)	Static tensile stress (MPa)	Macro-phenomena
4		0		0	0	36.7	(a)
5	1.0ms	2.13	5.0	10.8	10.8	36.7	(a)
7	1.0ms	2.63	5.0	13.4	13.4	36.7	(b)
6	1.0ms	3.13	5.0	15.9	15.9	36.7	(c)
8	1.0ms	3.38	5.0	17.2	17.2	36.7	(d)
9	250 μs	2.75	5.0	14.0	56.0	36.7	(a)
10	250 μs	2.88	5.0	14.6	58.6	36.7	(e)
11	250 μs	3.50	5.0	17.8	71.3	36.7	(f)
12	250 μs	4.40	5.0	22.3	89.1	36.7	(d)

(a) No observable macro-damage; (b) In the notched-tip region, the voids appear which occur in the form of interfacial debonding between the particles and the matrix; (c) Macro-crack is observed in the notched-tip region and the macro-crack is 0.13mm in the length; (d) The sample is totally fractured; (e) The laser irradiated region is away from the notched-tip region. The radial cracks are observed around the periphery of the laser beam. (f) Macrocracking is observed in the notched-tip region and macrocrack is 0.1mm in length.

The macro-phenomena and threshold intensity are given in Table I for the thermal failure of SiC particulate reinforced 6061 aluminium alloy composites induced by both laser thermal shock and far-field mechanical load. According to the test data given in

Table I, when only the mechanical loading exists, no visible macroscopic damage was observed on the rear surface of the specimen. On the other hand, when the laser energy density is respectively lower than $10.8J/cm^2$ and $14J/cm^2$ for 1.0ms and $250\mu s$ FWHM laser beam and the mechanical tensile stress is $36.7\,MPa$, we do not observe the macroscopic damage on the rear surface of the specimen. When the laser energy density is gradually increased, the damage is more and more serious. The specimen is totally fractured as the laser energy density is increased to $17.2J/cm^2$ and $22.3J/cm^2$ for 1.0ms and $250\mu s$ FWHM laser beam, respectively. It is noted that the laser intensity threshold value I_{cr} for the specimen fractured is different for 1.0ms and $250\mu s$ FWHM laser beam. In this case, the other parameters, such as specimen dimensions, laser spot and mechanical loading, are the same.

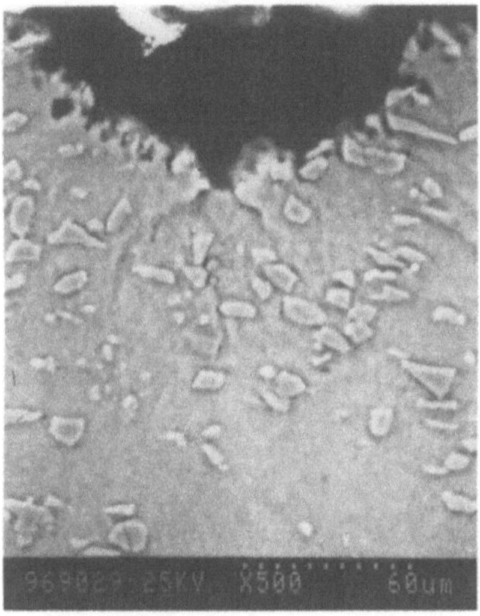

Figure 5. SEM of micro-voids in the notched-tip region

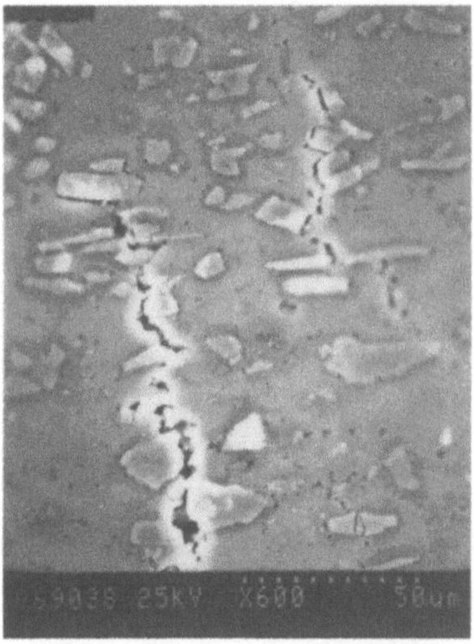

Figure 6. Radial crack showing voids in the matrix and separation of the SiC particle matrix interface for sample No.10

Figure 5 shows the SEM of micro-voids in the notched-tip region with a laser energy density of $13.4J/cm^2$ and 1.0ms FWHM. It is observed that the voids occur in the form of interfacial debonding between the particles and the matrix. In order to understand the initial damage behavior of SiC-particle-reinforced 6061-Al matrix composite induced by laser beam thermal shock, the laser beam irradiated region is moved away from the notched-tip region as shown in Fig.4(c). In this case, the radial crack appeared around the periphery of the laser beam as shown in Fig.6. The radial crack occurs by the void nucleation and growth within the matrix and by the decohesion at the interface between the particle and the matrix. Little SiC particle fracture is observed in Fig.6. When the reinforcement SiC particle is at the crack-tip, the SiC particle does not fracture but resists the crack propagation. One can conclude that the initial damage may be produced in the form of the separation of the SiC particle-matrix interface or in the form of void nucleation and growth within the matrix. The onset and progression of this ductile matrix failure are influenced by the development of local plastic strains and hydrostatic stresses during the coupled loading with both the laser thermal shock and the far-field mechanical load. The initial damage is similar to the fatigue fracture behavior.

When the energy density is increased to 15.9J/cm^2 or 17.8J/cm^2 for 1.0ms and 250μs FWHM laser beam, respectively, the microcracks formed in the notched-region grow into macroscopic cracks. The higher magnification SEM micrographs of macroscopic cracks is shown in Fig.7. It is shown that reinforcement SiC particle fracture is the dominant damage mechanism. The reinforcements are broken by cracks perpendicular to the loading axis and the fraction of broken reinforcements increases near the crack tip zone as shown in Fig.8. Note that the SEM in Fig.8 is distinct from the SEM in Fig.5 and Fig.6 in damage mechanism.

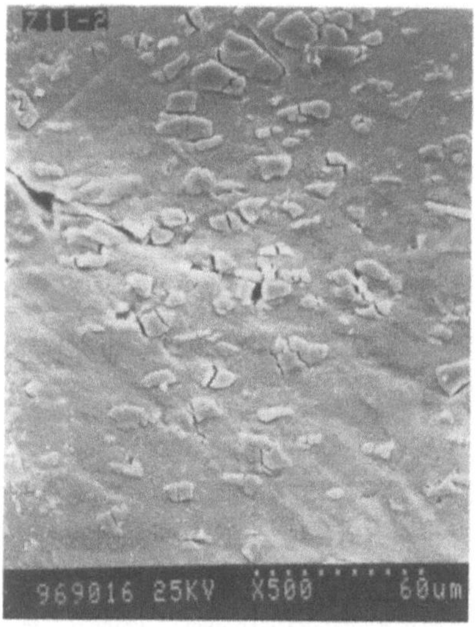

Figure 7. Photograph of macro-crack showing the cracked SiC particles

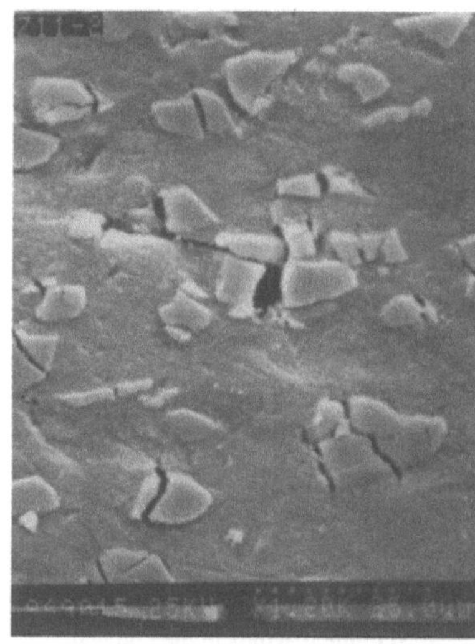

Figure 8. Phenomenon of SiC particles broken in the crack-tip region

3.3 Temperature rise and thermal stress

Surface and internal temperatures in the composite materials can be determined by the diffusion equation

$$\rho C_p \frac{\partial \theta}{\partial t} = k\left(\frac{\partial^2 \theta}{\partial r^2} + \frac{1}{r}\frac{\partial \theta}{\partial r}\right) + \frac{I(r,t)(1.0 - R_0)}{h} \tag{7}$$

where $I(r,t)$ is the intensity of incident laser beam, r is the coordinate in radial direction and the origin is taken at the center of the laser spot, t is time, $\theta(r,t)$ is the temperature rise and the reflection coefficient R_0 is experimentally determined and equals 0.4 approximately. In the above formulas k, ρ, C_p and h are the thermal conductivity, density, specific heat capacity of materials and the thickness of the plate, respectively. During the period of laser irradiation, thermal conductivity effect can be ignored. Therefore, the temperature rise is approximately

$$\theta = \frac{E_J(1.0 - R_0)}{\pi\rho C_p h a^2} \qquad 0 \le r \le a_0 \tag{8}$$

and $\theta = 0$ for $r \rangle a_0$. The temperature at the center of the laser spot is much high than that at the outer rim. The induced higher thermal expansion in the sample center is constrained by the cooler edge, causing tensile hoop stresses at the edge and tangential and radial compressive stresses in the center. For simplicity the sample is assumed to be a disk the thermal stresses is calculated under plane stress condition. The linear elastic solution for radial and tangential thermal stresses $\sigma_{rr}(r,t)$ and $\sigma_{\theta\theta}(r,t)$ has the following form

$$\sigma_{rr}(r,t) = \alpha E[\frac{1}{R^2}\int_0^R \theta(\xi,t)d\xi - \frac{1}{r^2}\int_0^r \theta(\xi,t)\xi d\xi] \tag{9}$$

$$\sigma_{\theta\theta}(r,t) = \alpha E[\frac{1}{R^2}\int_0^R \theta(\xi,t)d\xi + \frac{1}{r^2}\int_0^r \theta(\xi,t)d\xi - \theta(r,t)] \tag{10}$$

where E and α are the Young's modulus and thermal expansion coefficient respectively, R the radius of the disk. Only tangential thermal stress $\sigma_{\theta\theta}(r,t)$ may have contribution to crack opening. From equations (8) and (10), we have

$$\sigma_{\theta\theta} = -\frac{1}{2}\alpha E \frac{E_J(1.0 - R_0)}{\pi\rho C_p h a_0^2}\left[1 - \left(\frac{a_0}{R}\right)^2\right] \qquad 0 \langle r \langle a_0 \tag{11}$$

$$\sigma_{\theta\theta} = \frac{1}{2}\alpha E \frac{E_J(1.0 - R_0)}{\pi\rho C_p h a_0^2}\left[\left(\frac{a_0}{r}\right)^2 + \left(\frac{a_0}{R}\right)^2\right] \qquad a_0 \langle r \langle R \tag{12}$$

where a_0 is the radius of laser spot. We see that the tangential thermal stresses $\sigma_{\theta\theta}(r,t)$ remain negative values within laser spot region and positive values around laser spot region, respectively.

3.4 " Thermal shock " fracture toughness

The weight function method is used to calculate the stress intensity factor of the single edge notched sample. For simplicity we assume the sample to be a disk with a radially orientated single edge crack. Starting from the known solution of a constant stress σ_0 on the crack faces, it was shown by Oliveira and Wu[10,13] that the stress intensity factor of an arbitrary stress distribution has the following solution

$$K_I^T = \Phi\sigma_0\sqrt{\pi c} \tag{13}$$

$$\Phi = \frac{1}{\sqrt{2\pi c\psi_0(\bar{c})}}\int_0^{\bar{c}} \frac{\sigma_{\theta\theta}(x,t)}{\sigma_0}\left[\psi_1(\bar{c})(\bar{c}-x)^{-1/2} + \psi_2(\bar{c})(\bar{c}-x)^{1/2} + \psi_3(\bar{c})(\bar{c}-x)^{3/2}\right]dx \tag{14}$$

where c is the length of notch and $R = c + a_0$, $\bar{c} = c/R$. The solutions for the coefficient functions $\psi_0(\bar{c})$, $\psi_1(\bar{c})$, $\psi_2(\bar{c})$, and $\psi_3(\bar{c})$ for a polynomial of 4th order are given in [10]. To use this resuit, abscissa x in the tangential hoop stresses $\sigma_{\theta\theta}(r,t)$ of equation (10) have to be replaced by $x = 1 - r/R$. Integrating the equation (14), we obtain

$$K_I^T = \sqrt{\pi c}\,\alpha E \theta \phi \tag{15}$$

where

$$\phi = \frac{1}{\sqrt{2\pi \bar{c}\psi_0(\bar{c})}} \frac{1}{2} \bar{a}^{-2} [\psi_1(\bar{c}) \left(\frac{\bar{c}}{\bar{a}} + \frac{\arctan\left(\frac{\bar{c}}{\bar{a}}\right)}{\bar{a}\sqrt{\bar{a}}} + 2\sqrt{\bar{c}} \right) + \psi_2(\bar{c}) \left(\frac{\bar{c}^{-3/2}}{\bar{a}} - \frac{\sqrt{\bar{c}}}{\bar{a}} + \frac{2}{3}\bar{c}^{-3/2} + \frac{\arctan\left(\frac{\bar{c}}{\bar{a}}\right)}{\sqrt{\bar{a}}} \right)$$

$$+ \psi_3(\bar{c}) \left(3\sqrt{\bar{c}} - \bar{c}^{-3/2} + \frac{2}{5}\bar{c}^{-5/2} - \frac{3}{2}\sqrt{\bar{a}} \arctan\left(\frac{\bar{c}}{\bar{a}}\right) \right)] \tag{16}$$

and $\bar{a} = a_0 / R$. Consequently, the specimen subjected to thermal shock shows crack propagation with the minimum laser energy E_J. At this critical condition, the stress intensity factor representing thermal shock fracture toughness K_{Ic}^{TS} is:

$$K_{Ic}^{TS} = K_I^T + K_I^m \tag{17}$$

where K_I^m is the stress intensity factor induced by the far-field mechanical loading and it is given by the following expression[14]

$$K_I^m = F_{Iso}(\lambda)\sigma\sqrt{\pi c} \tag{18}$$

where

$$F_{Iso}(\lambda) = \sqrt{\frac{2}{\pi\lambda}\tan\left(\frac{1}{2}\pi\lambda\right)} \, \frac{0.752 + 2.02\lambda + 0.37\left[1 - \sin\left(\frac{1}{2}\pi\lambda\right)\right]^3}{\cos\left(\frac{1}{2}\pi\lambda\right)} \tag{19}$$

and $\lambda = c/w$, w is the width of the specimen.

The results of thermal shock fracture toughness K_{Ic}^{TS} are listed in Table III with the related parameters listed in Table II. The fracture toughness K_{Ic} at room temperature is also listed in Table III[15]. From the results listed in Table III we can clearly see that the " thermal shock " toughness is larger than that at room temperature. Also as the thermal shock becomes more intense, the thermal shock fracture toughness is larger. This result is the same as that obtained by Schneider and Petzow[6]. Schneider and Petzow measured the thermal shock fracture toughness of Si_3N_4 from room temperature up to 1000°C. Their results showed a constant fracture toughness of $6.1 MPa\sqrt{m}$ up to 750°C and a strong increase at higher temperatures. The dependence of thermal shock fracture toughness on the level of thermal shock is due to the difference of the mechanical properties of the composite for different levels of thermal shock. When the heating is slow and the temperature is not high, the deformation of the material is elastic.

However, at rapid heating rate and at high temperature, the deformation is viscoelastic or viscoplastic. This means that the rheological feature of the materials under intense thermal shock has a strong influence on the fracture toughness. On the other hand, the fracture toughness of materials is highly sensitive to the level of thermal shock.

Table II. Material Parameters and Other Constants

E=96.7GPa	$\alpha = 20.7\times10^{-6}/^0C$	$\sigma = 36.7MPa$	$c = 0.2cm$	$a_0 = 0.25cm$
$R = 0.45cm$	$w = 0.8cm$	$R_0 = 0.4$	$\phi = 1.37$	$K_I^m = 4.35MPa\sqrt{m}$

Table III. Thermal Shock Fracture Toughness

$K_{Ic}(MPa\sqrt{m})$ [15]	$K_{Ic}^{TS}(MPa\sqrt{m})$ $(FWHM = 1.0ms)$	$K_{Ic}^{TS}(MPa\sqrt{m})$ $(FWHM = 250\mu s)$
19.0	22.4	27.8

3.5 Mechanism of crack formation and propagation

As previously described, the initial crack is produced by the mechanisms of void formation in the matrix and separation of the SiC particle-matrix interface, while the crack propagation is dominated by SiC particle fracture. Why is there the difference between the damage mechanism between the crack formation and crack propagation? It is well known that the reinforcement of a hard ceramic in a soft metallic matrix produces composites with substantially higher yield strength compared to that of the matrix. The strengthening effect is primarily attributed to two factors in various earlier investigations. An alternative method of loading the particle is through the misfit strain generated during plastic flow as a result of the difference in elastic moduli between the particle and the matrix. A radial tensile stress σ_{rr} is developed across the interface which is given by $\sigma_{rr} \approx \sigma(\varepsilon_p)+\sigma_T$, where $\sigma(\varepsilon_p)$ is the matrix yield stress at the plastic strain locally attained adjacent to the particle and σ_T is any tensile hydrostatic stress developed in the neighborhood of the particle. Consequently, it is proposed that at early stage of laser irradiation the low yield stress of the matrix alloys causes the particle loading to be low and the voids appear in the matrix or within the interface of particle/matrix. Once the microcracks grow into macroscopic cracks, the hardening of the matrix due to the high strain rate in the case of macrocrack propagation causes high particle loading that exceeds the particle strength and causes the particle to fracture.

4. Conclusion

In this paper, the characteristics of rheological thermal fracture induced by laser beam thermal shock are theoretically investigated. The thermal shock results obtained from laser beam heating are markedly different from those obtained from mechanical loading. For thermal shock loading the maximum in the stress intensity factor over the crack length curves separates the regions of stable and unstable crack growth. The experimental results of rheological thermal failure of metal matrix composites induced by laser beam thermal shock show that the fracture toughness of the materials is highly sensitive to the level of thermal shock and the rheological behavior of the materials under intense thermal shock has a strong influence on the fracture toughness.

References

[1] Boustie, M and Cottet, F. (1991) Experimental and numerical study of laser induced spallation into aluminium and copper targets, *J. Appl. Phys.* **69**, 7533-7538

[2] Kar, A and Mazumder, J. (1990) Two-dimensional model for material damage due to melting and vaporization during laser irradiation, *J. Appl. Phys.* **68**, 3884-3891

[3] Hector Jr, L. G. and Hetnarski, R. B. (1996) Thermal stresses due to a laser pulse: elastic solution, *J. Appl. Mech.* **63**, 38-46

[4] Zhou, Y. C. , Duan, Z. P., Xie, B. M. (1997) Long pulsed laser induced reverse bulging and plugging, *Int. J. Engng Sci.*,

[5] Gupta, T. K. (1972) Strength degradation and crack propagation in thermally shocked Al_2O_3, *J. Am. Ceram. Soc.*, **55**, 249-253

[6] Schneider, G. A. and Petzow, G. (1991) Thermal shock testing of ceramics a new testing method, *J. Am. Ceram. Soc.* **74**, 98-102

[7] Bueckner, H. F. (1970) A novel principle for the computation of stress intensity factors, *ZAMM*, **50**. 529-546

[8] Rice, J. R. (1972) Some remarks on elastic crack-tip stress field, *Int. J. Solids Stuct.*, **8**, 751-758

[9] Zhou, Y. C., Zhu, Z. M., Duan, Z. P. and Yang, Q. B., (1997), Stress intensity factors for a center crack in an infinite plate subjected to laser beam thermal shock, *Engng. Fract. Mech.*, (accepted for publication)

[10] Oliveira, R. and Wu, X. R., (1987) Stress intensity factors for axial cracks in hollow cylinders subjected to thermal shock, *Engng Fract Mech.*, **27**, 185-197

[11] Song, S. G., Vaidya, R. U., Zurek, A. K. and Gray III G. T. (1996) Stacking faults in SiC particles and their effect on the fracture behavior of a 15 vol pct SiC/6061-Al matrix composite, *Metall. Mat. Trans. A*, **27**, 459-465

[12] Liu, G., Zhang, Z. H. and Shang, J. K. (1994) Interfacial microstructure and fracture of Al_2O_3 particulate reinforced Al-Cu composite, *Acta Metall.*, **42**,271-282

[13] Wu, X. R. (1984) Approximate weight functions for center and edge cracks in finite bodies, *Engng. Fract. Mech.*, **20**, 25-49

[14] Kagegama, K., (1989), in Application of fracture mechanics to composites, Composite Materials Series, Eds by Friedrich, K., Elserier, 327

[15] Zhao, D. and Tuler, F. R., (1994) Effect of particle size on fracture toughness in metal matrix composites, *Engng. Fract. Mech.*, **47**, 303-308

A CONSTITUTIVE MODEL OF A PARTICLE REINFORCED VISCOELASTIC COMPOSITE MATERIAL WITH DEBONDED MICROVOIDS*

ZHU-PING HUANG JIAN-KANG CHEN HUI-LING LI YI LIU
(Peking University, Beijing 100871)

Abstract

The statistical behavior of microvoids' evolution in a linear viscoelastic material which contains well bonded second phase particles is investigated. The particle size distribution is assumed to obey a logarithmic normal distribution. Because of the difference in mechanical properties between the matrix and the second phase particles, the debonding damage of particle-matrix interface may occur under the action of external loads. This kind of damage will lead to microvoids' nucleation and growth. In this paper, the reinforcing effect due to rigid particles and the weakening effect due to microvoids produced from the debonding on the overall mechanical property of the particle reinforced composite material are studied. By virtue of Eshelby's equivalent inclusion method and Mori-Tanaka theory, the average normal stress on the particle-matrix interface which governs the void nucleation and the dilational rate of void volume which governs the void growth are calculated. Then, based on the kinetic conditions for the microvoids' nucleation and growth as well as the balance law of voids' number, the distribution functions of the number densities both for perfectly bonded particles and for microvoids are obtained. Thus, a macroscopic constitutive relation of the considered composite material is derived. It is shown that the macroscopic strain rate,

*This project is supported by the National Natural Science Foundation of China (19632030) and the Doctoral Program Foundation of the State Education Commission of China

R. Wang (ed.), IUTAM Symposium on Rheology of Bodies with Defects, 133–144.

the particle-size dispersity and the interface adhesive strength play key roles in describing the overall mechanical property of such a composite material.

1. Introduction

A polymer composite comprised of a polymeric matrix material reinforced by dispersed second phase rigid particles has been used as an advanced engineering plastics in recent years [1,2]. Because the mechanical property of the dispersed phase differs from that of the matrix, any of the following microdamage nucleation mechanisms may be observed during the deformation process : namely, cracking of particles, debonding at the interface and fracture of the matrix adjacent to stronger particles. The nucleated microdamage will subsequently grow into microvoid under continued loading. Obviously, the mechanical property of the composite is not only influenced by the reinforcing effect due to rigid particles, but also influenced by the weakening effect due to microvoids. Hence, in order to obtain the macroscopic constitutive relation of the said composite, the evolutions of number densities both for the well bonded particles and the microvoids have to be considered.

Recently, the statistical evolution of microvoids in a rheological material was studied in Ref.[3]. The effective elastic-plastic behavior of a particle-reinforced composite including debonding damage was also discussed by Zhao and Weng [4], Tohgo and Chou [5]. There the matrix materials are taken to be elasto-plastic. However, the effects of the void growth and the particle-size distribution on the macroscopic constitutive relation were not considered, and as is shown below, these effects could be quite important in some cases.

In this paper, we propose a constitutive model of a rheological material filled with second phase particles. The constitutive relation of the matrix material is linear viscoelastic. The particle-size distribution is assumed to be log-normal distribution. It may be further assumed that during the deformation process, when the average normal stress on the interface exceeds the nucleation threshold, debonding damage at the interface will occur with its probability obeying the Weibull's statistical function. Then, by means of Eshelby's equivalent inclusion method and the Mori-Tanaka theory, the kinetic equations for the void nucleation and the void growth are established. From these kinetic equations together with the balance law of the voids' number, evolutions of the number densities for both the particles and the microvoids are obtained. The corresponding constitutive relation of the composite material is then derived. The influences of the loading rate, the particle-size dispersity and the interface adhesive strength are also examined under plane strain condition. It is found that all these factors will play significant roles in predicting the overall mechanical property of a rheological composite material with progressively debonded microdamage.

2. Nucleation of microvoids in a particle-reinforced composite

Now consider a particle-reinforced composite. The experimental observation [2],[6] indicates that the particle-size distribution in several polymer composites can be described by the logarithmic normal distribution

$$n_p(c_1) = \frac{N_p}{\sqrt{2\pi} c_1 \ln \omega} \cdot \exp\left[-\frac{(\ln c_1 - \ln \bar{c}_1)^2}{2 \ln^2 \omega} \right] \tag{1}$$

where c_1 denotes the radius of a spherical particle, $\bar{c}_1$ is the average radius of particles, ω is the particle-size dispersity and N_p is the total number of particles per unit volume. So, the initial volume fraction of particles is

$$f_{po} = \frac{4}{3} \pi \bar{c}_1^{\,3} N_p \tag{2}$$

As the applied load increases, the nucleation of microdamage in the composite may take place. A commonly observed nucleation mechanism is the debonding at particle-matrix interface. Two necessary critical conditions for the debonding at interface between the particle and the matrix are

(1) The critical energy criterion, i.e. the elastic energy released by forming the stress free surface is at least equal to the surface energy created.

(2) The mechanical debonding criterion, i.e. the normal stress σ at the interface must exceed the threshold bond strength σ_o between the particle and the matrix material.

It was pointed out by Argon[7], Goods and Brown[8] that for metallic composites, when particle size is larger than $100\,\mathring{A}$, the critical energy criterion is always satisfied if the normal stress σ reaches σ_o. We shall use it here, so in the following, only the mechanical debonding criterion will be considered. Extensive investigations on the nucleation of microdamage in ductile materials have been conducted by many researchers (e.g., Ref. [7-11]). However, the kinetics of nucleation process is complex and is not quite clear so far. Hence the expression of the nucleation rate is usually given empirically. Suppose the probability of debonding at the interface may be described by Weibull's distribution function

$$P(c_1,\sigma)=1-\exp\left[-\left(\frac{c_1}{c_o}\right)^2\left(\frac{\sigma-\sigma_o}{\sigma_u}\right)^m\right], \qquad (\sigma\geq\sigma_o) \tag{3}$$

where σ is the average normal stress at the interface, and c_o, σ_o, σ_u, m are material parameters, which should be determined by experiment. Then for $\sigma>\sigma_o$, the number of debonded particles per unit volume with radii between c_1 and $c_1+\mathrm{d}\,c_1$ may be given by

$$n_N(c_1,\sigma)\,\mathrm{d}\,c_1=n_p(c_1)P(c_1,\sigma)\,\mathrm{d}\,c_1 \tag{4}$$

It follows that nucleation rate of microvoids can be expressed by

$$\dot{n}_N=n_p(c_1)\cdot\frac{\partial P(c_1,\sigma)}{\partial\sigma}\cdot\frac{\mathrm{d}\,\sigma}{\mathrm{d}\,t}, \quad \left(\sigma>\sigma_o, \frac{\mathrm{d}\,\sigma}{\mathrm{d}\,t}>0\right) \tag{5}$$

where the average normal stress σ will be determined in the next section.

3. Kinetic equations of microvoids' nucleation and growth

Now suppose the matrix material of the composite is a linear viscoelastic one, its constitutive relation may be expressed by

$$\underset{\sim}{\sigma}_o(x,t)=\underset{\sim}{L}_o(t):\underset{\sim}{\varepsilon}_o(\underset{\sim}{x},0)+\int_0^t\underset{\sim}{L}_o(t-\tau):\dot{\underset{\sim}{\varepsilon}}_o(\underset{\sim}{x},\tau)\,\mathrm{d}\,\tau \tag{6}$$

where $\underset{\sim}{x}$ is the position vector, $\underset{\sim}{\sigma}_o$ and $\underset{\sim}{\varepsilon}_o$ are stress and strain tensors in the matrix respectively, $\underset{\sim}{L}_o$ is the fourth-order relaxation tensor. When the matrix material is isotropic, the components of the relaxation tensor in rectangular Cartesian coordinate may be written as

$$L_{ijkl}(t)=\frac{\nu\phi(t)}{(1+\nu)(1-2\nu)}\delta_{ij}\delta_{kl}+\frac{\phi(t)}{2(1+\nu)}\left(\delta_{ik}\delta_{jl}+\delta_{il}\delta_{jk}\right) \tag{7}$$

where ν is Poisson's ratio and $\phi(t)$ for a Maxwell fluid is an exponential function

$$\phi(t)=E\,\exp\left(-\frac{E}{\eta}t\right) \tag{8}$$

with E and η being Young's modulus and a coefficient of viscosity, respectively.

When the matrix material is filled with second phase rigid particles, the correspondence principle between an elastic and a linear viscoelastic solution in a heterogeneous body can be utilized. By taking the Laplace transform to all the field equations on such a equivalent inclusion problem, we can obtain the stress in rigid particles $\underset{\sim}{\sigma}_p$ and the strain of voids $\underset{\sim}{\varepsilon}_v$ as follows

$$\begin{aligned} \underset{\sim}{\sigma}_p &= -\underset{\sim}{L}_o(t):\underset{\sim}{\varepsilon}_p^*(0) - \int_o^t \underset{\sim}{L}_o(t-\tau):\dot{\underset{\sim}{\varepsilon}}_p^*(\tau)\,\mathrm{d}\tau \\ \underset{\sim}{\varepsilon}_v &= \underset{\sim}{\varepsilon}_v^*(t) \end{aligned} \tag{9}$$

where $\underset{\sim}{\varepsilon}_p^*$ and $\underset{\sim}{\varepsilon}_v^*$ denote eigenstrains of the particles and the voids respectively (see Ref. [12]) . When Poisson's ratio of the matrix ν remains constant, $\underset{\sim}{\varepsilon}_p^*$ and $\underset{\sim}{\varepsilon}_v^*$ can be expressed in terms of the average strain of the matrix $<\underset{\sim}{\varepsilon}_o>$ by

$$\begin{aligned} \underset{\sim}{\varepsilon}_p^* &= -\underset{\sim}{S}^{-1}:<\underset{\sim}{\varepsilon}_o> \\ \underset{\sim}{\varepsilon}_v &= \left(\underset{\sim}{I}-\underset{\sim}{S}\right)^{-1}:<\underset{\sim}{\varepsilon}_o> \end{aligned} \tag{10}$$

where $\underset{\sim}{I}$ is the fourth-order identity tensor, $\underset{\sim}{S}$ is Eshelby's tensor. For spherical particles, the components of $\underset{\sim}{S}$ in a rectangular Cartesian coordinate are

$$S_{ijkl}(t) = \frac{5\nu-1}{15(1-\nu)}\delta_{ij}\delta_{kl} + \frac{4-5\nu}{15(1-\nu)}\left(\delta_{ik}\delta_{jl}+\delta_{il}\delta_{jk}\right) \tag{11}$$

It should be noted that in Eqs. (9) and (10), the inverse transform has already been worked out.

Let $f_p(t)$ and $f_v(t)$ be the volume fraction of particles and the porosity of the composite at time t respectively. Then the average stress and the average strain may be written as

$$\begin{aligned} <\underset{\sim}{\sigma}> &= \left(1-f_p-f_v\right)<\underset{\sim}{\sigma}_o> + f_p<\underset{\sim}{\sigma}_p> \\ <\underset{\sim}{\varepsilon}> &= \left(1-f_p-f_v\right)<\underset{\sim}{\varepsilon}_o> + f_v<\underset{\sim}{\varepsilon}_v> \end{aligned} \tag{12}$$

Hence the macroscopic constitutive relation of the composite (the relationship between $<\underset{\sim}{\sigma}>$ and $<\underset{\sim}{\varepsilon}>$) may be obtained from Eqs. (6)-(12) so long as f_p and f_v are determined.

From eqs.(10)-(12), $\underset{\sim}{\sigma}_p$ and $\underset{\sim}{\varepsilon}_v$ in eq.(9) may be expressed in terms of the specified average strain $<\underset{\sim}{\varepsilon}>=\underset{\sim}{\varepsilon}^\infty(t)$ by

$$\underset{\sim}{\sigma}_p = \underset{\sim}{L}_o(0):\underset{\sim}{S}^{-1}:\underset{\sim}{A}_o(t):\underset{\sim}{\varepsilon}^{\infty}(t) - \int_o^t \frac{\partial}{\partial \tau}\left[\underset{\sim}{L}_o(t-\tau)\right]:\underset{\sim}{S}^{-1}:\underset{\sim}{A}_o(\tau):\underset{\sim}{\varepsilon}^{\infty}(\tau)\,\mathrm{d}\tau \tag{13}$$

$$\underset{\sim}{\varepsilon}_v = \left(\underset{\sim}{I} - \underset{\sim}{S}\right)^{-1}:\underset{\sim}{A}_o(t):\underset{\sim}{\varepsilon}^{\infty}(t) \tag{14}$$

where $\underset{\sim}{A}_o$ is a fourth-order tensor:

$$\underset{\sim}{A}_o = \left[\left(1 - f_p - f_v\right)\underset{\sim}{I} + f_v\left(\underset{\sim}{I} - \underset{\sim}{S}\right)^{-1}\right]^{-1} \tag{15}$$

The first invariants of $\underset{\sim}{\sigma}_p$ and $\underset{\sim}{\varepsilon}_v$ (denoted by σ_{pkk} and ε_{vkk} respectively), and hence the relative growth rate of the void volume $\frac{\mathrm{d}}{\mathrm{d}t}\varepsilon_{vkk}$ as well as the rate of the average normal stress on the interface $\frac{\mathrm{d}\sigma}{\mathrm{d}t} = \frac{1}{3}\frac{\mathrm{d}}{\mathrm{d}t}\sigma_{pkk}$ can easily be obtained directly from eqs.(13) and (14). So the kinetic equation for the void nucleation may be given by Eq.(5) with $\frac{\mathrm{d}\sigma}{\mathrm{d}t}$ replaced by $\frac{1}{3}\frac{\mathrm{d}\sigma_{pkk}}{\mathrm{d}t}$.

In general, during the deformation process, the void shape may no longer remain spherical. However, in the case of high stress triaxiality (for instance, in the case of plane strain deformation), the notation c can still be used as the average radius of a void. Thus the kinetic equation for the void growth may be written as

$$\dot{c} = \frac{1}{3}c \cdot \frac{\mathrm{d}}{\mathrm{d}t}\varepsilon_{vkk} = c \cdot g(t) \tag{16}$$

where

$$g(t) = \frac{\mathrm{d}}{\mathrm{d}t}\varepsilon_{vkk} \tag{17}$$

4. The statistical evolution of microvoids and the corresponding macroscopic constitutive relation

Suppose at the instant t_o, the average normal stress at the particle-matrix interface σ reaches σ_o, and the microvoids nucleation takes place. If the number density of the microvoids is denoted by $n(c,t)$, i.e., the number of the voids per unit volume with radius from c to $c + \mathrm{d}c$ at time t is $n(c,t)\mathrm{d}c$, then the balance law of the void number can be written as

$$\dot{n}(c,t) = -g(t) \cdot n + \dot{n}_N \tag{18}$$

where $\dot{n}_N$ and $g(t)$ are given by Eqs. (5) and (17).

The corresponding initial and boundary conditions are

$$n(c,t_o)=0 \quad , \quad n(\infty,t)=0 \tag{19}$$

In order to solve Eqs.(18) and (19), f_p and f_v in Eqs.(12)-(15) should be determined. Evolution equations of f_p and f_v are

$$\begin{aligned} \dot{f}_p &= -\frac{4}{3}\pi\int_o^\infty \dot{n}_N c_1^3\,\mathrm{d}c_1 \\ \dot{f}_v &= \frac{4}{3}\pi\int_o^\infty \dot{n}c^3\,\mathrm{d}c \end{aligned} \tag{20}$$

It follows that

$$\begin{aligned} f_p &= f_{po} + \int_{t_o}^t \dot{f}_p(\tau)\,\mathrm{d}\tau \\ f_v &= \frac{4\pi}{3}\int_o^\infty nc^3\,\mathrm{d}c \end{aligned} \tag{21}$$

Obviously, Eqs.(5), (13)-(21) are coupled to each other. So, these equations have to be solved simultaneously to obtain $n(c,t)$, f_p and f_v.

The number density of perfectly bonded particles is given by

$$n_p(c_1)\left[1-\int_o^t \frac{\partial P(c,\sigma)}{\partial\sigma}\frac{\mathrm{d}\sigma}{\mathrm{d}\tau}\mathrm{d}\tau\right], \quad \left(\sigma \ge \sigma_{o,}\ \frac{\mathrm{d}\sigma}{\mathrm{d}\tau}>0\right) \tag{22}$$

and the statistical evolution of microvoids is described by the distribution function $n(c,t)$, which are closely related to the determination of f_p and f_v. The derivation of the macroscopic constitutive relation of the composite is straightforward and comes directly from Eqs.(6)-(12) as soon as f_p and f_v are known

$$<\underset{\sim}{\sigma}>=\underset{\sim}{L}_o(t):\underset{\sim}{B}(t):\underset{\sim}{A}_o(0):<\underset{\sim}{\dot{\varepsilon}}(0)>+\int_o^t \underset{\sim}{\dot{L}}_o(t-\tau):\underset{\sim}{B}(t):\underset{\sim}{A}_o(\tau):<\underset{\sim}{\dot{\varepsilon}}(\tau)>\mathrm{d}\tau \tag{23}$$

where the fourth-order tensor $\underset{\sim}{B}$ is defined by

$$\underset{\sim}{B}=\left(1-f_p-f_v\right)\underset{\sim}{I}+f_p\underset{\sim}{S}^{-1}$$

5. An illustrative example and discussions

In order to simplify the calculation, here a Maxwell fluid model is used for the constitutive equation of the matrix material(see eqs.(7) and (8)), and only the one-dimensional plane strain deformation with constant strain rate is considered

$$\varepsilon_{11}^{\infty}=\varepsilon_{11}^{o\,\infty}\left(\frac{t}{t_c}\right)\quad,\quad \varepsilon_{22}^{\infty}=\varepsilon_{33}^{\infty}=0$$

where t_c is the reference time and $\varepsilon_{11}^{o\,\infty}$ is the macroscopic strain at $t=t_c$. Then, from eqs.(13) and (14), the first invariants of $\underset{\sim}{\sigma}_p$ and $\underset{\sim}{\varepsilon}_v$ may be given by

$$\sigma_{pkk}=-\left(\frac{E}{1-2\nu}\right)\left[\varepsilon_{pkk}^{*}(t)+\frac{E}{\eta}\int_0^t\exp\left(-\frac{E}{\eta}\tau\right)\varepsilon_{pkk}^{*}(t-\tau)\,\mathrm{d}\tau\right]$$

$$\varepsilon_{vkk}=\frac{3(1-\nu)}{2(1-2\nu)}\cdot\frac{\varepsilon_{11}^{\infty}}{h(t)}$$

where
$$\varepsilon_{pkk}^{*}=-\frac{3(1-\nu)}{(1+\nu)}\cdot\frac{\varepsilon_{11}^{\infty}}{h(t)}$$

$$h(t)=\left(1-f_p-f_v\right)+\frac{3(1-\nu)}{2(1-2\nu)}\cdot f_v$$

moreover, the expression of in eqs.(16)-(18) may be written as

$$g(t)=\frac{1-\nu}{2(1-2\nu)}\left[\frac{\dot{\varepsilon}_{11}^{\infty}}{h(t)}-\frac{\dot{h}(t)}{h^2(t)}\varepsilon_{11}^{\infty}\right]$$

After introducing the following non-dimensional parameters

$$\tilde{t}=\frac{t}{t_c}\ ,\ \tilde{t}_o=\frac{t_o}{t_c}\ ,\ \tilde{c}_1=\frac{c_1}{\bar{c}_1}\ ,\ \tilde{c}=\frac{c}{\bar{c}_1}\ ,\ \tilde{\sigma}_\tau=\frac{\sigma_o}{E}\ ,S_u=\frac{\sigma_o}{\sigma_u}\ ,$$

$$\tilde{\sigma}=\frac{\sigma}{\sigma_o}\ ,\tilde{\mu}_o=\frac{E}{\eta}t_c\ ,\ \tilde{n}_p=\bar{c}^4 n_p\ ,\ \tilde{n}=\bar{c}^4 n\ ,\ \tilde{g}=t_c g$$

we can rewrite all the governing equations in the non-dimensional forms. The material parameters are taken to be

$$\nu=\frac{1}{3}\ ,f_{po}=0.2\ ,\ \tilde{c}_o=1.5\ ,\ \frac{\sigma_u}{\sigma_o}=3\ ,\ \tilde{\mu}_o=0.5\ ,\ \omega=2.0\ ,\ m=4\ \text{and}\ t_c=t_o$$

The number density of microvoids $n(c,t)$ is calculated numerically, and depicted in Fig.1 and Fig.2. It is shown that at the early stage of the deformation ($t\le 3t_o$), the distribution curve of the number density gradually moves toward the direction of larger size with the peak value increases rapidly indicating that void nucleation is dominant (Fig.1). However, as time goes on, the void growth will become more and more noticeable. Hence at the later stage ($t>3t_o$), the curve of the number density moves in the direction of larger void size with a decreased peak value, and as a whole the curve becomes smoother and flatter.

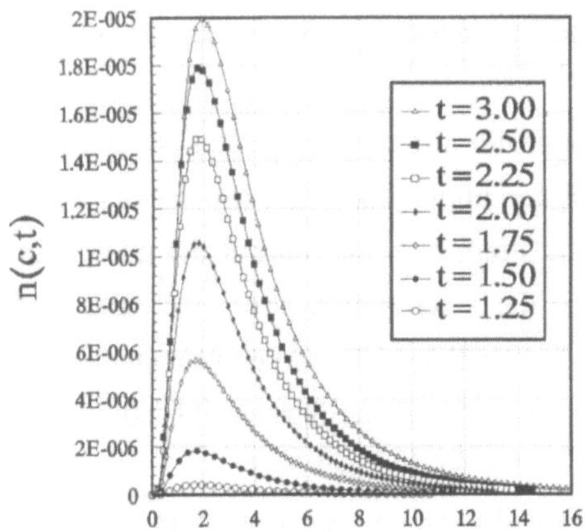

Fig.1 The evolution of microvoids' number density for $t \leq 3t_c$

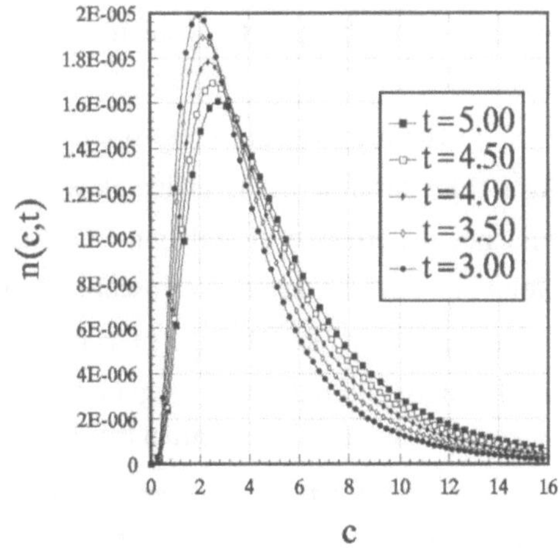

Fig.2 The evolution of the microvoids' number density for $t \geq 3t_c$

As the macroscopic strain $\varepsilon_{11}^{\infty}$ monotonically increases, the progressive debonding and microvoids' growth take place. The evolutions of the volume concentration of the well bonded particles f_p and porosity f_v are shown in Fig.3. At the beginning, f_p decreases more and more rapidly, and then decreases slowly and tends to a saturation value. On the other hand, the porosity f_v results from both the

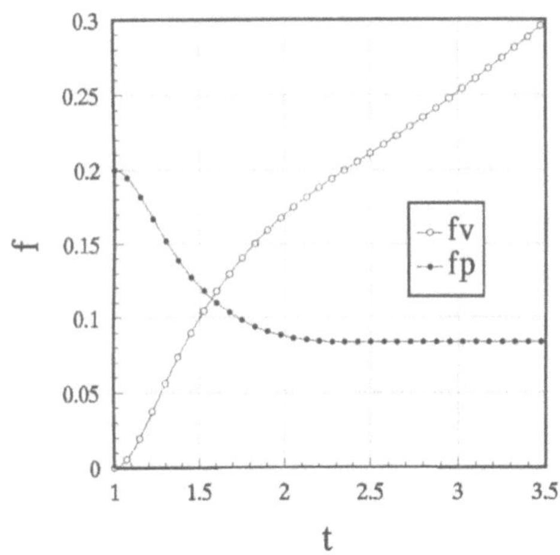

Fig.3 Evolution of the volume fraction of perfectly bonded particles f_p and porosity f_v. (m=2)

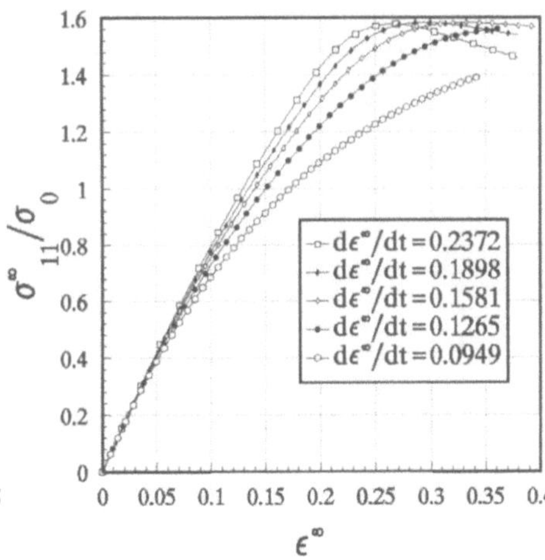

Fig.4 The macroscopic sress-strain relations with different strain rates

nucleation and the growth of microvoids, and continues to increase until the material fails.

The macroscopic stress-strain relations with different strain rates are shown in Fig.4. The deformation is under plane strain condition. There are two factors which influence the stress-strain relation : One is the strain rate, the other is the microvoids evolution. For the same macroscopic strain, the macroscopic stress increases with increasing strain rate. But the higher strain rate will lead to earlier void nucleation, which in turn will reduce the magnitude of the stress. The latter factor will become more pronounced at much higher strain rate (e.g. $\dot{\varepsilon}_{11}^{\infty} = 0.2372$ in Fig.4).

The influence of the particle-size dispersity ω on the stress-strain relation is shown in Fig.5. It is found that a higher value of ω is unfavorable for improving the mechanical property of the composite. For lower value of ω (<2.0), microvoids evolve gradually, and the stress-strain curve increases continuously. On the contrary, for higher value of ω (>2.0), microvoids evolve rapidly, and the softening effect becomes notable, leading to a decrease of the material strength. As is displayed in Fig.5, when $\omega > 2.0$, the stress-strain curve has a maximum, after which the stress decreases continuously, and the carrying capacity is reduced. This result is also consistent with the experimental observations for PVC/NBR blends[13]. Thus, a lower value of ω is needed for attaining a better quality of the composite material.

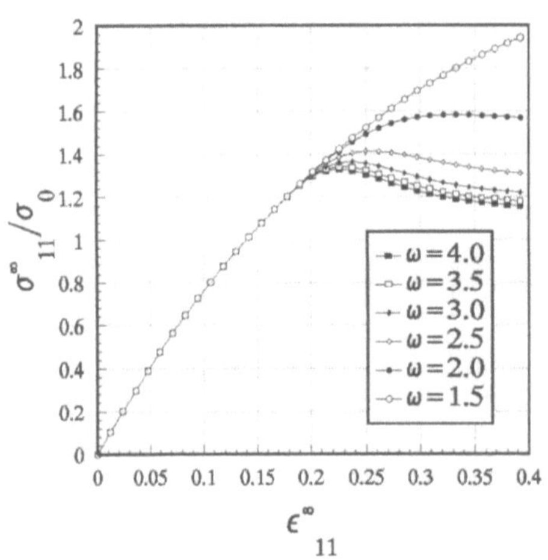

Fig.5 The stress-strain relations with different ω

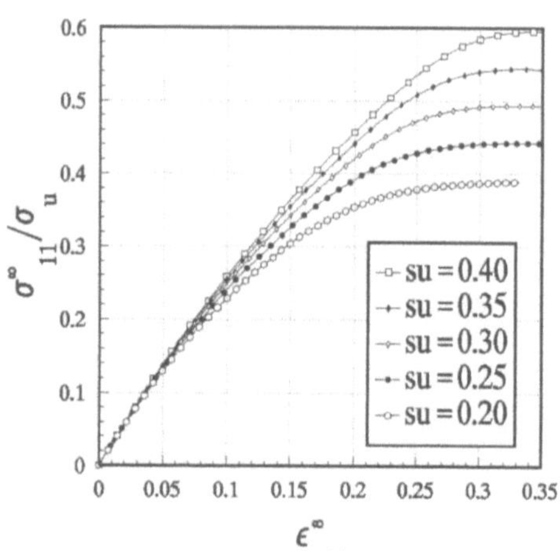

Fig.6 The influence of the interfacial strength on the stress-strain relation

The interphase adhesive strength also has a remarkable influence on the macroscopic constitutive relation as indicated in Fig. 6. Higher interfacial strength will reinforce the composite more effectively. However, in case the interfacial strength is extremely high, other damage mechanisms, such as cracking of particles and craze-fracture of the matrix, are possible. Thus in order to improve the mechanical property of the composite, the effect of the matrix toughness should also be taken into consideration[14].

Acknowledgment The authors would like to thank Professor R. Wang for his helpful discussion on this subject.

References

[1] Zhu X. G., Liu C. K., Qi Z. N., Wang G. H., Choy C. L. (1992) Toughening mechanism of HDPE/$CaCO_3$ ultra-tough composite, *Proc. of Int. Sym. on Polym. Alloys & Comp. (Hong Kong)*, 328-330

[2] Fu Q., Shen J., and Wang G. (1992) Factors affecting the toughness of HDPE/$CaCO_3$ blends, *Polym. Mater. Sci. Eng. (in Chinese)* Vol. 1, 107-112

[3] Li H. L. (1996) Statistical behavior of microdamage's evolution in ductile materials and its stochastic model, Ph.D. Thesis (in Chinese with English abstract), Peking University

[4] Zhao Y. H., Weng G. J. (1996) Plasticity of a two-phase composite with partially debonded inclusions, *Int. J. Plasticity*, Vol.12(6), 781-804

[5] Tohgo K., Chou T. (1996) Incremental theory of particulate-reinforced composites including debonding damage, *JSME Int. J. Series A*, Vol.39, 389-397

[6] Wu S. H. (1985) Phase structure and adhesion in polymer blends: A criterion for rubber toughening, *Polymer*, Vol. 26, 1855-1863

[7] Argon A. S., Im J., Safoglu R. (1975) Cavity formation from inclusions in ductile fracture, *Metall. Trans.*, Vol.6 A, 825-838

[8] Goods S. H., Brown L. M. (1979) The nucleation of cavities by plastic deformation, *Acta Metall.*, Vol.27, 1-15

[9] Fisher J. R., Gurland J., (1981) Void nucleation in spheroidized carbon steels, *Metal. Sci.*, Vol.15 ,185-202

[10] Needleman A. J., (1987) A continuum model for void nucleation by inclusion debonding, *J. Appl. Mech.*, Vol.54, 525-531

[11] Curran D. R., Seaman L., Shockey D. A.,(1987) Dynamic failure of solids, *Phys. Rep.*, Vol.147, 253-388

[12]Toshio Mura, (1987) *Micromechanics of Defects in Solids,* Martinus Nijhoff Publishers, Dordrecht

[13] Liu Z. H., Zhu X. G., et al., (1996) Brittle-ductile transition in polymer blends, *Acta Polymerica Sinica (in Chinese)*, No. 4, 468-473

[14]Fu Q., Zhang Y. L., and Wang G. H. (1994) The effect of matrix toughness on the brittle- ductile transition of HDPE/$CaCO_3$ blends, *Chinese J. of Polymer Science,* Vol. 12, No. 4 ,309-315

DYNAMIC DEBONDING BETWEEN FIBERS AND MATRIX IN FIBER-REINFORCED COMPOSITES

NAI-CHIEN HUANG
Department of Aerospace and Mechanical Engineering
University of Notre Dame
Notre Dame, Indiana 46556

Abstract

In this paper, dynamic debonding between fibers and matrix is studied based on an axisymmetrical model of a single fiber surrounded by a cylindrical matrix. Asymptotic fields of stress and particle velocity are found in the vicinity of the crack front based on the concept of local plane strain for an interfacial crack between two dissimilar materials. The method of modified material stiffness constants and conversion relations between stress intensity factors for a propagating interfacial crack and a stationary interfacial crack are employed in the study.

Introduction

In the problem of crack bridging in fiber-reinforced composites, fracture can usually be induced by the dynamic loading through an interfacial debonding between fibers and matrix. In order to study the dynamic strength of the composite, it is necessary for us to determine the asymptotic stress fields in the vicinity of the front of a propagating interfacial crack and the rate of energy release for interfacial crack propagation. The information thus obtained can then be applied to the measurement of dynamic interfacial crack toughness. It can also be used for the investigation of breakage of fibers resulting from crack kinking at the tip of the crack.

Basic Equations

Our investigation is based on an axisymmetric model of a single fiber surrounded by a hollow cylindrical matrix with an interfacial crack which propagates with a prescribed speed. Both the fiber and the matrix are considered to be homogeneous, isotropic and linearly elastic. We are interested in the determination of the asymptotic fields of stresses and particle velocities in the vicinity of the front of the propagating interfacial crack.

When the body forces are absent, Navier's equations of motion for linearly elastic material can be written as

$$\mu\nabla^2 u_i + (\lambda+\mu)\frac{\partial e}{\partial x_i} = \rho\frac{\partial^2 u_i}{\partial t^2} \quad \text{for} \quad i = 1, 2, 3, \tag{1}$$

R. Wang (ed.), IUTAM Symposium on Rheology of Bodies with Defects, 145–152.

where u_i are displacement components; λ and μ are Lamè's constants; ∇^2 is the Laplacian operator, ρ is the mass per unit volume; t is time, e is the dilatation of deformation. In the cylindrical coordinates system (r,θ,z) with axisymmetry, all field quantities are independent of θ. Thus

$$\nabla^2 = \frac{\partial^2}{\partial r^2} + \frac{1}{r}\frac{\partial}{\partial r} - \frac{1}{r^2} + \frac{\partial^2}{\partial z^2} \tag{2}$$

and

$$e = \frac{\partial u_r}{\partial r} + \frac{u_r}{r} + \frac{\partial u_z}{\partial z} . \tag{3}$$

Stresses and strains are singular at the crack front. For the interest of asymptotic analysis in the vicinity of the crack front, the terms of lower order of differentiations with respect to r and z in Navier's equations may be neglected. Thus for any point with a distance from the crack front much smaller than the radius of the fiber, equations (2) and (3) may be written approximately as

$$\nabla^2 \simeq \frac{\partial^2}{\partial r^2} + \frac{\partial^2}{\partial z^2} , \tag{4}$$

$$e \simeq \frac{\partial u_r}{\partial r} + \frac{\partial u_z}{\partial z} . \tag{5}$$

Hence, Navier's equations are reduced to

$$\mu\left(\frac{\partial^2}{\partial r^2} + \frac{\partial^2}{\partial z^2}\right)u_r + (\lambda + \mu)\frac{\partial}{\partial r}\left(\frac{\partial u_r}{\partial r} + \frac{\partial u_z}{\partial z}\right) = \rho\,\frac{\partial^2 u_r}{\partial t^2} , \tag{6}$$

$$\mu\left(\frac{\partial^2}{\partial r^2} + \frac{\partial^2}{\partial z^2}\right)u_\theta = \rho\,\frac{\partial^2 u_\theta}{\partial t^2} , \tag{7}$$

$$\mu\left(\frac{\partial^2}{\partial r^2} + \frac{\partial^2}{\partial z^2}\right)u_z + (\lambda + \mu)\frac{\partial}{\partial z}\left(\frac{\partial u_r}{\partial r} + \frac{\partial u_z}{\partial z}\right) = \rho\,\frac{\partial^2 u_z}{\partial t^2} . \tag{8}$$

In the following, a moving coordinate system with the origin at any point on the crack front is employed. Let a be the radius of the fiber and $\ell(t)$ the length of the crack. The crack propagates in the z-direction with speed $v(t) = \frac{\partial \ell}{\partial t}$. The local moving coordinates are

$$\xi_1 = z - \ell(t) , \quad \xi_2 = r - a , \tag{9}$$

and the ξ_3 - axis in the θ-direction is perpendicular to the ξ_1 and ξ_2 axes. Put

$$\tau = t \tag{10}$$

and denote

$$u_1 = u_z, \qquad u_2 = u_r, \qquad u_3 = u_\theta \, . \tag{11}$$

Note that

$$\frac{\partial^2}{\partial t^2} = v^2 \frac{\partial^2}{\partial \xi_1^2} - 2v \frac{\partial^2}{\partial \xi_1 \partial \tau} + \frac{\partial^2}{\partial \tau^2} - \frac{\partial v}{\partial \tau} \frac{\partial}{\partial \xi_1} \, . \tag{12}$$

For the interest of the field quantities in the vicinity of the crack front, we may write

$$\frac{\partial^2}{\partial t^2} \simeq v^2 \frac{\partial^2}{\partial \xi_1^2} \, . \tag{13}$$

Thus, for the asymptotic analysis of field quantities, equations (6), (7) and (8) can be expressed as

$$\mu \left(\frac{\partial^2}{\partial \xi_1^2} + \frac{\partial^2}{\partial \xi_2^2} \right) u_1 + (\lambda + \mu) \frac{\partial}{\partial \xi_1} \left(\frac{\partial u_1}{\partial \xi_1} + \frac{\partial u_2}{\partial \xi_2} \right) = \rho \, v^2 \frac{\partial^2 u_1}{\partial \xi_1^2} \, , \tag{14}$$

$$\mu \left(\frac{\partial^2}{\partial \xi_1^2} + \frac{\partial^2}{\partial \xi_2^2} \right) u_2 + (\lambda + \mu) \frac{\partial}{\partial \xi_2} \left(\frac{\partial u_1}{\partial \xi_1} + \frac{\partial u_2}{\partial \xi_2} \right) = \rho \, v^2 \frac{\partial^2 u_2}{\partial \xi_1^2} \, , \tag{15}$$

$$\mu \left(\frac{\partial^2}{\partial \xi_1^2} + \frac{\partial^2}{\partial \xi_2^2} \right) u_3 = \rho v^2 \frac{\partial^2 u_3}{\partial \xi_1^2} \, . \tag{16}$$

Equations (14) (15) and (16) are identical to the governing equations for the dynamic displacement field in the vicinity of the crack tip in a plane strain problem where an interfacial crack propagates with speed v(t) in the ξ_1-direction. The in-plane displacement components u_1 and u_2 are generated by the in-plane axisymmetrical loading in r and z directions while the anti-plane displacement component u_3 is generated by the twist of the fiber relative to the matrix. For the plane strain problem, Eshelby, Read and Shockley (1953) suggested that the displacement components can be expressed as

$$u_k = 2a_k f(z) \qquad \text{for } k = 1, 2, 3, \tag{17}$$

where

$$z = \xi_1 + p\xi_2 \, , \tag{18}$$

and p and a_k are respectively eigenvalue and eigenvector to be determined. By substitution, it is found that

$$\begin{bmatrix} c_d^2 - v^2 + p^2 c_s^2 & p\left(c_d^2 - c_s^2\right) & 0 \\ p\left(c_d^2 - c_s^2\right) & c_s^2 - v^2 + p^2 c_d^2 & 0 \\ 0 & 0 & c_s^2 - v^2 + p^2 c_s^2 \end{bmatrix} \begin{bmatrix} a_1 \\ a_2 \\ a_3 \end{bmatrix} = 0, \tag{19}$$

where

$$c_s = \left(\frac{\mu}{\rho}\right)^{1/2}, \quad c_d = \left(\frac{\lambda+2\mu}{\rho}\right)^{1/2} \tag{20}$$

are respectively the speed of shear wave and the speed of dilatational wave.

Plane Strain Problem

The plane strain problem of interfacial crack propagation between two elastic materials has been studied by Yang, Suo and Shih (1991). It was also investigated independently by Huang (1992) based on the method of modified stiffness constants using the conversion of stress intensity factors from the case of stationary crack to the case of propagating crack. In the study, the stress intensity factors in the problem of stationary crack are defined according to the work by Suo (1990).

In the following, the results obtained by Huang (1992) will be summarized. First, we assume that the speed of the interfacial crack is sufficiently low that it is in the subsonic range $v < c_s$ and also there is no configuration instability in the crack front. Thus the crack front is always in the plane perpendicular to the fiber. Denote

$$\alpha_s = \left(1 + v^2/c_s^2\right)^{1/2}, \quad \alpha_d = \left(1 - v^2/c_d^2\right)^{1/2}, \quad D = 4\alpha_s\alpha_d - \left(1 + \alpha_s^2\right)^{1/2}. \tag{21}$$

We shall use the subscript 1 to refer to the matrix and the subscript 2 to refer to the fiber. Put

$$a = \sum_{j=1}^{2} \frac{1}{\mu_j D_j} \alpha_{sj} \left(1 - \alpha_{sj}^2\right), \tag{22}$$

$$b = \sum_{j=1}^{2} (-1)^{j-1} \frac{1}{\mu_j D_j} \left(1 + \alpha_{sj}^2 - 2\alpha_{sj}\alpha_{dj}\right), \tag{23}$$

$$c = \sum_{j=1}^{2} \frac{1}{\mu_j D_j} \alpha_{dj} \left(1 - \alpha_{sj}^2\right) \tag{24}$$

and

$$\omega = (a/c)^{1/2}. \tag{25}$$

The oscillation index is given by

$$\varepsilon = \frac{1}{2\pi} \ln \left[\frac{(ac)^{1/2} - b}{(ac)^{1/2} + b} \right] . \tag{26}$$

Asymptotic Fields in the Vicinity of the Crack Front

Let K_{D1}, K_{D2}, and K_{D3} be the dynamic stress intensity factors for mode I, mode II and mode III deformations respectively. In the following, the asymptotic fields for stress and particle velocity are found for the region of matrix only. The corresponding equations for the region of the fiber can be obtained by replacing α_{s1} by α_{s2} , replacing α_{d1} by α_{d2}, replacing $e^{\pi\varepsilon}$ by $e^{-\pi\varepsilon}$ and replacing $e^{-\pi\varepsilon}$ by $e^{\pi\varepsilon}$. A branch cut is made on the line of the crack. If we omit the subscript 1 in α_{s1} and α_{d1}, we have the following asymptotic fields for stress and particle velocity in the matrix near the front of the crack in the case of in-plane deformation.

$$\begin{aligned}\sigma_{zz} = \sigma_{11} = K_{D1}\left[Q_1F_{11}\left(R_s^+,R_d^+,\omega\right) + Q_2F_{11}\left(R_s^-,R_d^-,\omega\right)\right] \\ -\omega K_{D2}\left[Q_1F_{11}\left(I_s^+,I_d^+,\omega\right) - Q_2F_{11}\left(I_s^-,I_d^-,\omega\right)\right],\end{aligned} \tag{27}$$

$$\begin{aligned}\sigma_{rr} = \sigma_{22} = K_{D1}\left[Q_1F_{22}\left(R_s^+,R_d^+,\omega\right) + Q_2F_{22}\left(R_s^-,R_d^-,-\omega\right)\right] \\ -\omega K_{D2}\left[Q_1F_{22}\left(I_s^+,I_d^+,\omega\right) - Q_2F_{22}\left(I_s^-,I_d^-,-\omega\right)\right],\end{aligned} \tag{28}$$

$$\begin{aligned}\sigma_{rz} = \sigma_{12} = K_{D1}\left[Q_1F_{12}\left(I_s^+,I_d^+,\omega\right) + Q_2F_{12}\left(I_s^-,I_d^-,-\omega\right)\right] \\ +\omega K_{D2}\left[Q_1F_{12}\left(R_s^+,R_d^+,\omega\right) - Q_2F_{12}\left(R_s^-,R_d^-,-\omega\right)\right],\end{aligned} \tag{29}$$

$$\begin{aligned}\dot{u}_z = \dot{u}_1 = -\frac{2v}{\mu} K_{D1}\left[Q_1F_1\left(R_s^+,R_d^-,\omega\right) + Q_2F_1\left(R_s^-,R_d^-,-\omega\right)\right] \\ +\frac{2v}{\mu}\omega K_{D2}\left[Q_1F_1\left(I_s^+,I_d^-,\omega\right) - Q_2F_1\left(I_s^-,I_d^-,-\omega\right)\right],\end{aligned} \tag{30}$$

$$\begin{aligned}\dot{u}_r = \dot{u}_2 = -\frac{2v}{\mu} K_{D1}\left[Q_1F_2\left(I_s^+,I_d^-,\omega\right) + Q_2F_2\left(I_s^-,I_d^-,-\omega\right)\right] \\ -\frac{2v}{\mu}\omega K_{D2}\left[Q_1F_2\left(R_s^+,R_d^-,\omega\right) - Q_2F_2\left(R_s^-,R_d^-,-\omega\right)\right],\end{aligned} \tag{31}$$

where

$$F_{11}(M, N, \omega) = -2\alpha_s g_1(\omega)M + \left(2\alpha_d^2 - \alpha_s^2 + 1\right) g_2(\omega)N , \tag{32}$$

$$F_{22}(M, N, \omega) = 2\alpha_s g_1(\omega)M - \left(1 + \alpha_s^2\right)g_2(\omega)N , \tag{33}$$

$$F_{12}(M, N, \omega) = \left(1 + \alpha_s^2\right)g_1(\omega)M - 2\alpha_d g_2(\omega)N , \tag{34}$$

$$F_1(M, N, \omega) = -\alpha_s g_1(\omega)M + g_2(\omega)N , \tag{35}$$

$$F_2(M, N, \omega) = g_1(\omega)M - \alpha_d g_2(\omega)N \,, \tag{36}$$

$$g_1(\omega) = 2\alpha_d - \frac{1}{\omega}\left(1 + \alpha_s^2\right) \,, \tag{37}$$

$$g_2(\omega) = 1 + \alpha_s^2 - \frac{2}{\omega}\alpha_s \,, \tag{38}$$

$$R^+ = r^{-1/2}e^{-\varepsilon\theta} \cos\left(\frac{\theta}{2} - \varepsilon \ln r\right) , \tag{39}$$

$$R^- = r^{-1/2}e^{\varepsilon\theta} \cos\left(\frac{\theta}{2} + \varepsilon \ln r\right) , \tag{40}$$

$$I^+ = -r^{-1/2}e^{-\varepsilon\theta} \sin\left(\frac{\theta}{2} - \varepsilon \ln r\right) , \tag{41}$$

$$I^- = -r^{-1/2}e^{\varepsilon\theta} \sin\left(\frac{\theta}{2} + \varepsilon \ln r\right) , \tag{42}$$

$$Q_1 = \frac{e^{\pi\varepsilon}}{4(2\pi)^{1/2}D \cosh \pi\varepsilon} \,, \tag{43}$$

$$Q_2 = \frac{e^{-\pi\varepsilon}}{4(2\pi)^{1/2}D \cosh \pi\varepsilon} \,, \tag{44}$$

$$r_s = \left(\xi_1^2 + \alpha_s^2 \xi_2^2\right)^{1/2} , \tag{45}$$

$$r_d = \left(\xi_1^2 + \alpha_d^2 \xi_2^2\right)^{1/2} , \tag{46}$$

$$\theta_s = \tan^{-1}\left(\frac{\alpha_s \xi_2}{\xi_1}\right) , \tag{47}$$

$$\theta_d = \tan^{-1}\left(\frac{\alpha_s \xi_2}{\xi_1}\right) . \tag{48}$$

Subscripts s and d can be added to R^+, R^-, I^+, I^-, r and θ. It is seen that there are oscillatory singularities in stresses and particle velocities in the case of in-plane deformation.

For the anti-plane deformation, the asymptotic fields of stress and particle velocity in the matrix are

$$\sigma_{\theta z} = \sigma_{31} = -\frac{K_{D3}}{(2\pi r_s)^{1/2}\alpha_s}\sin\frac{\theta_s}{2}, \tag{49}$$

$$\sigma_{rz} = \sigma_{32} = \frac{K_{D3}}{(2\pi r_s)^{1/2}}\cos\frac{\theta_s}{2}, \tag{50}$$

$$\dot{u}_\theta = \dot{u}_3 = \frac{vK_{D3}}{\mu\alpha_s(2\pi r_s)^{1/2}}\sin\frac{\theta_s}{2}. \tag{51}$$

Hence, there is no oscillation in the case of anti-plane deformation.

When the two materials are identical, all asymptotic field equations are reduced to the classical expressions given in Freund's text book (1990).

Energy Release Rates

The energy release rate for the in-plane deformation is found to be

$$G = \frac{v^2\left(c\,K_{D1}^2 + a\,K_{D2}^2\right)}{8ac\cosh^2\pi\varepsilon}\sum_{j=1}^{2}\frac{c\alpha_{sj} + a\alpha_{dj}}{\mu_j\,c_{sj}^2\,D_j}. \tag{52}$$

For the anti-plane deformation, the energy release rate is found to be

$$G = \frac{K_{D3}^2}{4}\sum_{j=1}^{2}\frac{1}{\mu_j\,\alpha_{sj}}. \tag{53}$$

Again, when the two materials are identical, equations (52) and (53) are reduced to the classical expressions given in the text book by Freund (1990).

Dynamic Stress Intensity Factors

In the above expressions of asymptotic fields of stress and particle velocity and energy release rates, the dynamic stress intensity factors are involved in the equations. Expressions of dynamic stress intensity factors for a propagating interfacial crack can be derived from the equations of the corresponding static stress intensity factors for a stationary crack through the following conversion relations

$$K_1 = \frac{K_{D1}}{\omega},\quad K_2 = K_{D2},\quad K_3 = K_{D3}. \tag{54}$$

For example, in the problem of a semi-infinite stationary crack on the negative ξ_1-axis, if the following concentrated tractions are prescribed on the crack surface to be

$$\sigma_{12}(\xi_1) = -\sigma_{12}^{*}\,\delta(\xi_1 + d),\quad \sigma_{22}(\xi_1) = -\sigma_{22}^{*}\,\delta(\xi_1 + d), \tag{55}$$

where $\delta(\xi_1)$ is the Dirac delta function and d is a distance, the static complex stress intensity factor is

$$K = K_1 + iK_2$$

$$= -\left(\frac{2}{\pi}\right)^{1/2} \cosh \pi\varepsilon \int_{-\infty}^{0} (-\xi_1)^{-1/2-i\varepsilon} \left[\left(\frac{c}{a}\right)^{1/2} \sigma_{22}(\xi_1) + i\sigma_{12}(\xi_1)\right] d\xi_1$$

$$= \left(\frac{2}{\pi}\right)^{1/2} \cosh \pi\varepsilon \; d^{-1/2-i\varepsilon} \left(\frac{\sigma_{22}^*}{\omega} + i\,\sigma_{12}^*\right). \tag{56}$$

The corresponding expression for the dynamic stress intensity factors is

$$\frac{K_{D1}}{\omega} + iK_{D2} = \left(\frac{2}{\pi}\right)^{1/2} \cosh \pi\varepsilon \; d^{-1/2-i\varepsilon} \left(\frac{\sigma_{22}^*}{\omega} + i\,\sigma_{12}^*\right). \tag{57}$$

By the method of superposition, equation (57) can be used to determine the dynamic stress intensity factors due to the distributed force applied on the boundary of a plastic zone of Dugdale type ahead of the front of a propagating interfacial crack.

Conclusions

In this paper, the asymptotic analysis of stresses and particle velocities in the vicinity of the front of a propagating interfacial crack between the fiber and the matrix is presented. All field equations and energy release rates per unit crack propagation are expressed in terms of dynamic stress intensity factors. Our results can be used for further studies on dynamic debonding between fibers and matrix and the breakage of the fiber at the crack front caused by local dynamic stresses in a fiber-reinforced composite.

References

Eshelby, J.D., Read, W.T. and Shockley, W. (1953) Anisotropic Elasticity with Applications to Dislocation Theory, *Acta Metallurgia* 1, 251-259.

Freund, L.B. (1990) *Dynamic Fracture Mechanics*, Cambridge University Press, 155-171 and 221-235.

Huang, N.C. (1992) Interfacial Crack Propagation Between Two Isotropic Elastic Media, Technical Report, Department of Aerospace and Mechanical Engineering, University of Notre Dame, Notre Dame, Indiana 46556.

Suo, Z.G. (1990) Singularities, Interfaces and Cracks in Dissimilar Anisotropic Media, *Proceedings of Royal Society, London* A427, 331-358.

Yang, W., Suo, Z. and Shih, C.F. (1991) Mechanics of Dynamic Debonding, *Proceedings of Royal Society, London* A433, 679-697.

A MODEL FOR SHEAR STRESS RELAXATION AROUND A FIBER BREAK IN UNIDIRECTIONAL COMPOSITES AND CREEP RUPTURE ANALYSIS

N. OHNO*, H. KAWABE*, T. MIYAKE** and M. MIZUNO*
** Department of Micro System Engineering, Nagoya University*
Chikusa-ku, Nagoya 464-01, JAPAN
*** Nagoya Municipal Industrial Research Institute*
Rokuban, Atsuta-ku, Nagoya 456, JAPAN

Abstract This paper describes a model of stress relaxation in broken fibers in unidirectional metal matrix composites reinforced with long brittle fibers. A cylindrical cell with a broken fiber embedded in a power law creeping matrix is employed, and the broken fiber is assumed to have a bilinear distribution of axial stress. Then, on the basis of energy balance in the cell, a relaxation equation of interfacial shear stress acting on stress recovery segments is derived in a simple form. Under constant overall strain the relaxation equation is approximated rationally and integrated to obtain an analytical solution, which is shown to agree fairly well with the numerical analysis of Du and McMeeking (1995). Moreover, the relaxation equation is combined with Curtin's (1991) model, so that rupture time in long term creep is evaluated analytically and explicitly on the assumption of global load sharing. It is shown that the resulting relation represents well the dependence of creep rupture time on applied stress observed experimentally on a unidirectional metal matrix composite.

1. Introduction

Unidirectional polymer matrix and metal matrix composites reinforced with long brittle fibers may suffer from fiber breaks under initial loading and subsequent stress holding in creep. Since it is probable that fiber breaks induce creep rupture of such composites, the time-dependent evolution of stress profiles in broken fibers and neighboring intact fibers in creeping matrices has been studied in several works so far.

This kind of study can be traced back to that of Lifshitz and Rotem (1970). They studied creep rupture of GFRPs by assuming a linear viscoelastic theory; they analyzed the stress relaxation in broken fibers resulting from matrix viscosity, and then they evaluated creep rupture time by extending the rupture model of Rosen (1964). More detailed analysis was made by Lagoudas et al. (1989) with respect to the time-dependent evolution of stress profiles in neighboring intact fibers and by Mason et al. (1992)

R. Wang (ed.), IUTAM Symposium on Rheology of Bodies with Defects, 153–165.

concerning the effect of nonlinear, matrix viscosity on stress profiles around a fiber break. Phoenix et al. (1988) and Otani et al. (1991), on the other hand, performed creep experiments of carbon fiber/epoxy microcomposites. They thus suggested that the extension of stress recovery parts in broken fibers with time as well as the time-dependent development of interfacial debonding can be main reasons for the creep rupture.

For metal matrix composites, creep experiments of continuous SiC fiber/Ti alloy systems have been done in recent works (Schwenker et al., 1993; Ohno et al. 1994a, 1994b, 1996; Weber et al., 1996). It was thus observed that such material systems exhibit creep rupture even at stress levels fairly lower than tensile strength, and that fiber breaks occur with the increase of creep strain. Numerical simulations to elucidate the effect of fiber breaks on creep and creep rupture of unidirectional metal matrix composites were done by Goda (1993), Kelly and Barbero (1993), Du and McMeeking (1995), and Song et al. (1995). The simulations confirmed that the broken fibers in metal matrix composites also can have the time-dependent evolution of stress profiles resulting in the extension of stress recovery parts with time. Especially Du and McMeeking (1995) ascertained that the stress relaxation in broken fibers becomes significant in long term creep, and that the model of Curtin (1991) in combination with that of McLean (1985) is effective in predicting analytically the creep rupture time in short term creep. Weber et al. (1996) showed the validity of the Curtin-McLean model by performing short term creep experiments of a SiC/Ti alloy composite.

Now a problem of interest is to formulate simply the stress relaxation in broken fibers for the purpose of estimating analytically the rupture time in long term creep of unidirectional metal matrix composites. Ohno and Yamakawa (1996) presented a model for the stress relaxation by approximating bilinearly the stress distribution in broken fibers and by imposing the deformation compatibility at broken fiber ends, and they showed that the model leads to an analytical solution which simulates well the stress relaxation computed by Du and McMeeking. Iyengar and Curtin (1997) also presented a simple model for the stress relaxation in broken fibers, and for a few integers of the stress exponent of matrix creep they obtained an analytical but implicit solution to estimate the rupture time in long term creep.

In this paper, the stress relaxation in broken fibers in unidirectional metal matrix composites is formulated in a simple form from the view point of energy balance, and the resulting relation is applied to estimating analytically the creep rupture time in long term creep. To begin with, using an overall balance equation of energy in a cylindrical cell consisting of a broken fiber and a power-law creeping matrix, we derive a relaxation equation of interfacial shear stress acting on broken fibers, in which axial stress profiles are approximated bilinearly. Then, the relaxation equation is approximated rationally and integrated to obtain an analytical solution for the interfacial shear stress relaxation under constant overall strain, and the solution is combined with Curtin's (1991) model based on the assumption of global load sharing, so that the creep rupture time in long term creep of composites is evaluated explicitly in an analytical form. Finally, the analytical relations derived are discussed on the basis of the creep experiment on a unidirectional SCS-6/Beta21S metal matrix composite at 500°C done by Ohno et al. (1996).

2. Model for Stress Relaxation in Broken Fibers

We consider a unidirectional composite consisting of long brittle fibers and a power-law creeping matrix with weak interface. Let the diameter and volume fraction of fibers be D and V_f, respectively. We suppose that the fibers have a scatter of tensile strength, and that fiber breaks occur dispersedly in the composite under tensile loads as a consequence of weak interface.

To simplify the modeling of stress relaxation in broken fibers, we assume that the intact fibers and the matrix have uniform distributions of longitudinal normal stresses σ_f and σ_m, respectively. For the intact fibers and the matrix, then, we have the following equations concerning longitudinal composite strain ε:

$$\varepsilon = \sigma_f / E_f, \quad \dot{\varepsilon} = \dot{\sigma}_m / E_m + B\sigma_m^n, \tag{1), (2}$$

where E_f and E_m indicate Young's moduli of the fibers and the matrix, respectively, B and n are the creep constant and exponent of the matrix, and $(\dot{\ })$ denotes the differentiation with respect to time t.

For the broken fibers, on the other hand, let us approximate bilinearly the distribution of σ_f in the broken fibers, as shown in Fig. 1. Then, since the interfacial shear stress τ becomes z-independent in the stress recovery region of $0 \le z \le \delta$, the distribution of σ_f and the stress recovery length δ are expressed in the following forms, which were derived for short fibers by Kelly and Tyson (1965):

$$\sigma_f = \begin{cases} 4\tau z / D, & 0 \le z \le \delta, \\ E_f \varepsilon, & \delta < z, \end{cases} \qquad \delta = \frac{D E_f \varepsilon}{4\tau}. \tag{3), (4}$$

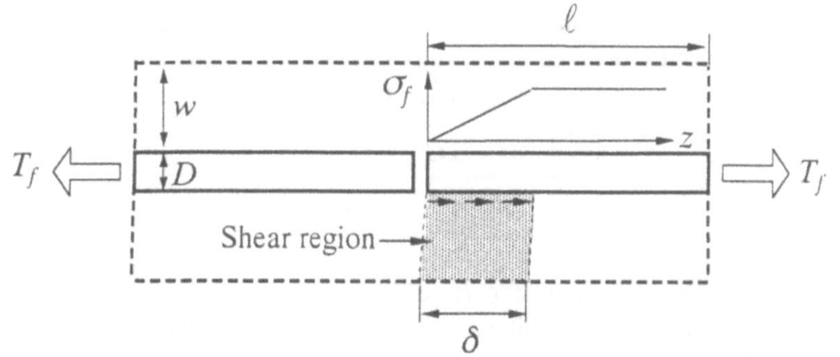

Figure 1. Cylindrical cell containing a broken fiber.

Now, in order to formulate the stress relaxation in broken fibers resulting from matrix creep, we consider a cylindrical cell of length 2ℓ in which a broken fiber is embedded (Fig. 1). Let us denote the cross areas of the fiber and matrix phases in the cell by A_f and A_m, respectively. We assume further that the cell is sufficiently long in the axial direction, so that at $z = \ell$ the broken fiber in the cell is subjected to the same load T_f and displacement u_ℓ as the intact fibers:

$$T_f = A_f E_f \varepsilon, \quad u_\ell = \ell\varepsilon. \tag{5), (6}$$

Let us suppose that the work done by T_f is converted in the cell to the elastic tensile strain energy in the broken fiber, the elastic shear strain energy in the matrix and the energy dissipation due to shear creep in the matrix, because we have assumed that the fiber break does not disturb the uniform distributions of matrix normal stress σ_m in the cell and fiber stress σ_f in the neighboring intact fibers. This energy balance can be expressed in the following rate form if we notice that the matrix in the cell is subjected to z-independent τ in the region of $0 \le z \le \delta$, and if we ignore the radial variation of shear stress in the matrix:

$$\frac{d}{dt}\left(\int_0^{\ell} \frac{\sigma_f^2}{2E_f} A_f dz + \frac{\tau^2}{2G_m} A_m \delta\right) + \tau\dot{\gamma}_m^c A_m \delta = T_f \dot{u}_\ell, \tag{7}$$

where G_m and $\dot{\gamma}_m^c$ denote the rigidity and shear creep rate of the matrix, respectively. Using eqns $(3)_2$, (5) and (6), eqn (7) is rewritten as

$$\frac{d}{dt}\left(\frac{E_f \varepsilon^2}{2} A_f \delta - \int_0^{\delta} \frac{\sigma_f^2}{2E_f} A_f dz\right) = \frac{d}{dt}\left(\frac{\tau^2}{2G_m} A_m \delta\right) + \tau\dot{\gamma}_m^c A_m \delta. \tag{8}$$

This equation means that the fiber strain energy released due to the stress relaxation in the broken fiber is used to shear the matrix around the fiber break elastically and in creep in the cell. Substitution of eqns $(3)_1$ and (4) into eqn (8) leads to

$$\dot{\tau} = \frac{\left(A_f E_f \varepsilon^2 - \dfrac{A_m \tau^2}{2G_m}\right)\dfrac{\tau\dot{\varepsilon}}{\varepsilon} - A_m \tau^2 \dot{\gamma}_m^c}{\dfrac{A_f E_f \varepsilon^2}{3} + \dfrac{A_m \tau^2}{2G_m}}. \tag{9}$$

As seen from the above equation, if ε is constant or increases slowly, matrix shear creep causes the relaxation of τ, which induces time-dependent extension of the stress recovery length δ. Thus, the above equation can be regarded as an evolution equation of τ which describes the stress relaxation in broken fibers. On the other hand, if ε increases rapidly, eqn (9) allows τ to develop. The value of τ, however, is limited by interfacial sliding stress τ_0, i.e., $\tau \le \tau_0$. Here and from now on τ is taken to be positive since $\dot{\varepsilon} \ge 0$ in the present work.

If the matrix is a material of the Mises type, $\dot{\gamma}_m^c$ in eqn (9) is written as

$$\dot{\gamma}_m^c = 3B\sigma_e^{n-1}\tau, \tag{10}$$

where $\sigma_e = (\sigma_m^2 + 3\tau^2)^{1/2}$. If σ_m is negligibly small, eqn (10) is reduced to

$$\dot{\gamma}_m^c = 3^{(n+1)/2} B\tau^n. \tag{11}$$

3. Stress Relaxation under Constant Overall Strain

This section deals with the relaxation of τ under constant overall strain ε_0. Let us suppose that the cell shown in Fig. 1 is subjected to $\varepsilon = \varepsilon_0$ applied instantaneously at $t = 0$, and that at $t = 0$ the fiber break occurs and shear sliding takes place at the interface in the stress recovery region of $0 \le z \le \delta$.

At $t \ge 0$, $\varepsilon(t) = \varepsilon_0$, so that eqn (9) is reduced to

$$\dot{\tau} = \frac{-A_m\tau^2\dot{\gamma}_m^c}{\dfrac{A_f E_f \varepsilon_0^2}{3} + \dfrac{A_m\tau^2}{2G_m}}. \tag{12}$$

Since the interfacial sliding stress τ_0 is usually much lower than the elastic shear stress generated at perfectly bonded interface around fiber breaks, it is appropriate to assume that while τ relaxes from τ_0, τ satisfies

$$\tau << E_f\varepsilon_0\left(2A_f G_m / 3A_m E_f\right)^{1/2}, \quad \text{i.e., } A_m\tau^2/2G_m << A_f E_f \varepsilon_0^2/3. \tag{13), (14}$$

Then eqn (12) has an approximation

$$\dot{\tau} = -3A_m\tau^2\dot{\gamma}_m^c / A_f E_f \varepsilon_0^2. \tag{15}$$

Substitution of eqn (10) into (15) gives

$$\dot{\tau} = -9A_m B\tau^3(\sigma_m^2 + 3\tau^2)^{(n-1)/2} / A_f E_f \varepsilon_0^2. \tag{16}$$

Now let us assume that σ_m relaxes much more quickly than τ. This assumption, which was ascertained numerically by Du and McMeeking (1995), is usually valid, as discussed later in this section. Then, since the effect of σ_m on the relaxation of τ is negligible, the above equation becomes

$$\dot{\tau} = -3^{(n+3)/2} A_m B\tau^{n+2} / A_f E_f \varepsilon_0^2. \tag{17}$$

Hence, by integrating this equation and taking the initial value of τ to be equal to the sliding stress τ_0, the relaxation of τ is expressed analytically as

$$\tau(t) = \tau_0\left[1 + (n+1)t/t_\tau\right]^{-1/(n+1)}, \quad t_\tau = A_f E_f \varepsilon_0^2 / 3^{(n+3)/2} A_m B\tau_0^{n+1}. \tag{18), (19}$$

Here t_τ denotes the relaxation time of τ defined as $t_\tau = -\tau_0/\dot{\tau}(0)$. When τ relaxes as eqn (18), the stress recovery length δ expressed as eqn (4) has the time-dependent extension

$$\delta(t) = \frac{DE_f\varepsilon_0}{4\tau_0}\left[1 + (n+1)\frac{t}{t_\tau}\right]^{1/(n+1)} \tag{20}$$

At $t >> t_\tau$, eqns (18) and (20) take the forms

$$\tau(t) = \left[\frac{A_f E_f \varepsilon_0^2}{(n+1)\cdot 3^{(n+3)/2} A_m B t}\right]^{1/(n+1)}, \quad \delta(t) = \frac{DE_f\varepsilon_0}{4}\left[\frac{(n+1)\cdot 3^{(n+3)/2} A_m B t}{A_f E_f \varepsilon_0^2}\right]^{1/(n+1)} \tag{21), (22}$$

The relaxation of σ_m under constant overall strain, on the other hand, is expressed by eqn (2) with $\varepsilon = \varepsilon_0$:

$$\dot{\sigma}_m = -BE_m\sigma_m^n. \tag{23}$$

If elastic stress $E_m\varepsilon_0$ is taken to be the initial value of σ_m, the above equation has the solution

$$\sigma_m(t) = E_m\varepsilon_0\left[1 + (n-1)t/t_\sigma\right]^{-1/(n-1)}, \quad t_\sigma = 1/BE_m^n\varepsilon_0^{n-1}, \tag{24), (25}$$

where t_σ denotes the relaxation time of σ_m defined as $t_\sigma = -E_m\varepsilon_0/\dot{\sigma}_m(0)$.

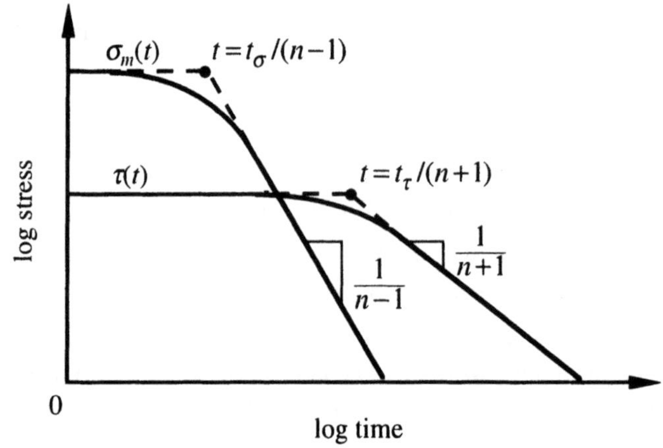

Figure 2. Illustration of the relaxation functions $\tau(t)$ and $\sigma_m(t)$ under constant overall strain.

Now we compare the relaxation functions $\tau(t)$ and $\sigma_m(t)$ given by eqns (18) and (24), respectively, (Fig. 2). The functions have different exponents $-1/(n+1)$ and

$-1/(n-1)$ as a result of eqns (17) and (23), in which $\dot{\tau}$ and $\dot{\sigma}_m$ under constant overall strain are proportional to $-\tau^{n+2}$ and $-\sigma_m^n$, respectively. Consequently, as illustrated schematically in Fig. 2, usually τ relaxes much more slowly than σ_m; in other words, the stress relaxation in broken fibers becomes significant in long term creep.

4. Model for Short Term Creep

The previous sections have dealt with a model which describes the stress relaxation in broken fibers. Before applying the model to estimating the rupture time of unidirectional composites in long term creep, we describe the Curtin-McLean model discussed by Du and McMeeking (1995). The Curtin-McLean model is effective for short term creep, because it is based on the assumption of no stress relaxation in broken fibers.

To analyze the longitudinal creep of unidirectional composites with no fiber breaks, McLean (1985) solved eqns (1) and (2), along with

$$\sigma = (1-V_f)\sigma_m + V_f\sigma_f, \tag{26}$$

where σ denotes applied stress and is constant here. He thus obtained

$$\varepsilon(t) = \frac{\sigma}{V_f E_f} - \frac{1-V_f}{V_f E_f}\left[\left(\frac{E_m \sigma}{E_c}\right)^{1-n} + \frac{(n-1)V_f E_f E_m B t}{E_c}\right]^{1/(1-n)}, \tag{27}$$

where E_c denotes the modulus based on the rule of mixtures; i.e., $E_c = (1-V_f)E_m + V_f E_f$.

Curtin (1991) presented a model for the load carrying capacity of fibers in unidirectional composites with weak interface. In his model, the broken fibers have the stress recovery parts subjected to constant interfacial shear stress τ_0, so that the bilinear distribution of σ_f expressed by eqns (3) and (4) with $\tau = \tau_0$ prevails in the broken fibers. On the assumption of global load sharing, then, he showed that overall strain ε induces the following fiber stress averaged on a cross section perpendicular to the fiber direction:

$$\overline{\sigma}_f = (1-q)E_f\varepsilon + qE_f\varepsilon/2, \tag{28}$$

where q denotes the probability that a fiber is broken within the stress recovery distance δ_0 $(= DE_f\varepsilon/4\tau_0)$ from the cross section. In the above equation, the first and second terms in the right hand side are concerned with the intact and broken fibers within δ_0 from the cross section, and it is noted that such broken fibers have the average stress equal to $E_f\varepsilon/2$ on the cross section because breaks are located randomly. If δ_0 is much smaller than fiber length, the Weibull statistics provides q with

$$q = \frac{2\delta_0}{L_0}\left(\frac{E_f\varepsilon}{\sigma_0}\right)^m, \tag{29}$$

where L_0, σ_0 and m are the Weibull parameters. Then, eqn (28) becomes

$$\overline{\sigma}_f = E_f\varepsilon\left[1 - \frac{1}{2}\left(\frac{E_f\varepsilon}{S_c}\right)^{m+1}\right], \tag{30}$$

where S_c indicates a characteristic strength such that

$$S_c = \left(2\sigma_0^m \tau_0 L_0 / D\right)^{1/(m+1)}. \tag{31}$$

It is shown from eqn (30) that when ε increases $\overline{\sigma}_f$ takes the maximum

$$S_{max} = S_c \frac{m+1}{m+2}\left(\frac{2}{m+2}\right)^{1/(m+1)}. \tag{32}$$

Consequently S_{max} can be regarded as the load carrying capacity of fibers in composites.

Du and McMeeking (1995) combined the McLean and the Curtin model mentioned above so as to predict the creep rupture time which they computed by discretizing a shear lag equation of creep of unidirectional metal matrix composites with weak interface. They thus found that the creep rupture time in short term creep is estimated fairly well as the time at which the intact fiber stress expressed by McLean's model reaches the load carrying capacity of fibers, S_{max}, derived by Curitn. This creep rupture time t_r is obtained by substituting eqn (27) into $E_f\varepsilon(t_r) = S_{max}$:

$$t_r = \frac{E_c}{(n-1)V_f E_f E_m B}\left[\left(\frac{1-V_f}{\sigma - V_f S_{max}}\right)^{n-1} - \left(\frac{E_c}{E_m \sigma}\right)^{n-1}\right]. \tag{33}$$

It is obvious that eqn (33) is effective for $\sigma > V_f S_{max}$.

5. Estimation of Rupture Time in Long Term Creep

Since eqn (33) was derived by taking into account only the relaxation of matrix stress σ_m, this equation is not effective for long term creep in which the stress relaxation in broken fibers becomes significant. In this section, creep rupture time t_r in such long term creep is estimated analytically by employing the relaxation model of interfacial shear stress τ derived in Sections 2 and 3.

The problem of estimating t_r in long term creep, however, is complicated, because the fiber breaks which occurred at different times in the progress of creep have different amounts of relaxation of τ; i.e., τ relaxes more significantly at older breaks, and if a break has just occurred τ is equal to τ_0 there. Now, to estimate t_r analytically, let us simplify the problem by introducing the following assumptions (Iyengar and Curtin, 1997):

(1) All the fiber breaks have the same amount of relaxation of τ as that around an initial break.

(2) Longitudinal normal stress in the matrix, σ_m, is completely relaxed, i.e., $\sigma_m = 0$.

The first assumption permits us to employ Curtin's model described in the preceding section, because all the fiber breaks have the same stress recovery length

$$\delta(t) = DE_f\varepsilon / 4\tau(t). \tag{34}$$

Here τ is not constant but time-dependent. Then, eqn (29) is modified as

$$q(t) = \frac{2\delta(t)}{L_0}\left[\frac{E_f\varepsilon(t)}{\sigma_0}\right]^m, \tag{35}$$

and consequently eqn (30) takes the form

$$\bar{\sigma}_f(t) = E_f\varepsilon(t)\left\{1 - \frac{\tau_0}{2\tau(t)}\left[\frac{E_f\varepsilon(t)}{S_c}\right]^{m+1}\right\}. \tag{36}$$

The second assumption, on the other hand, allows applied stress σ to satisfy $\sigma = V_f\bar{\sigma}_f$. Therefore, eqn (36) becomes

$$\frac{\sigma}{V_f} = E_f\varepsilon(t)\left\{1 - \frac{\tau_0}{2\tau(t)}\left[\frac{E_f\varepsilon(t)}{S_c}\right]^{m+1}\right\}. \tag{37}$$

Now let us approximate $\tau(t)$ in the above equation. Since we have assumed that σ_m is completely relaxed, the following strain is yielded at $t = 0$ except for the effect of fiber breaks, as seen from eqn (27):

$$\varepsilon' = \sigma / V_f E_f. \tag{38}$$

Creep strain then takes place due to the fiber breaks and the stress relaxation in broken fibers. Such strain, however, is usually much smaller than ε', as shown later. It leads us to approximate $\tau(t)$ in eqn (37) using the relaxation function of τ at constant overall strain ε'. The function is given by eqns (18) and (19) with ε_0 replaced by ε':

$$\tau(t) = \tau_0\left[1 + (n+1)t / t'_\tau\right]^{-1/(n+1)}, \qquad t'_\tau = \sigma^2 / 4 \cdot 3^{(n+3)/2}\alpha(\alpha - V_f^{1/2})V_f E_f B\tau_0^{n+1}, \tag{39), (40}$$

where $\alpha = \left(\pi / 2\sqrt{3}\right)^{1/2}$ for the hexagonal array of fibers.

Substituting eqn (39) into (37), and solving the resulting equation for time t, we have

$$t = \frac{t'_\tau}{n+1}\left\{\left[2\left(1-\frac{\sigma}{V_f E_f \varepsilon}\right)\left(\frac{S_c}{E_f \varepsilon}\right)^{m+1}\right]^{n+1} - 1\right\}. \tag{41}$$

Since $\dot{\varepsilon} = \infty$ or $dt/d\varepsilon = 0$ at rupture, the above equation gives the strain and time at rupture as follows:

$$\varepsilon_r = \frac{m+2}{m+1}\cdot\frac{\sigma}{V_f E_f}, \qquad t_r = \frac{t'_\tau}{n+1}\left[\left(\frac{V_f S_{max}}{\sigma}\right)^{(m+1)(n+1)} - 1\right], \tag{42), (43}$$

where S_{max} indicates the load carrying capacity of fibers in composites with $\tau = \tau_0$, and was given by eqn (32).

Since $t_r > 0$, eqn (43) is effective for $\sigma < V_f S_{max}$. For σ satisfying $(\sigma / V_f S_{max})^{(m+1)(n+1)} \ll 1$, eqn (43) is reduced to

$$t_r = \frac{t'_\tau}{n+1}\left(\frac{V_f S_{max}}{\sigma}\right)^{(m+1)(n+1)} \tag{44}$$

Eqns (38) and (42) give

$$\varepsilon_r / \varepsilon' = (m+2)/(m+1). \tag{45}$$

Hence, if $m > 5$, the difference between ε_r and ε' is less than about 15 percent. We therefore can say that creep strain taking place before rupture is fairly small if ε' is taken to be the initial instantaneous strain by assuming the complete relaxation of σ_m in long term creep. This validates the use of the relaxation function $\tau(t)$ at constant overall strain ε' in approximating $\tau(t)$ in eqn (37).

6. Discussion

Let us discuss the analytical predictions of creep rupture time on the basis of the creep experiment on a 6-ply unidirectional SCS-6/Beta21S metal matrix composite at 500°C done by Ohno et al. (1996). The composite consists of continuous SiC fibers, SCS-6, and a metastable beta titanium alloy, Beta21S.

The analytical predictions (33) and (43) give the σ versus t_r relations shown by the solid lines in Fig. 3. It is obvious that the predictions overestimate the experiments especially in long term creep. The disagreement can be ascribed to some extent to the degradation of fibers near the edges of specimens due to oxidation. We may consider that such fibers, which had little bonding with the matrix and could carry in effect no load if broken once, caused a decrease in the load carrying capacity of fibers. The decrease may be taken into account by simply reducing V_f from 0.35 to 0.31, because about ten percent of fibers in each ply were damaged due to oxidation. Since the oxidation induced damage was significant in long term creep, the long term prediction

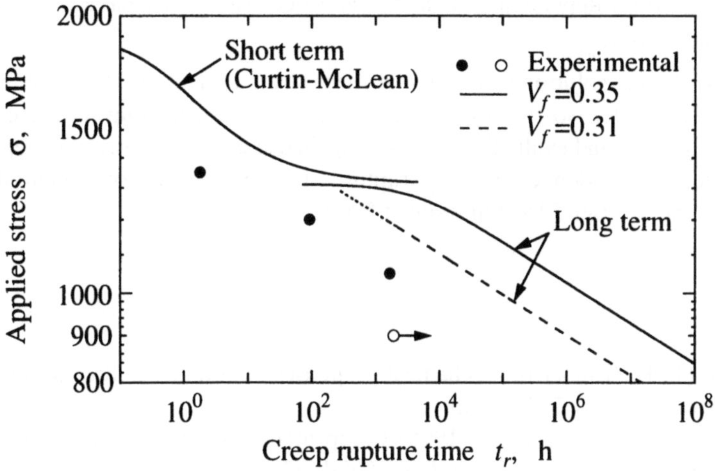

Figure 3. Creep rupture time of SCS-6/Beta21S at 500°C; experiments (Ohno et al., 1996), short term prediction by eqn (33) for $V_f = 0.35$, and long term predictions by eqn (43) for $V_f = 0.35$ and by eqn (44) for $V_f = 0.31$.

(44) with $V_f = 0.31$ is shown by the dashed line in the figure. It is seen that such a long term prediction gets closer to the experiments.

7. Concluding Remarks

In the present work, the stress relaxation in broken fibers in unidirectional metal matrix composites was formulated using a simple model, and the rupture time in long term creep was estimated explicitly in an analytical form. The results obtained in this work are summarized as follows.

(1) A relaxation equation of interfacial shear stress acting on broken fibers in composites was derived by considering the overall balance of energy in a cell consisting of a power-law creeping matrix and a broken fiber, in which fiber stress was assumed to distribute bilinearly.

(2) For the cell with low sliding resistance at interface and negligible normal stress in the matrix, the relaxation equation mentioned above was approximated rationally and integrated to obtain an analytical solution under constant overall strain ε_0.

(3) Interfacial shear stress τ around fiber breaks usually relaxes much more slowly than matrix normal stress σ_m, because the relaxation equation of τ has stronger

nonlinearity in stress dependence than σ_m, and because τ can have a much longer relaxation time than σ_m.

(4) The analytical solution of interfacial shear stress relaxation was combined with Curtin's (1991) model, so that the rupture time and strain in long term creep were estimated analytically and explicitly in terms of applied stress σ. This relation represented well the stress dependence of creep rupture time observed experimentally on a unidirectional SCS-6/Beta21S composite at 500°C.

References

Curtin, W.A. (1991) Theory of mechanical properties of ceramic-matrix composites. *J. Am. Ceram. Soc.* **74**, 2837-2845.

Du, Z.-Z. and McMeeking, R.M. (1995) Creep models for metal matrix composites with long brittle fibers. *J. Mech. Phys. Solids* **43**, 701-726.

Goda, K. (1993) Estimation of creep rupture life of a B/Al composite using Monte Carlo simulation method. *Proc. 12th Symposium on Material and Structural Reliability* (in Japanese), p. 45-49. The Society of Materials Science, Japan.

Iyengar, N. and Curtin, W.A. (1997) Time-dependent failure in fiber-reinforced composites by matrix and interface shear creep. *Acta Mater.* (to appear).

Kelly, A. and Tyson, W.R. (1965) Tensile properties of fibre-reinforced metals: copper/tungsten and copper/molybdenum. *J. Mech. Phys. Solids* **13**, 329-350.

Kelly, K.W. and Barbero, E. (1993) The effect of fiber damage on the longitudinal creep of a CFMMC. *Int. J. Solids Struct.* **30**, 3417-3429.

Lagoudas, D.C., Hui, C.-Y. and Phoenix, S.L. (1989) Time evolution of overstress profiles near broken fibers in a composite with a viscoelastic matrix. *Int. J. Solids Struct.* **25**, 45-66.

Lifshitz, J.M. and Rotem, A. (1970) Time-dependent longitudinal strength of unidirectional fibrous composites. *Fibre Sci. & Technol.* **3**, 1-20.

Mason, D.D., Hui, C.-Y. and Phoenix, S.L. (1992) Stress profiles around a fiber break in a composite with a nonlinear, power law creeping matrix. *Int. J. Solids Struct.* **29**, 2829-2854.

McLean, M. (1985) Creep deformation of metal-matrix composites. *Compos. Sci. & Technol.* **23**, 37-52.

Ohno, N., Fujita, T., Miyake, T., Nakatani, H. and Imuta, M. (1996) Creep rupture of a unidirectional SCS-6/Beta21S metal matrix composite at 450, 500 and 550°C. *Mater. Sci. Res. Int.* **2**, 199-205.

Ohno, N., Okamoto, N., Miyake, T., Nishide, S. and Masaki Jr., S. (1994a) Acoustic emission and fiber damage in creep of a unidirectional SCS-6/Ti-15-3 metal matrix composite at 450°C. *Scripta Metall. Mater.* **31**, 1549-1554.

Ohno, N., Toyoda, K., Okamoto, N., Miyake, T. and Nishide, S. (1994b) Creep behavior of a unidirectional SCS-6/Ti-15-3 metal matrix composite at 450°C. *ASME J.*

Engng Mater. Technol. **116**, 208-214.

Ohno, N. and Yamakawa, T. (1996) A model for shear stress relaxation around a fiber break in unidirectional metal matrix composites. *JSME Int. J., Ser. A* **39**, 517-525.

Otani, H., Phoenix, S.L. and Petrina, P. (1991) Matrix effects on lifetime statics for carbon fibre-epoxy microcomposites in creep rupture. *J. Mater. Sci.* **26**, 1955-1970.

Phoenix., S.L., Schwartz, P. and Robinson IV, H.H. (1988) Statics for the strength and lifetime in creep-rupture of model carbon/epoxy composites. *Compos. Sci. & Technol.* **32**, 81-120.

Rosen, B.W. (1964) Tensile failure of fibrous composites. *AIAA J.* **2**, 1985-1991.

Schwenker, S.W., Evans, D.J. and Eylon, D. (1993) Longitudinal creep behavior and damage in SCS-6/Ti-6Al-4V metal matrix composites. *Titanium '92, Science and Technology* (ed. F.H. Froes and I. Caplan), p. 2593-2600. The Minerals, Metals & Materials Society.

Song, Y., Bao, G. and Hui, C.Y. (1995) On creep of unidirectional fiber composites with fiber damage. *Acta Metall. Mater.* **43**, 2615-2623.

Weber, C.H., Du, Z.-Z. and Zok, F.W. (1996) High temperature deformation and fracture of a fiber reinforced titanium matrix composite. *Acta Mater.* **44**, 683-695.

STUDIES ON RHEOLOGICAL RELATION OF MATERIALS BY TAKING INTO ACCOUNT THE RATE-DEPENDENT EVOLUTION OF INTERNAL DEFECTS AT HIGH STRAIN RATES

L.L. WANG (WANG Lili)*, **, Z.B. JIANG*, J.Y. CHEN*
* *Mechanics and Materials Science Research Centre, Ningbo University Ningbo, Zhejiang, 315211, China*
** *Dept. of Modern Mechanics, Univ. of Science and Technology of China Hefei, Anhui, 230026, China*

Abstract. It was experimentally observed that a rheological process of material is usually accompanied with an evolution of internal defects or micro-damage. Based on the micro-observation of the twinning evolution for Zircaloy-4, the adiabatic shear banding evolution for titanium alloy TB2 and the micro-cracks evolution for cast magnesium alloy ZM5-T4, thermoplastics PMMA and cement mortar, a rate-dependent defect/damage evolution law is suggested on the basis of thermo-activated mechanism. Correspondingly, a damage-modified rate-dependent rheological relation, taking into account the damage-weakening effect, is proposed and discussed.

1. Introduction

It is well-recognized that what distinguished explosion/impact dynamics from static mechanics are mainly two so-called dynamic effects, i.e. the inertia effect and the strain-rate effect[1, 2]. The former is studied by wave propagation in various forms, and the later has promoted the study of dynamic mechanical behavior of materials under an extensive range of strain rates.

The study of the dynamic mechanical behavior of materials has been receiving greater attention by both mechanicians and material scientists in the recent one hundred years[2]. It is worthwhile to point out the following two common characteristics regardless of what specific material was studied.

Firstly, when the test range of strain rates is large enough, most materials display, more or less, the so-called rate(time)-dependent mechanical behavior. In other words, all the mechanical behavior observed in a certain range of strain rates are essentially rheological, containing both of rate independent deformation and rate dependent viscous flow, although different rheological relations may be suggested to describe the specific

R. Wang (ed.), IUTAM Symposium on Rheology of Bodies with Defects, 167–178.

rate dependent response for different material, respectively.

Secondly, when the micro-structure of material is observed in a range of observation scales, most materials reveal, more or less, a certain form of the so-called internal defects. It means that all the practical materials are rheological bodies with defects. In other words, the rheological process of any practical material is a process, more or less, accompanied with an internal defect evolution process.

The complexity of the problem lies in the interactive or coupled relation between the pure flow/deformation process and the pure defect (damage) evolution process. In fact, on the one hand, the defect/damage evolution is a stress-assisted process, which accompanies and is strongly dependent on the flow/deformation process. On the other hand, the influence of the defect/damage evolution on the material strength or the stress-flow/deformation response could not be neglected.

Thus, the study of rheological relation of materials by taking into account the internal defects evolution has become one of the research frontiers, receiving more and more attention by both mechanicians, rheologists and material scientists.

The dynamic mechanical behavior for different materials including metallic alloys, polymers and concrete were experimentally studied in the authors' laboratory from both the macroscopic and microscopic view-points, in a range of strain rates from 10^{-5} s^{-1} (quasi-static loading) up to 10^{3} s^{-1} (impulsive loading). In the present paper, the main experimental and theoretical results are summarized and discussed, and a class of damage-modified rate-dependent constitutive relation, taking into account the internal defect weakening effect, is finally proposed for the materials studied.

2. Experimental Results

2.1. TWINNING EVOLUTION FOR ZIRCALOY-4

The dynamic mechanical behavior for Zircaloy-4, a close-packed hexagonal (HCP) zirconium alloy, was experimentally investigated under a range of strain rates from 10^{-5} s^{-1} (static loading) up to 10^{3} s^{-1} (impulsive loading)[3,4]. It was found that the annealed Zircaloy-4 is highly sensitive to strain rates. The ratio of the dynamic flow stress and the static flow stress for a given strain is as high as 140 %, or the so-called dimensionless logarithmic strain-rate sensitivity $\bar{\lambda}$ is as high as 6-7 %, where $\bar{\lambda}$ is defined as

$$\bar{\lambda} = \frac{1}{\sigma_s}\frac{\partial\sigma}{\partial \log\dot{\varepsilon}} \approx \frac{\sigma_d - \sigma_s}{\sigma_s(\log\dot{\varepsilon}_d - \log\dot{\varepsilon}_s)}, \tag{1}$$

σ, ε and $\dot{\varepsilon}$ denote stress, strain and strain rate respectively, and the subscripts d and s denote the quantities at dynamic/high strain rate and at static strain rate, respectively. The corresponding microstructures observed after different tests are shown in Figure 1, which reveals the following three essential facts.

(1) The rate-dependent macroscopic elastic visco-plastic deformation for the Zircaloy-4 is mainly corresponding to the microscopic development of twinning, a kind of plane defect in lattice. In other words, the macroscopic rate-dependent flow/deformation is accompanied by a certain form of microscopic evolution of internal defects.

(2) By comparing Fig.1(b) and (d), it can be seen that the twinning density increases with increase of strain-rate for a given strain, showing the strain-rate strengthening effect.

(3) On the other hand, by comparing Fig.1(a), (b) and (c), it can be seen that for a given strain-rate, the twinning density increases with increasing strain, showing the strain hardening effect.

Thus, the evolution of twinning is dependent on both the strain-rate and the strain.

(a) $\dot{\varepsilon} = 10^3 \ s^{-1}$, $\varepsilon_r = 3.8\ \%$.

(b) $\dot{\varepsilon} = 10^3 \ s^{-1}$, $\varepsilon_r = 7.8\ \%$

(c) $\dot{\varepsilon} = 10^3 \ s^{-1}$, $\varepsilon_r = 12.6\ \%$.

(d) $\dot{\varepsilon} = 5x10^{-4} \ s^{-1}$, $\varepsilon_r = 8.6\ \%$

Fig. 1. Microstructures(500X) of Zircaloy-4 specimens, showing the dependency of twin evolution on both strain-rate $\dot{\varepsilon}$ and residual strain ε_f .

2.2. ADIABATIC SHEAR BANDING EVOLUTION FOR TB2

For TB2, a body-centered cubic (β phase) titanium alloy, our experimental results also show that its mechanical behavior is sensitive to strain rates, $\bar{\lambda}$=3.5-5.6 % at different given strain[5, 6]. However, the micro-observation reveals that, as shown in Fig. 2 and Fig. 3, the macroscopic visco-plastic flow/deformation for TB2 at high strain rates is microscopically characterized by the development of another basic, internal damage, i.e. in the form of the so-called adiabatic shear band.

Fig. 2 clearly shows that for a given strain rate (if high enough), with increase of strain, the shear band develops in a sequence of (a) trace of the localized shear deformation, as shown in Fig.2a, (b) formation of the deformed shear band, characterized by highly localized shear deformation without micro-structure change, as shown in Fig.2b, (c) transformation from the deformed shear band into the transformed shear band, wherein the material has undergone a transformation of micro-structure, as shown in Fig.2c, and (d) crack propagating along the shear band, as shown in Fig.2d.

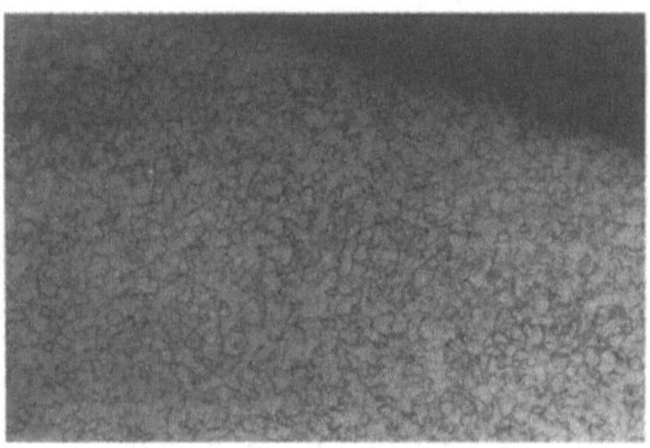

(a) $\dot{\varepsilon} = 1.5x10^3\ s^{-1}$, $\varepsilon_r = 8.8$ %.

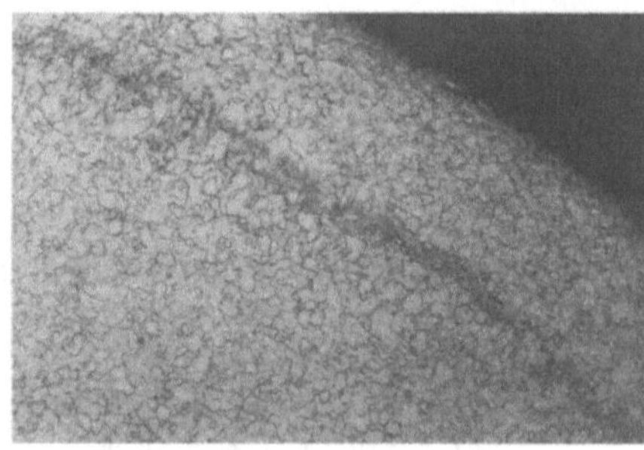

(b) $\dot{\varepsilon} = 1.5x10^3\ s^{-1}$, $\varepsilon_r = 16.3$ %

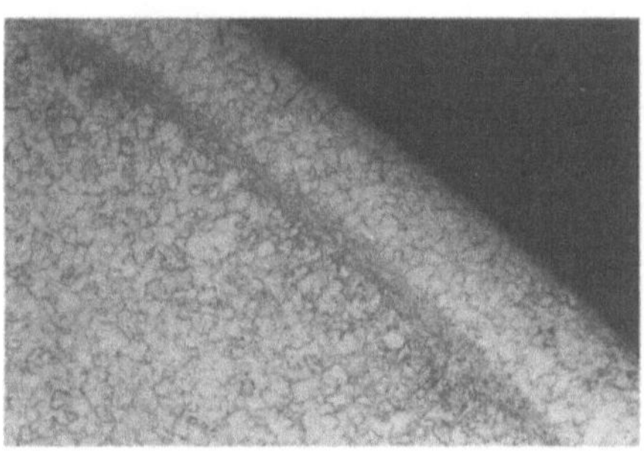

(c) $\dot{\varepsilon} = 1.5x10^3\ s^{-1}$, $\varepsilon_r = 27.4$ %.

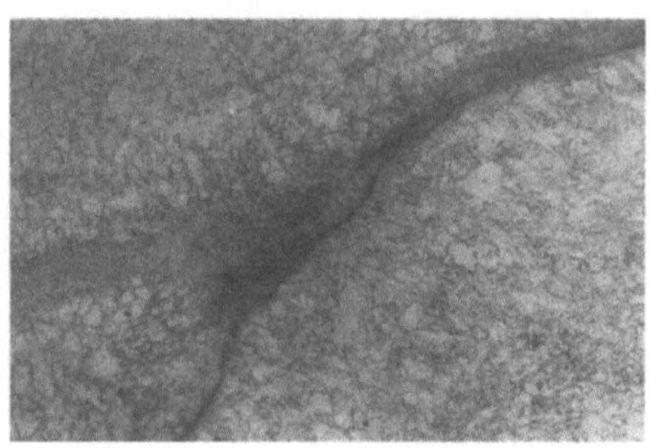

(d) $\dot{\varepsilon} = 1.5x10^3\ s^{-1}$, $\varepsilon_r = 42.3$ %

Fig.2. Micrographs(100X) of TB2 specimens after impact tests, showing the development of adiabatic shear band with increase of strain under strain rate of $1.5x10^3\ s^{-1}$.

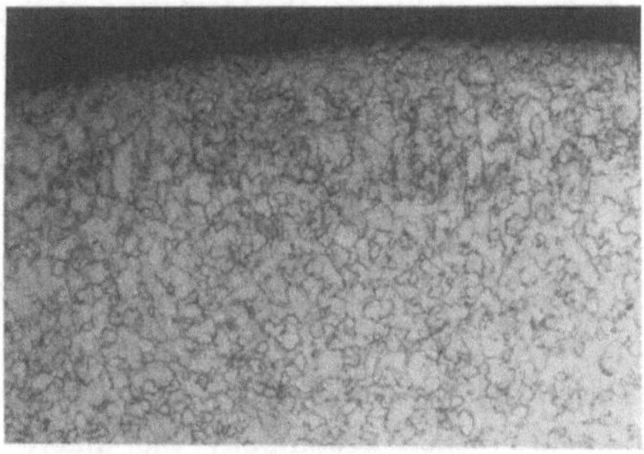

(a) $\dot{\varepsilon} = 0.88x10^3 \ \ s^{-1}, \ \ \varepsilon_r = 16\ \%.$

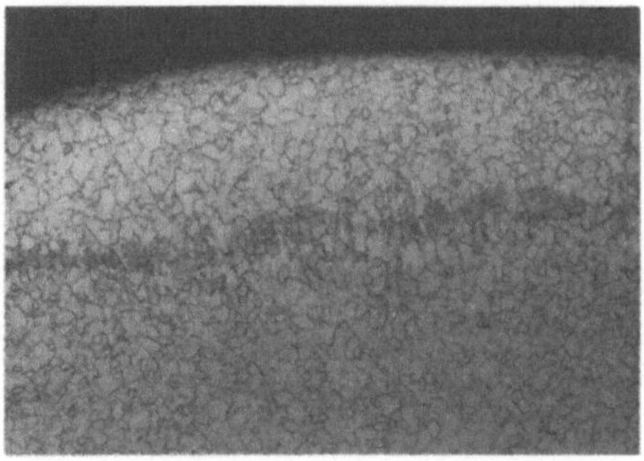

(b) $\dot{\varepsilon} = 1.45x10^3 \ \ s^{-1}, \ \ \varepsilon_r = 16\ \%..$

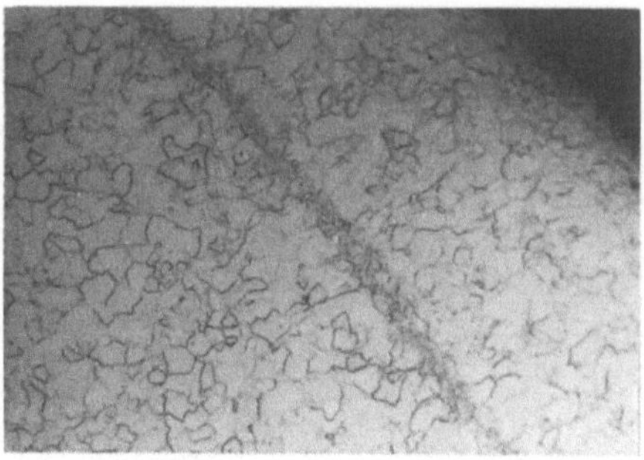

(c) $\dot{\varepsilon} = 2.21x10^3 \ \ s^{-1}, \ \ \varepsilon_r = 16\ \%..$

Fig.3. Micrographs of TB2 specimens after impact tests, showing the development of adiabatic shear band with increase of strain rate at a given strain of 16 %.

Furthermore, Fig. 3 shows that for a given strain (if large enough), with increase of strain rate the shear band also develops in a similar sequence, that is, (a) nucleation of the deformed shear band, as shown in Fig.3a, (b) formation of the deformed shear band, as shown in Fig.3b, (c) transformation of the deformed shear band into the transformed shear band, as shown in Fig.3c, and so on.

Thus, the evolution of the internal damage in the form of adiabatic shear band is also dependent on both strain rate and strain.

2.3. MICRO-CRACK EVOLUTION FOR ZM5-T4, PMMA AND CONCRETE

For ZM5-T4, a cast magnesium alloy, it was microscopically found that the dominant form of internal defect/damage which accompany the macroscopic deformation of material, is the micro-cracks along the grain boundaries[7], as shown in Fig. 4.

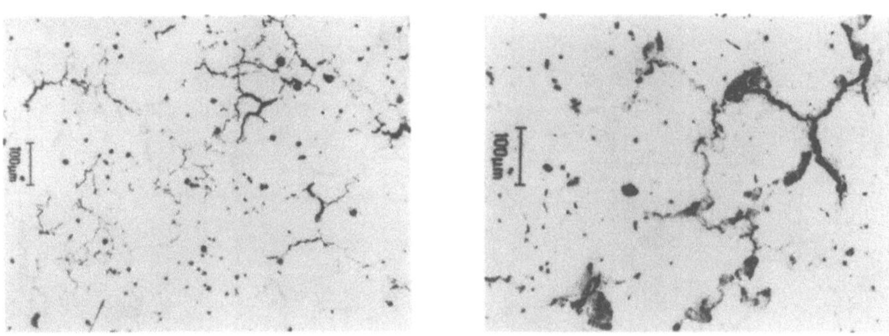

(a) (b)

Fig. 4. Micro-cracks along grain boundaries, accompanying the deformation of ZM5-T4 specimen

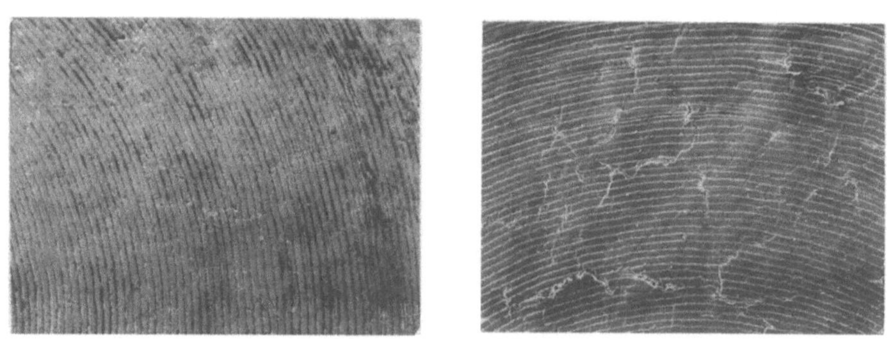

(a) $\dot{\varepsilon} = 10^{-4}\ s^{-1},\ \varepsilon_r = 10.4\ \%$. (b) $\dot{\varepsilon} = 10^{3}\ s^{-1},\ \varepsilon_r = 10.0\ \%$

Fig. 5. Surface cracks observed for two ZM5-T4 specimens tested at different strain-rates while subjected to almost the same strain.

Moreover, it was shown that the micro-cracks increase not only with increasing strain, but also with increasing strain rate, as shown in Fig. 5. In other words, the evolution of micro-cracks for ZM5-T4 is also dependent on both strain rate and strain.

For polymethylmethacrylate(PMMA), a thermoplastics, it was found[8] that before the final fracture of the specimen tested, more or less micro-cracks can be seen within the transparent PMMA specimen as a precursor of fracture. Moreover, the cracks, or more generally speaking the internal damage in the form of cracks for PMMA was found to increase with increasing strain, and to decrease with decreasing strain rate until no crack can be seen at low enough strain rate. It means that the evolution of internal cracks for PMMA is also dependent on both strain rate and strain.

For cement mortar, a sand reinforced cement composite material, a similar phenomena of crack-evolution accompanying the deformation process at different strain rates were experimentally observed by the authors. Although the cracks within the opaque concrete specimens is not as easily visible and detectable as within the transparent PMMA specimens, experimental evidences do reveal that the micro-cracks certainly develop with the deformation process for a given strain rate, and also increase with increase of strain rate for a given strain. In fact, the fraction sizes of the broken specimen become smaller and finer with increasing strain rate, as can be seen in Fig.6,

(a) $\dot{\varepsilon} = 1.5x10^{-5}$ s^{-1}.

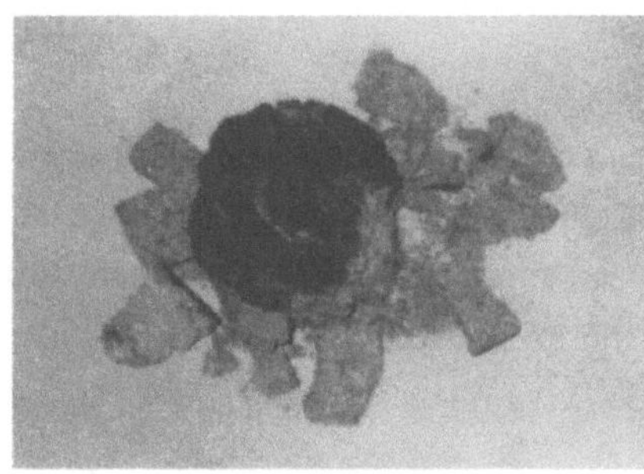

(b) $\dot{\varepsilon} = 2.0x10^{-3}$ s^{-1}.

(c) $\dot{\varepsilon} = 1.0x10^{0}$ s^{-1}.

(d) $\dot{\varepsilon} = 1.7x10$ s^{-1}.

Fig.6. Fracture appearances of cement mortar specimens tested at different strain rates.

implying that the higher the strain rate is, the more cracks within the specimen have developed before the final fracture.

Thus, two common, basic characteristics can be drawn from the experimental results.

(1) From the combined macroscopic and microscopic view-point, a macro-flow/deformation process of material is microscopically accompanied with an evolution process of internal defects/damage in different forms.

(2) The defect/damage evolution is dependent on both strain and strain rate, regardless whether the form of defect/damage is twin, adiabatic band or micro-crack.

Thus, when the rate-dependent constitutive relation of rheological material is studied, the related rate-dependent evolution of defect/damage should be taken into account.

3. Constitutive Relation of Materials by Taking Account of Damage Evolution

It is well-known that the macroscopic thermo-viscoplastic behavior of materials can be microscopically explained by the thermal activated motion of dislocation, a kind of lattice effect, and described by the following Arrhenius type equation

$$\dot{\varepsilon}^p = \dot{\varepsilon}_0 \exp(-\frac{U_s(\sigma)}{kT}), \quad (2)$$

where $\dot{\varepsilon}^p$ is the viscoplastic strain rate, $\dot{\varepsilon}_0$ is a frequency coefficient, k the Boltzmann's constant, T the absolute temperature, and the activation energy U_s is a function of the applied stress σ .In the thermal activation theory, a key problem is how the U_s depends on the σ , or what form of the function $U_s(\sigma)$ should be. The simplest, linear relation of $U_s(\sigma)$ was proposed by Seeger[9], which gives

$$\sigma = \sigma_L + \sigma_S = \sigma_L + \frac{U_0}{V} + \frac{kT}{V}\ln(\frac{\dot{\varepsilon}}{\dot{\varepsilon}_0}) = \sigma_A + \lambda_T \ln(\frac{\dot{\varepsilon}}{\dot{\varepsilon}_0}), \quad (3)$$

where σ_L measures the long-range resistance to dislocation motion (athermal component of stress), σ_S the short-range resistance (thermal component of stress), U_0 the activation energy in the absence of stress, V the activation volume, and in the Eq.(3) the total strain rate $\dot{\varepsilon}$ has been used instead of $\dot{\varepsilon}^p$ by assuming that the elastic strain rate component $\dot{\varepsilon}^e$ is negligible small. Eq.(3) means that a linear function exists between σ and $\log\dot{\varepsilon}$, and its slop characterizes the strain rate sensitivity of materials as defined in Eq.(1).

To describe the nonlinear function between σ and $\log\dot{\varepsilon}$, which is frequently observed in a wide range of strain rates, a thermo-viscoplastic constitutive equation as follows was proposed by Wang[10] on the basis of thermo-activated mechanism with a spectrum of hyperbolic-shape barriers,

$$\dot{\varepsilon} = \dot{\varepsilon}_0 \sum_{i=1}^{n} \psi_i \exp\left(\frac{(1+\sigma/\sigma_0)^{1-m_i} - 2^{1-m_i}}{1-m_i} \frac{\sigma_0 V}{kT}\right), \qquad \text{with } \sum_{i=1}^{n} \psi_i = 1, \tag{4}$$

where ψ_i is the strain rate weight function, σ_0 the characteristic stress, and $m(\geq 0)$ is a material parameter characterizing the thermo-activated barriers. Obviously, when m=o and n=1, Eq.(4) reduces to Seeger's model, i.e. Eq.(3).

To take into account the weakening influence of damage evolution upon the material strength, introduce the following macro-parameter of damage, as usual,

$$D = \frac{\sigma - \sigma_{app}}{\sigma}, \quad o \leq D \leq 1, \qquad \text{or, } \sigma_{app} = (1-D)\sigma, \tag{5}$$

where σ is the stress for material without damage, the same as that in Eqs. (2)-(4), and σ_{app} is the apparent stress for material with damage, i.e. the decreased stress due to damage.

In principle, the rheological constitutive relation taking into account the damage evolution, i.e. the damage-modified rheological constitutive relation can be obtained by combining Eqs. (4) and (5), if the additional equation for damage evolution is given.

Since the experimental results mentioned above reveal that the damage evolution process is also a rate process similar to the rate process for viscoplastic deformation, it is not unreasonable to consider the damage evolution as a thermal activated process. In other words, similar to Eq. (2), the rate-dependent damage evolution can be expressed by

$$\dot{D} = \frac{\partial D}{\partial t} = \dot{D}_0 \exp\left(-\frac{U_D}{kT}\right), \tag{6}$$

where $\dot{D}$ is the rate of damage evolution, $\dot{D}_0$ is the frequency coefficient and U_D is the activation energy for damage evolution, respectively. Instead of analyzing the relation between the U_D and the applied stress σ in detail, assume that a proportional relation exists between the U_D in Eq. (6) and the U_s in Eq. (2), that is

$$U_D = \alpha U_s, \tag{7}$$

where α is a material parameter. Then, from Eqs. (2) and (6), after some mathematics calculations, we have

$$\frac{\dot{D}}{\dot{D}_0} = \left(\frac{\dot{\varepsilon}}{\dot{\varepsilon}_0}\right)^{\alpha}, \qquad \text{or, } \dot{D} = K_D \dot{\varepsilon}^{\alpha}, \quad K_D = \frac{\dot{D}_0}{\dot{\varepsilon}_0^{\alpha}}, \tag{8}$$

or by integrating,

$$D = K_D \Psi_D(\dot{\varepsilon}), \quad \Psi_D(\dot{\varepsilon}) = \int_{t_0}^{t} \dot{\varepsilon}^{\alpha} dt, \tag{9}$$

which formulates a rate dependent damage evolution law in a simple form for engineering application.

It is worthwhile discussing the following three special cases of Eq. (9).

(1) In the case of constant strain rate, it is easy to integrate the Eq. (8), and if assume that a strain-threshold, ε_{th} , exists for damage evolution, then the Eq. (9) reduces to

$$D = K_D \dot{\varepsilon}^{\alpha-1}(\varepsilon - \varepsilon_{th}), \qquad \text{for } \varepsilon > \varepsilon_{th}, \tag{10}$$

which explicitly describes a both strain and strain-rate dependent damage evolution process. It was found that at high strain rates and large deformation, the damage-modified nonlinear rheological behavior of PMMA, wherein the damage evolution should be taken into consideration as mentioned above, can be satisfactorily described by the so-called ZWT viscoelastic equation modified by the Eqs. (5) and (10), as shown in [8,11]. Moreover, it was also found that based on the above Eq. (10) and by introducing a critical damage D_f, the dynamic fracture criterion for PMMA at high strain rates can be simply expressed as $D \geq D_f$, which is equivalent to a two control variables (in terms of strain rate and strain) dynamic fracture criterion[8,11].

(2) When a proportional relation exists between $\Psi_D(\dot{\varepsilon})$ and $\ln(\dot{\varepsilon}/\dot{\varepsilon}_0)$, Eq. (9) reduces to a simpler form similar to Eq. (3),

$$D = D_0 + \lambda_D \ln(\frac{\dot{\varepsilon}}{\dot{\varepsilon}_0}), \tag{11}$$

where $\lambda_D (> 0)$ is the strain rate sensitivity of damage evolution. Without any loss of generality, if the Seeger's model is discussed as an example, then from Eqs. (3), (5) and (11) the damage evolution modified Seeger's viscoplastic constitutive equation can be derived[7], that is,

$$\sigma = \sigma_A(1 - D_0) + (\lambda'_T - \lambda'_D)\ln(\frac{\dot{\varepsilon}}{\dot{\varepsilon}_0}), \tag{12}$$

where, $\lambda'_T = \lambda_T(1 - D_0) > 0$, and $\lambda'_D = \sigma_A \lambda_D + \lambda_T \lambda_D \ln(\frac{\dot{\varepsilon}}{\dot{\varepsilon}_0}) > 0$. Correspondingly, the similar damage evolution modified equation can be derived for Eq. (4) or other forms of rate dependent constitutive relations. It is worthwhile noting that the sign before $\lambda'_D (>0)$ is negative, which means that an inverse (negative) strain rate effect accompanies with a damage evolution process. Thus, depending on the value of the ratio $\bar{\lambda} = \lambda'_T / \lambda'_D$, three cases may exist:

(a) When $\bar{\lambda} > 1$, it shows an apparent strain-rate strengthening (positive sensitivity).

(b) When $\bar{\lambda} = 1$, it shows an apparent strain-rate independence (non-sensitivity).

(c) When $\bar{\lambda} < 1$, it shows an apparent strain-rate weakening (negative sensitivity).

Thus, it provides a new possible explanation for inverse strain rate effect based on damage weakening mechanism.

(3) The thermal softening induced by the adiabatic temperature rise at high strain rates. The adiabatic shear banding mentioned above is generally attributed to the so-called thermo-plastic instability, or more strictly speaking, the so-called thermo-viscoplastic constitutive instability[12]. The thermal softening induced by the adiabatic heating during the adiabatic shearing process, actually, can be considered as a weakening induced by a generalized damage. In fact, the adiabatic temperature rise dT and the corresponding stress decrease $d\sigma$ during the adiabatic shearing can be expressed, respectively, as

$$dT = \frac{\beta\sigma_r\, d\varepsilon}{\rho C_V}, \qquad \text{and} \quad d\sigma = (\frac{\partial\sigma}{\partial T})dT = -\varphi\, dT, \tag{13}$$

where β is the Taylor-Quinney coefficient of visco-plastic work converted to heat, ρ the material density, C_V the specific heat. If $\varphi = -\partial\sigma / \partial T$ (>0) can be regard as a constant, then we have

$$dD = -\frac{d\sigma}{\sigma_r} = \frac{\varphi\beta}{\rho C_V} d\varepsilon, \qquad \text{or,} \quad \dot{D} = \frac{\varphi\beta}{\rho C_V}\dot{\varepsilon}, \tag{14}$$

which is actually the special case of Eq. (9) when $K_D = \varphi\beta / (\rho C_V)$ and $\alpha = 1$. Thus, the adiabatic temperature rise can be indeed regarded as a kind of generalized damage.

4. Conclusions

The following conclusions can be drawn from the above experimental evidence and theoretical analyses.

(1) A macroscopic rheological flow/deformation process is actually accompanied with a microscopic/mesoscopic defect/damage evolution process, which should be taken into account when the macroscopic rheological constitutive relation is to be thoroughly studied.

(2) The dominant form of internal defect/damage may be different for different material studied. However, their evolution processes are all dependent on both strain and strain rate. Thus, it should be emphasized that the rate dependent constitutive response for the material studied actually contains the rate dependent evolution response of internal defect/damage too. Moreover, the "adiabatic temperature rise" can be regarded as a kind of generalized damage.

(3) Assuming a thermo-activation mechanism for damage evolution, a rate dependent damage evolution law, and consequently a corresponding damage modified rate dependent constitutive relation is proposed in the present paper. Both the positive strain rate sensitivity and the negative strain rate sensitivity can be described by the

proposed constitutive relation.

5. Acknowledgments

The research work was supported by the Natural Science Foundation of Zhejiang.

6. References

[1] Wang Lili (1985) *Foundations of Stress Waves,* National Defense Industry Press, Beijing*.

[2] Zukas, J., Nicholas, T., Swift, H., Grezczuk, L. and Curran, D.(1982) *Impact Dynamics,* Wiley-Interscience, New York.

[3] Wang, L.L., Kobayashi, A. et al.(1986) Dynamic Compressive Behavior of Zircaloy at High Strain Rates, in Gu Haicheng and He Jiawen (eds.), *Microstructural and Mechanical Behavior of Materials,* Engineering Materials Advisory Services Ltd., London, pp. 807-814.

[4] Kobayashi, A., Wang, L.L., Lu, W.X., Hu, S.S., Tang, Z.P. and Hashimoto, S.(1988) Macro/ Microscopic Approach to Dynamic Mechanical Behavior of Zircaloy-4 Subjected to Tension and Compression, in C.Y. Chiem, H.D. Kunze and L.W. Meyer(eds.), *Impact Loading and Dynamic Behavior of Materials,* DGM Informationsgesellschaft Verlag, pp. 815-822.

[5] Wang, L.L., Lu, W.X., Hu, S.S., and Tang, Z.P.(1987) Study of the Initiation and Development of Adiabatic Shear Bands for a Titanium Alloy under High Strain Rates, in K. Kawata and J. Shioiri(eds.), *Macro- and Micro-Mechanics of High Velocity Deformation and Fracture,* IUTAM Symposium on MMMHVDF, Tokyo, Japan, August 12-15, 1985, Springer-Verlag, Berlin, pp. 395-406.

[6] Wang L.L., Bao H.S. and Lu Weixian(1988) The Dependence of Adiabatic Shear Banding on Strain-rate, Strain and Temperature, *Journal de Physique* **49**, Colloque C3, 207-214.

[7] Wang Lili, Bao Hesheng and Lu Weixian (1993) The Inverse Strain-Rate Effect Induced by Damage and Its Influence on Constitutive Relation and Thermo-Viscoplastic Instability, *Explosion and Shock Waves* **13**, 1-8**.

[8] Zhou Fenghua, Wang Lili and Hu Shisheng (1992) A Damage-Modified Nonlinear Visco-Elastic Constitutive Relation and Failure Criterion of PMMA at High Strain Rates, *Explosion and Shock Waves* **12**, 333-342**.

[9] Seeger A.(1955) The Generation of Lattice Defects by Moving Dislocation, and Its Application to the Temperature Dependence of the Flow Stress of Face-Centered Cubic Crystals, *Phil. Mag.* **46**, 1194-1217.

[10] Wang, Li-Lih(Wang Lili) (1984) A Thermo-Viscoplastic Constitutive Equation Based on Hyperbolic-Shape Thermo-Activated Barriers, *Tran. ASME, J. Engrg. Mater. Tech.* **106**, 331-336.

[11] Wang Lili and Yang Liming (1992) The Nonlinear Viscoelastic Constitutive Relation of Solid Polymeric Materials, in Wang Lili, Yu Tongxi and Li Yongchi(eds.), *Progress in Impact Dynamics,* The Press of the University of Science and Technology of China, Hefei, pp. 88-116*.

[12] Wang lili (1992) Adiabatic Shearing, the Constitutive Instability of Materials under Impact Loading, in Wang Lili, Yu Tongxi and Li Yongchi(eds.), *Progress in Impact Dynamics,* The Press of the University of Science and Technology of China, Hefei, pp. 3-33*.

* In Chinese.

** In Chinese with English abstract.

DAMAGE WAVE PROPAGATION IN ELASTIC-BRITTLE MATERIALS

XI ZHANG and YIU-WING MAI
Center for Advanced Materials Technology
Department of Mechanical and Mechatronic Engineering
The University of Sydney
Sydney, NSW 2006, Australia

Abstract

A theory of damage wave propagation in elastic-brittle materials is developed within the framework of thermodynamics. Because the local extent of damage is a result of microscopic movement of its neighborhood, we include the gradient of damage and the additional kinetic energy in the construction of thermodynamic functions. A specific elastic-brittle material model is presented. The governing equations of the coupled thermo-damage-mechanism are derived. It is shown that the equation of the damage evaluation is a non-linear wave equation and has a solitonic solution of the kink type. The propagation speed is determined using energy analysis. Dissipative mechanisms, like internal friction, irreversible phase transformation and chemical reactions, reduce the speed of damage wave. More detailed discussions are presented in the paper.

1. Introduction

The dynamic responses of elastic-brittle materials, such as glasses, concretes and ceramics, have received much attention over the last few years. The evolution of damage in these elastic-brittle materials under impact loading is explained in terms of so-called 'failure wave' or 'damage wave'. Recent experimental investigations of compressive failure in glasses conducted by Brar et al (1991), Clifton (1993) and Bourne et al (1995) reported a loss of spall strength behind the shock wave, which they describe as the 'failure wave'. The failure wave appeared to start at discrete nucleation sites until coalescence has occurred, and then spread to form a continuous failure front. Therefore, it depends very much on the kinetics of changing microstructures. The wave front velocity is found to be less than the shock wave speed and a function of the applied compressive stress. Also, it varies with the material composition, i.e. the micro-damage evolution. It is found by Bourne et al (1995) that the wave speed is less in soda-lime float glass than in borosilicate glass. The precise nature of the failure

R. Wang (ed.), IUTAM Symposium on Rheology of Bodies with Defects, 179–190.

wave is, as yet, uncertain, because it is possibly associated with the shear stress level in the shock wave, the evolution of temperature and the micro-defects. However, we believe that the evolution of these microscopic motions must be accounted for in any predictive model or theory. Attention here is focused on the influence of material damage evolution on the damage wave motion, but not on its interactions with thermal effect and stress field.

The dissipative mechanisms in glasses may be attributed to the formation of microcracks, voids and phase transformation. All these mechanisms evolve within the framework of thermodynamics. Based on continuum damage theory, see Leimatre and Cabcache (1990) and Maigun(1990), the damage quantities which appear in the expression of the free energy density or the evaluation equations are the scalar internal variable and its gradient. For our purpose, it is neither necessary nor helpful to identify the tensorial character of the internal and external state variables. In some cases, however, such a description is appropriate, as emphasized by Onat and Leckie (1988). The gradient of damage is introduced to account for the influence of damage at a material point on the damage of its neighborhood, i.e. nonlocal averaging scheme. Of course, a material characteristic size is introduced. Introduction of the gradient of state variables is attributed to the works of Lehman(1989) and Maugin(1990). A constitutive model, which depends on the damage and gradient of damage, as well as temperature and its gradient, is presented here. The governing equations are derived from a sensible assumption of the free energy density. Motivated by the work of Clifton(1993), the failure wave is considered as the propagation of localized damage caused by thermally activated phase transition in glasses. It follows that a form of free energy density similar to that in phase transformation should be adopted in this paper.

The objective of the present work is to derive the equation of damage waves and to characterise the features of its solutions based on observations of the mechanical behavior of damaged materials. It is expected that this non-linear equation has solitonic structures, stemming from its reversible physical and irreversible dissipative processes. This characteristic of damage wave has been pointed out and considered in general by Maugin (1990). But results for specific materials cannot be found to our knowledge.

This paper is organized as follows. In section 2, a statement of wave propagation in an infinite medium is presented. The basic principles of thermodynamics based on internal variables is briefly reviewed and the corresponding equations of motion is derived. In section 3, a specific case is studied for elastic-brittle materials with specified free energy density. Indeed, the resultant damage wave is a solitary wave due to energy dispersion. The analytic solution in one-dimensional cases is given in section 4 and the results are discussed in detail. Finally, some useful conclusions are given in section 5.

2. Problem Formulation

Let us consider an infinite solid, whether glass, concrete or ceramics, with a changing microstructure. In each material element, its location is decided by the coordinate x_i, and its motion by independent displacement $u_i(x_i, t)$ at time t. To describe the effect of micro defects and their neighborhood on the material degradation, we introduce the scalar $\beta(x,t)$ as an additional independent variable with a range [0,1]. Like the deformation field in classical wave theory, its distribution in the medium is called the damage field and its motion the damage wave. The parameter β may also serve to define the changes relative to some reference state in the micro- and macro-cracks, such as, geometry of individual cracks or statistical averages pertaining to distributed microcracks, void volume, and the degree of molecular enlargements, crosslinking or crystallinity. Rice (1971) also discussed the significance of β for describing the dissipative mechanisms of crystalline slip, diffusion and phase transformation. With these structure-related variations in mind, we shall call β the structural parameter that is interchangeable with the internal variable.

The particular feature of the classical damage theory based only on $\beta(x,t)$, is that finite element approximation depends much on mesh sensitivity. This spurious mesh dependency is unacceptable. Many remedial methods have been put forward, such as, non-locality, strain gradient and Correscat medium. In our work, we introduce the gradient of damage as an additional internal variable in the thermodynamic description of the dissipative process. This quantity is clearly related to the micro-movements of micro defects and their interactions, as pointed out by Bazant(1994).

Under impact loading, a major feature of the material behavior is the heat generated with the deformation. The resultant non-uniform temperature is controlled by the coupled thermomechanical behavior, i.e. elastic deformation, damage and heat flux.

The reversible energy I is defined by:

$$I = \int_{0}^{t}\int_{V} \{\frac{1}{2}\rho v_i v_i + \frac{1}{2}\iota\dot{\beta}\dot{\beta} - \rho u\} dV dt \tag{1}$$

in which ρ is mass density, ι the acceleration energy of the microscopic links which is proportional to the mass density; u is the free energy density; v_i is the velocity vector; $\dot{\beta}$ denotes the rate of damage. The first term inside the bracket is the usual kinetic energy and the second term is caused by the additional kinetic energy due to the degree of freedom of β. The latter is a direct result that damage is not considered here as a local quantity, but is defined as an averaging quantity over a finite volume. The microscopic links in this finite domain also yield changes in the kinetic energy of this local material. It is analogous to the fact that relative rotation of a small rigid domain would impart additional kinetic energy. So, ι also depends on the characteristic length l_c of this finite volume.

Within the framework of thermodynamics, the free energy provides a new equation to describe the evolution of the damage quantity. It is assumed that the free energy density is a function of the strain tensor ε_{ij}, entropy s and entropy flux s_i, the damage field and its gradient $\beta, \beta_{,i}$. It can be expressed as a function of a set of thermodynamic state variables:

$$u = u(\varepsilon_{ij}, s, s_i, \beta, \beta_{,i}) \tag{2}$$

If the heat flux is q_i, then the flux of entropy is given by:

$$s_i = q_i / T \tag{3}$$

in which T is the absolute temperature.

For simplicity, the damage work is only associated with β except for the thermal effect. We choose the dissipative energy as:

$$W^* = \int_0^t \int_V \{T\dot{s} + T_{,i} s_i + A\beta\} dV dt \tag{4}$$

in which A denotes the damage work per unit volume. It can be caused by the breakage of links at the atomic level and/or the formation of voids. The body force is neglected without loss of generality.

The problem we study here is related to the deformation and evolution of damage in an infinite medium. It is reasonable not to include the external forces, heat flux and damage source on the boundary in deriving the equations of motion and constitutive relations.

To obtain the motion and damage evaluation equations for thermal and mechanical state variables, we can construct the energy functional $\Pi = I + W^* + W$, in which W is external work caused by the applied force, heat flux and damage source on the boundary. The Lagrangian equation of the function $\delta\Pi = 0$ are:

(1) Equations of motion

$$\begin{aligned} \rho\dot{v}_i &= \sigma_{ij.j} \\ \iota\ddot{\beta} &= H_{i,i} - B + A \end{aligned} \tag{5}$$

(2) Evolution equations of state variables

$$\begin{aligned} &\sigma_{ij} = \rho\frac{\partial u}{\partial \varepsilon_{ij}}, \quad B = \rho\frac{\partial u}{\partial \beta}, \quad H_i = \rho\frac{\partial u}{\partial \beta_{,i}} \\ &T = \frac{\partial u}{\partial s}, \quad T_{,i} = \frac{\partial u}{\partial s_i} \end{aligned} \tag{6}$$

in which B and H_i are generalized forces associated with β and $\beta_{,i}$ respectively and related to the reversible physical processes. A is related to irreversible processes, such as cracking, crystalline slip and chemical reactions.

For simplicity, we assume small deformation and let ε_{ij} be small too. Also, because the time derivative of β is not necessary continuous, we can choose:

$$\dot{\beta} = \lim_{\Delta t \to 0} \frac{\beta(t) - \beta(t - \Delta t)}{\Delta t} \tag{7}$$

The time derivative of the functional Π is the constitutive law. So, the constitutive structure is related to the history of the deformation. The time-related variation of Π leads to:

$$-\rho\dot{u}+\sigma_{ij}\dot{\varepsilon}_{ij}-(Ts_i)_{,i}+(H_i\dot{\beta})_{,i}+$$
$$T(\rho\dot{s}+s_{i,i})+T_{,i}s_i+2A\dot{\beta}-B\dot{\beta}-H_i\dot{\beta}_{,i}=0 \tag{8}$$

To obtain the constitutive equations, we resort to the basic laws of thermodynamics. The first law states that :

$$\dot{u}=\sigma_{ij}\varepsilon_{ij}-[Ts_i-H_i\dot{\beta})]_{,i} \tag{9}$$

in which the extra entropy flux is $H_i\dot{\beta}$ and the internal energy sources vanish due to no plastic dissipation in elastic-brittle materials. Besides, the constitutive law must also obey the second law of thermodynamics, i.e. Clasussis-Duhem inequality,

$$T(\rho\dot{s}+s_{i,i})\geq 0 \tag{10}$$

Combining (8), (9) and (10), the following inequality is obtained

$$-T_{,i}s_i-2A\dot{\beta}+B\dot{\beta}+H_i\dot{\beta}_{,i}\geq 0 \tag{11}$$

Very often this is split into three parts for three different sources of dissipated energies: thermal effect, intrinsic dissipation and irreversible external process, such as phase transformation, chemical reactions, etc. Hence:

$$-T_{,i}s_i\geq 0$$
$$B\dot{\beta}+H_i\dot{\beta}_i\geq 0 \tag{12}$$
$$-A\dot{\beta}\geq 0$$

Furthermore, the inequalities in (12) can be satisfied by assuming:

$$A=-\tau\dot{\beta} \quad \text{and} \quad s_i=-\kappa_{ij}T_{,j} \tag{13}$$

in which τ is a positive constant and κ_{ij} a positive definite tensor.

In addition, normality is assumed for the evolution of damage variables, i.e. there exists a convex free energy density u, see Moreau (1970), so that (10) holds. Similarly, the dissipative force A must be a convex function of $\dot{\beta}$ and the state variables mentioned above. It is expected that there is a convex pseudo-potential of dissipation $\varpi(\dot{\beta})$, so that:

$$A=\frac{\partial\varpi(\dot{\beta})}{\partial\dot{\beta}} \tag{14}$$

The same results can be found in the work of Fremond and Nedjar(1996) and its references. They obtained this result by imposing the condition $0\leq\beta\leq 1$.Equation (13) is similar to the Fourier law and thus the tensor κ_{ij} is called the thermal conductivity coefficient. Because we do not deal with coupling mechanisms in this paper, the governing equation for the temperature field is not necessary.

3. Governing Equation

Damage in elastic-brittle materials takes the form of microcracks and voids when a certain critical condition is reached. Its direct result is the reduction in material properties, such as, strength and elastic modulus. Clifton (1993) interpreted the microcracking in soda-lime glass under impact loading as the propagation of localized regions of phase transformation. The phase change could be a transformation to a crystalline phase or an amorphous phase with a higher coordination number. Because of the large strains involved in phase transition, stress- induced microcracking can be expected to result from heterogeneous nucleation and growth of transformation region. Following the work of Leimatre and Chaboche (1985), a simple modeling of damage by a scalar variable β is introduced in the preceding section. Physically, it measures the decrease in internal surface which transmits internal forces in an isotropic material over a finite domain. In particular, it represents the microstructural rearrangement of particles due to the existence of phase transformation. For similar reasons its gradient is also introduced in the specific free energy. The idea of considering its gradient as an internal variable dates back to Maugin (1990) and Fremond and Nedjar (1996).

With respect to many applications, it is advantageous to replace the entropy s by the temperature T and its increment θ. This can be achieved by a corresponding Legendre transformation leading to the Helmholtz energy density:

$$\psi = u - Ts \tag{15}$$

ψ is also assumed to be convex and sub-differential with respect to the thermodynamic state variables. Its value must be positive in all sub-space.To be consistent wth the internal variable theory, a sensible expression of free energy is given by:

$$\psi = W_0 - \frac{1}{2}\beta^2 W_e + \frac{1}{4}\beta^4 W_1 + \frac{1}{2}\kappa^2 \beta_{,i}^2 \tag{16}$$

in which W_0 denotes the initial reference strain energy without damage, W_1 the minimum reference strain energy prior to failure, W_e the elastic strain energy of undamaged material at a certain state (ε_{ij}, θ); it measures the interaction of the strain and temperature fields with the material damage, i.e.

$$W_e(\varepsilon_{ij}, \theta) = \frac{1}{2}[\lambda(\theta)\varepsilon_{ii}^2 + 2\mu(\theta)\varepsilon_{ij}\varepsilon_{ij}] + K\alpha\theta\varepsilon_{ii} + \frac{1}{2}\chi\theta_{,i}^2 \tag{17}$$

in which λ, μ are elastic moduli, K is bulk modulus, $\theta = T - T_0$ in which T_0 is reference temperature, α is thermal expansion coefficient, and χ is heat flux. The quantity κ measures the non-local effect on the degradation of material strength. As no plastic deformation is allowed, the material strength is a function of the extent of phase transformation, i.e., β. For simplicity, it is assumed to be the momentum of shear strength of a material point after damage has occurred over a finite domain. It satisfies the requirement that the Helmholtz energy density function is positive, convex and sub-differential in the space of $(\varepsilon_{ij}, T, T_{,i}, \beta, \beta_{,i})$. And because of the occurrence of damage, $\psi \le W_0$.

Substituting (17) into (6), we obtain:

$$\begin{aligned} \sigma_{ij} &= \lambda\varepsilon_{ii} + 2\mu\varepsilon_{ij} + K\alpha\theta \\ B &= -W_e\beta + W_1\beta^3 \\ H_i &= \kappa\beta_{,i} \end{aligned} \tag{18}$$

To evaluate the dissipative force associated with β, we assume that the pseudo-potential of dissipation is expressed by:

$$\varpi = \frac{1}{2}\tau\dot{\beta}^2 \tag{19}$$

in which τ is the viscosity parameter of damage already defined in (13) but with an opposite sign.

The expression of pseudo-potential of dissipation is chosen so that the damage arises only from the additional viscous phenomenon. Thus, it follows:

$$A = \tau\dot{\beta} \tag{20}$$

Substituting (18) and (20) into (5), we can obtain the equation for the evolution of damage as:

$$\iota\ddot{\beta} = \kappa\beta_{,ii} + W_e\beta - W_1\beta^3 + \tau\dot{\beta} \tag{21}$$

4. One-Dimensional Solutions

As an example, let us consider the damage wave propagation in an semi-infinite rod. suddenly loaded by impact at its end. Equation (21) in this one-dimensional problem becomes:

$$\iota\ddot{\beta} = \kappa\beta_{,xx} + W_e\beta - W_1\beta^3 + \tau\dot{\beta} \tag{22}$$

To simplify the above equation, we introduce the following parameters:

$$c_0^2 = \frac{\kappa}{\iota}, c_1 = \frac{W_1}{\iota}, c_2 = \frac{W_e}{\iota}, \gamma = \frac{\tau}{\iota}, \tag{23}$$

in which c_0 denotes the limiting velocity of the material, c_1 describes the effect of the damage or the material inhomogeneity, $c_2(\varepsilon_{xx}, \theta)$ the effect of internal fields and γ the material damping effect. The governing equation can be rewritten as:

$$\ddot{\beta} = c_0^2\beta_{,xx} - c_1\beta^3 + c_2\beta + \gamma\dot{\beta} \tag{24}$$

Now, we proceed with the dimensional variables to simplify the governing equation for damage:

$$t' = \sqrt{c_1}t, x' = x/d, d = c_0/\sqrt{c_1} \tag{25}$$

Simple manipulation yields:

$$\ddot{\beta} - \beta_{,x'x'} - \beta + \beta^3 + \gamma'\dot{\beta} + V(x',t')\beta = 0 \tag{26}$$

in which, $\gamma' = \gamma/\sqrt{c_1}$ and $V(x',t') = c_2/\sqrt{c_1} - 1$.

It should be noted that $V(x',t')$, which represents the damage or inhomogeneity effects at time t′, is a function of the strain and temperature at a local material point. A coupled mechanism exists for the evaluation of these thermodynamic state variables. For the sake of simplicity, it is assumed to be known in the assessment of damage evaluations only.

The above equation of damage evolution is the Klein-Gordon cubic non-linear equation, or non-integrable φ^4 model. This appears in quantum field theory problems, see Rajaraman(1982), and in phase transition theory, see Krumhans and Schrieffer(1975). In addition, Clifton(1993) attributed the damage wave in glass to phase transformation. This explains our choice of the free energy in the form of eq. (16).

To study the behavior of non-linear equations, we will apply a version of the asymptotic method of small-parameter expansion. This weak perturbation method is suitable to the analysis of solitons, irrespective of whether the unperturbed system is intergrable or not, see Abdullaev (1994). So the corresponding equation can be rewritten as:

$$\ddot{\beta} - \beta_{,x'x'} - \beta + \beta^3 = \varepsilon R[\beta] \tag{27}$$

in which $0 < \varepsilon << 1$, $R[\beta] = -\gamma' \dot{\beta} - V(x', t')\beta$ and it approaches zero when $x' \to \pm\infty$ because $\beta \to 0$ when no damage occurs; or γ' and $V(x', t')$ approach zero when damage reaches a maximum.

First let us consider the evolution of the steady-state particular solution of (27), i.e. $\varepsilon = 0$. Steady-state solutions are sought in the form of $\beta = \beta(x' - vt')$, in which v is the propagation speed. It should be noted that v is a relative speed, i.e. $v = \mathrm{v} / c_0$, if v is absolute speed. It is determined by the wave form and considered as an unknown constant. From (27), we have:

$$(v^2 - 1)\beta_{\xi\xi} - \beta + \beta^3 = 0 \tag{28}$$

in which $\xi = x' - vt'$.

Multiplication of both sides by β_ξ and integration yields:

$$(v^2 - 1)(\beta_\xi)^2 - \beta^2 + \frac{1}{2}\beta^4 = C \tag{29}$$

in which C is a constant. For the case of $C = -0.5$, i. e. the minimum of the function $0.5\beta^4 - \beta^2$, the solitary solution can be obtained; otherwise, non-linear periodic solution exists. We consider the former case here. Thus,

$$\int_{\beta_0}^{\beta} \frac{d\beta}{\beta^2 - 1} = \pm \frac{\xi}{\sqrt{2(1 - v^2)}} \tag{30}$$

and the steady-state solution is given by:

$$\beta(x', t') = \pm \tanh\left[\frac{x' - vt' - x_0}{\sqrt{2(1 - v^2)}}\right] \tag{31}$$

in which the signs $\pm$ refer to two different types of solitons, kink and anti-kink, x_0 is the initial equilibrium position.

It is seen that $\beta \to 1$ for $\xi \to \infty$ and the initial value of the damage variable is zero. This is consistent with the evolution process of damage. And if there are no energy dissipation processes, the material point would reach the failure criterion sooner than if there is a damping effect, as illustrated in Fig.1. The sign + is assumed in (31) and this means that the solution is soliton of the kink type.

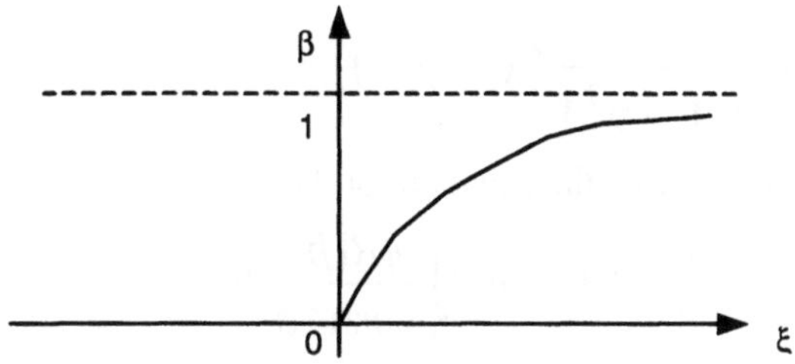

Fig.1 Kink wave of solitons

The steady-state Lagrangian density is:

$$L = \frac{1}{2}(\beta_{t'}^2 - \beta_{x'}^2 - \beta^2 + \frac{\beta^4}{2} + \frac{1}{2}) \tag{32}$$

then the energy of a moving soliton of the kink type is:

$$E = \int_0^\infty L dx = \frac{2}{3\sqrt{2}} \frac{1}{\sqrt{1-v^2}} \tag{33}$$

Expanding the energy in terms of Taylor's series, this gives

$$E = E_0 + \frac{1}{2} m v^2 \qquad \text{for} \quad v << 1 \tag{34}$$

in which the mass $m = \frac{2}{3\sqrt{2}}$ and the initial energy $E_0 = \frac{2}{3\sqrt{2}}$.

It is shown that the velocity of the solitary wave is an arbitrary constant. It can be predicted through the wave form or the stored energy measured experimentally because a part of the energy is dissipated in the form of damage evolution, especially the motion of localized phase boundary. In general, the speed of a solitonic wave is dependent on the initial condition. Different impact stress causes different propagation velocity of the boundary of the failure wave. But it should be pointed out that the maximum speed is less than its limit c_0.

Let us define the kink energy center by the expression:

$$z(t') = \frac{\int_0^\infty x' E dx'}{E} \tag{35}$$

Following the procedure of Rodrigez-Plaza and Vazquez(1990), we obtain equations for the kink velocity and kink energy center for the perturbed case in (27). Differentiating (33) with respect to t and using the perturbed nonlinear equation, we obtain the differential equation for the velocity of solitary wave of the kink type as:

$$\dot{v}(t') = -\frac{3}{4}\varepsilon(1-v^2)\int_0^\infty \frac{R[\beta]}{\cosh^2\phi}dx' \tag{36}$$

in which $\phi = \dfrac{1}{\sqrt{2}(1-v^2)}(x'-z(t'))$.

In the same way, we find the equation for the kink center as:

$$\dot{z}(t') = -\frac{3\sqrt{2}}{4}\varepsilon v(1-v^2)\int_0^\infty \frac{\phi R[\beta]}{\cosh^2\phi}dx' \tag{37}$$

For homogeneous deformation, that is, the magnitude of $V(x',t')$ is independent of the location of the material point, from (36) and (37), we obtain:

$$\begin{aligned} \dot{v}(t') &= -\varepsilon\gamma' v(1-v^2) \\ \dot{z}(t') &= v + \frac{3}{2}\varepsilon v(1-v^2)V(t') \end{aligned} \tag{38}$$

It is seen from (38) that the propagation speed decreases due to the existence of damping effects. In addition, irreversible damage can extend the position of the kink center of solitonic wave, and thus delay the time for material damage to reach the critical condition.

5. Discussions and Conclusions

Damage evolution in elastic-brittle materials has received much recent attention because of its significant effect on strength and ductility. A plausible method to describe the material degradation is the internal variable method within the framework of thermodynamics. For simplicity, we use the scalar internal variable $\beta(x',t')$ as the measure of the extent of damage at a material point. For practical prediction of the evolution of microstructures, we modify the expression for the power of internal force and inertia of the whole system, and assume that they are only dependent on the damage rate and its gradient which are related to microscopic movements. Physical and chemical processes, like phase transformation, microcracking and creep diffusion, can also be incorportated in this thermodynamic model.

Due to the heterogeneity of microstructures, two types of interactions, i.e., among various sites and orientations, exist and must be somehow represented in the macroscopic continuum description. The interaction at the different sites controls the localization of damage, while their relative rotation controls the kinetics of microstructures. As an analogy, this situation can be modeled as a lot of particles with spring connections between them. The interactions not only provide the whole system with potential energy, but also with kinetic energy, resulting from the relative movement of particles and the rotation of the springs. These interactions are ignored by the classical, local continuum damage models, but are taken into account in the non-local models. So, the damage measurement at a material point depends not only on the thermal field at the same point, but also on its neighborhood, see Bazant(1994). For dynamic cases, the effect of relative microstructural rotation, like grain rotations caused by phase transformation, becomes more obvious. In this paper, damage is not simply a measure of the decrease of net area in transport of internal forces, but an additional degree of freedom in describing the motion of a material point.

As stated above, the material damage at a point must be an average value over a finite area characterized by a length l_c. This length has a major influence on the results of analysis, such as size effect and wave velocity.

Dissipative mechanisms due to changing temperature, formation of microcracks, voids, chemical reactions and phase transition all absorb energies and are included in a scalar internal variable. They would have a definite influence on the elastic degradation in strength. Irreversible energy absorption causes a damping effect on the propagation of the failure wave and decreases the wave speed as indicated in equation (38).

As expected, the wave solution displays the feature of solitons of the kink type due to the non-linear terms existing in the governing equation. The results show that the wave velocity depends very much on the initial activation. With increasing impact stress, the absorbed energy by the damage processes will increase. It is seen from (33) that it would lead to an increase of the failure wave velocity. This is in good agreement with the experimental results of Brar et al (1991).

The maximum speed is less than $c_0 = \sqrt{\kappa / \iota}$, in which κ, ι measure the non-local effect of material damage. Using dimensional analysis, we can find that:

$$\begin{aligned} \kappa &= \mu l_c^2 \kappa(\text{material constants, } \beta) \\ \iota &= \rho l_c^2 \iota(\text{material constants, } \beta) \end{aligned} \tag{39}$$

With increasing damage, the non-local effect becomes intensive. It decreases the tensile strength and enlarges the influence area on this material point. This means that $\kappa \leq 1$, and $\iota \geq 1$, if we take the undamaged material as the reference states. In turn, it yields an important result that $c_0 \leq c_s$, in which c_s is the shear wave speed. As an extreme, if the material, like Pyrex glasses, is very perfect, the damage extent is very small, no damping effect is detected and no failure wave can be observed because its speed approaches the shear wave speed.

It is anticipated that the model based on our formulation is free of spurious mesh sensitivity due to the introduction of an internal material size. It can be employed to predict correctly the behavior of brittle materials and compare with experimental results. However, the theory is by no means complete and application of this model to numerical algorithms, such as Finite Element Method requires further work..

Due to much uncertainty in the formation of failure wave, especially its related physical processes, additional experimental and theoretical research is also needed. Of course, the lack of distinction between observable deformation and internal variable creates ambiguity. Whether or not a transformation shock occurs behind the elastic wave needs further experimental investigation too. In addition, for a complex thermomechanical system, different forms of internal variables, like vector and tensor, are needed for the description of orientational hardening materials. And how to extend this approach to elastic-plastic solids and surface wave is being investigated.

Acknowledgments

The authors would like to thank the Australian Research Council (ARC) for supporting this research. Xi Zhang is supported by an OPRS from the Australian Government and an ARC scholarship from the research grant.

References

Abdullaev, F. Kh.(1994) Theory of Solitons in Inhomogeneous Media, John Wiley & Sons Press.

Bazant, Z. P.(1994) Nonlocal damage theory based on micromechanics of crack interactions, *ASCE, J. Engng Mech.* 120, 593-617.

Bourne, N. K., Rosenberg, Z. and Field, J. E.(1995), High-speed photography of compressive failure wave in glasses, *J. Appl. Phys.* 78, 3736-3739.

Brar, N. S., Rosenberg, Z. and Bless, S. J., (1991) Impact-induced failure waves in glass bars and plates, *Appl. Phys. Lett.*, 59, pp3396-3398.

Clifton, R. J. (1993) Analysis of failure wave in glasses, *Appl. Mech. Rev.* 46, pp540-546.

Fremond, M. and Nedjar, B. (1996) Damage, gradient of damage and principle of virtual power, Int. J. Solids Structures, 33, pp1083-1103.

Krumhansl, J.L. and Schrieffer, J.R.,(1975) Phys. Rev. B11, 3565.

Lemaitre, J. and Chaboche, J. L., (1990), Mechanics of Solids, Cambridge University Press, Cambridge, UK.

Lehmann, Th.(1989) Some thermodynamic consideration on inelastic deformations including damage process, *Acta Mech.* 79,1-24.

Maugin, G. A. (1990) Internal variables and dissipative structures, *Int. J. Non-Equilib Thermodyn.* 15, 173-192.

Moreau, J. J.(1970) sur les lois de frottement de viscosite et de plasticite, C. R. Acad. Sci. Paris, 271, 608-611.

Onat, E. T., and Leckie, F. A. (1988) Representation of mechanical behavior in the presence of changing internal structures, *J. Appl. Mech.*, 55, 1-10.

Rajaraman, R. (1982) An introduction to solitons and quantum filed theory, North-Holland, Amsterdam.

Rodrigez-Plaza, M. J. and Vazquez, L.(1990) *Phys. Rev.* B41, 11437

EFFECT OF INITIAL FLAWS IN HIGH CYCLE FATIGUE OF SG CAST IRON

A.-S. BÉRANGER
Renault - Direction de la Recherche - Service 60152
F-92109 Boulogne Billancourt, France.

AND

R. BILLARDON, F. HILD AND H. YAACOUB AGHA
Laboratoire de Mécanique et Technologie
ENS de Cachan - CNRS - Université Paris 6
F-94235 Cachan, France.

Abstract – An expression of the cumulative failure probability of a structure is proposed for cyclic loading conditions. This expression is dependent on the initial flaw distribution and on the microcrack propagation law. Two independent sets of experiments were carried out on specimens made of Spheroidal Graphite cast iron. These specimens were tested under cyclic tension with different load ratios. The initial flaw distribution and the propagation law parameters are identified from the first set of experimental results. The expression of the failure probability is then used to predict the second set of experimental data. The effect of the load ratio is discussed and the influence of stress field heterogeneity is studied.

1. Introduction

The structural integrity of a component can either be assessed by using deterministic crack initiation criteria and by ignoring micro-inhomogeneities (e.g., flaws) within the material, or by modeling the presence of these inhomogeneities, their possible evolution with the number of cycles, and if needed their statistical distribution. For structures made of cast iron and subjected to high cycle fatigue, local failure is due to the presence of initial flaws randomly distributed within the material. Therefore, these flaws as well as their stable propagation with the number of cycles need to be accounted for.

R. Wang (ed.), IUTAM Symposium on Rheology of Bodies with Defects, 191–203.

The fatigue process in materials can be schematically divided into two stages. Macrocrack initiation, which is often due to initial flaws, has to be considered for both brittle and ductile materials. Macrocrack propagation is usually unstable for brittle materials, while first stable and then instable for ductile materials. In this paper, we will focus our attention on SG cast iron subjected to high cycle fatigue. The structure is therefore assumed to remain macroscopically elastic, whereas the microscopic evolution of the flaws is described according to a generalized Paris' law up to local failure.

Statistical methods applied to predicting failure under monotonic conditions have been extensively used. The first attempt was made by Weibull [1] and was based on a statistical treatment of failure. Monotonic and cyclic loading conditions were analyzed. Batdorf and Crose [2] modeled initial flaws by cracks whose sizes and orientations were randomly distributed. Lamon and Evans [3] derived another model based on similar assumptions. The aim of this paper is to study the influence of randomly distributed initial flaws on the failure of SG cast iron components. The effect of stress field heterogeneity is also studied.

2. Flaw propagation in high cycle fatigue

In heterogeneous materials, initial heterogeneities are sometimes sphere-like cavities (flaws due to cooling down in SG cast iron), or sphere-like brittle inclusions with low interfacial strength (e.g., graphite in cast iron). During microcrack propagation, it is assumed that the cracked surface increases with no morphological change. Therefore, the radius a of the surface is the only parameter to be accounted for.

In SG cast iron there are four different defect populations that may lead to high cycle fatigue failure. First, pinholes located at or close to the as-cast surface, second, graphite nodules ($a < 30$ μm), third macro-shrinkage defects ($a > 0.5$ mm) and fourth, micro-shrinkage pores ($a \leq 0.5$ mm) which are very difficult to detect but can be present in the components. This last class of defects is studied in this paper since they tend to be the main micro-propagation sites leading to final fracture. They are generally larger than the graphite nodules.

The flaw geometry is described by a dimensionless factor Y such that a general stress intensity factor K is given by

$$K = Y\sigma\sqrt{a} \tag{1}$$

where σ stands for an equivalent uniaxial tensile stress (e.g., maximum principal stress). It is worth noting that the values of the parameter Y depend on the geometry of the initial defect and on whether or not the flaw intersects a free surface.

To take into account the localized non-linear behavior of the material in the vicinity of the crack tip under cyclic loading conditions, the onset of microcrack propagation is described by a criterion postulated by Pellas et al. [4]

$$\Delta K_{\text{eff}} = g(R)K_{\max} - K_{\text{th}}(a) \geq 0 \tag{2}$$

where $K_{\text{th}}(a)$ refers to a so-called threshold stress intensity factor which is a function of the current crack size [5]. The function $g(R)$ models the influence of load ratio R. It is supposed that the load history is simple, such that the maximum principal stress direction is constant throughout the load history. Therefore bifurcation of the crack is not considered.

In many practical situations, it can be assumed that the flaw size is bounded by a maximum value a_{M}. A cyclic threshold stress can be defined below which no failure occurs (i.e., the failure probability is equal to zero). This cyclic threshold stress, S_{th}, is related to the threshold stress intensity factor K_{th}. Its expression, when $g(R) = 1$, can be derived from eqs. (1) and (2)

$$S_{\text{th}} = \frac{K_{\text{th}}(a_{\text{M}})}{Y\sqrt{a_{\text{M}}}} \tag{3}$$

The microcrack propagation law keeps the main features of the macrocrack propagation law based upon the generalized Paris' law proposed by Pellas et al. [4]

$$\frac{da}{dN} = C\left[\frac{K_{\max}g(R) - K_{\text{th}}(a)}{K_{\text{c}} - \frac{K_{\text{th}}(a)}{g(R)}}\right]^n \tag{4}$$

where C and n are material parameters. The failure criterion is supposed to be given by $K_{\max} = K_{\text{c}}$. In high cycle fatigue, the flaw size does not increase significantly over a major part of the loading domain up to fracture. Therefore it is assumed that the evolution of the threshold stress intensity factor depends only on the initial flaw size a_0. Furthermore, the flaw size varies usually between 60 μm and 500 μm where K_{th} is only weakly dependent on a_0 [5]. As a first approximation, the threshold stress intensity factor K_{th} will be taken as a constant. The following closed-form solution can be derived by integration of eq. (4)

$$\varphi\left(\sqrt{\frac{a_{\text{c}}}{a_{\text{M}}}}\right) - \varphi\left(\sqrt{\frac{a_{\text{c0}}}{a_{\text{M}}}}\right) = C^*\left[\frac{g(R)\,K_{\text{th}}}{K_{\text{c}} - \frac{K_{\text{th}}}{g(R)}}\right]^n \left(\frac{\sigma_{\max}}{S_{\text{th}}}\right)^n N_{\text{F}} \tag{5}$$

where a_{c} and a_{c0} denote the critical and initial flaw sizes, respectively, and $\sigma_{\max}$ the maximum tensile stress over one cycle. The size a_{c} can be calculated from eq. (1) when $K_{\max}$ is equal to K_{c}, and a_{c0} represents the initial flaw size that becomes critical (i.e., equal to a_{c}) after N_{F} cycles. The constant C^* is equal to $\frac{C}{a_M}$. The value of the function φ is given by (when $n \neq 1$ and $n \neq 2$)

$$\varphi(x) = 2\frac{(x - x_{th})^{1-n}[x_{th} - (n-1)x]}{(n-1)(n-2)} \tag{6}$$

where x_{th} is the normalized threshold defect size obtained by using eqs. (1) and (2)

$$x_{th} = \sqrt{\frac{a_{th}}{a_M}} = \frac{S_{th}}{\sigma_{max}\, g(R)} \tag{7}$$

The aim of the next section is to derive an expression for the cumulative failure probability for stable defect growth under cyclic loading conditions.

3. Cumulative failure probability

For sake of simplicity, only the case of constant (tensile) load level σ_{max} is discussed. Furthermore, it is assumed that the flaw population can be characterized by a single parameter, the flaw size a. Initial heterogeneities are usually randomly distributed within the material, and are modeled by a flaw size distribution function f. This function needs to be determined in order to assess the reliability of heterogeneous materials under given loading conditions. The cumulative failure probability P_{F0} of an element Ω_0 is the probability of finding initial flaws larger than the critical flaw size a_{c0} [6]

$$P_{F0} = \int_{a_{c0}}^{+\infty} f_0(a)da \tag{8}$$

A generalization can be found in Ref. [7]. Equation (8) is valid if we assume that the flaw size evolution is deterministic, and the only flaws to cause failure are those initially present in material body. The cumulative failure probability P_F of a structure Ω can be related to the cumulative failure probability P_{F0} in the framework of the weakest link theory [8] by

$$P_F = 1 - \exp\left[\frac{1}{V_0}\int_{\Omega} \ln(1 - P_{F0})dV\right] \tag{9}$$

Equation (9) holds when the flaw interaction can be neglected. In SG cast iron, the average distance between flaws is large compared to their maximum size so that this hypothesis is very often fulfilled.

4. Stress field heterogeneity

Hild et al. [9] have introduced stress heterogeneity factors to characterize the stress field. These factors are dependent on the load type and describe monotonic loadings. In the following, a modified version of these factors is introduced. Under the condition that the flaw size is bounded by a_M, the flaw size distribution can be approximated by

$$f_0(a) = \frac{W(a_M - a)^{\beta-1}}{a_M^\beta} \text{ with } W > 0 \text{ and } \beta > 0 \quad (10)$$

The cumulative failure probability of a volume element can be approximated by

$$P_{F0} = \frac{W}{\beta}\left(1 - \frac{a_{c0}}{a_M}\right)^\beta \quad (11)$$

In the following, the scatter of endurance limits (i.e., $N \to \infty$ and $a_{c0} \to a_{th}$) will be investigated

$$P_{F0} = \frac{W}{\beta}\left[1 - \left\{\frac{S_{th}}{\sigma_{max} g(R)}\right\}^2\right]^\beta \quad (12)$$

When the stress level σ_{max} is close to $\frac{S_{th}}{g(R)}$, eq. (12) can be rewritten as

$$P_{F0} = \left[\frac{\langle \sigma_{max} - \frac{S_{th}}{g(R)} \rangle}{\sigma_0}\right]^\beta \text{ with } \sigma_0 = \frac{S_{th}}{2g(R)}\left[\frac{W}{\beta}\right]^{\frac{1}{\beta}} \quad (13)$$

Equation (13) corresponds to a three-parameter Weibull law with $m = \beta$. The parameter β gives the tendency of the initial flaw distribution for large defects (i.e., $a \cong a_M$). By using eq. (9), the failure probability of a structure Ω of volume V can then be approximated by

$$P_F = 1 - \exp\left[-\frac{V}{V_0} H_m^* \left\{\frac{\langle \sigma_F - \frac{S_{th}}{g(R)} \rangle}{\sigma_0}\right\}^m\right] \quad , \quad \sigma_F = \max_{M \in \Omega} \{\sigma_{max}(M)\} \quad (14)$$

with

$$H_m^* = \frac{1}{V} \frac{\int_\Omega \langle \sigma_{max}(M) g(R) - S_{th} \rangle^m dV}{\langle \sigma_F g(R) - S_{th} \rangle^m} \quad , \quad \text{if } \sigma_F g(R) > S_{th} \quad (15)$$

where H_m^* is a modified stress heterogeneity factor which depends upon the load type and the load level. The corresponding effective volume V_{eff}^* is given by

$$V_{eff}^* = H_m^* V \quad (16)$$

The effective volume V_{eff}^* is also dependent on the load level. In simple cases such as pure tension, this volume is zero when $\sigma_{max} g(R) \leq S_{th}$, and is equal to V when $\sigma_{\max} g(R) > S_{\text{th}}$.

The previous results will be applied to rotating bending ($R = -1$). In this case, H_{m}^* can be expressed by

$$H_{\text{m}}^* = H_{\text{m}} \left(1 - \frac{S_{\text{th}}}{\sigma_{\text{F}}}\right) \left[1 + \frac{S_{\text{th}}}{\sigma_{\text{F}}(m+1)}\right] \quad , \quad \text{with } H_{\text{m}} = \frac{2}{m+2} \tag{17}$$

Figure 1 shows the change of the modified stress heterogeneity factor as a function of the stress level for the particular case of rotating bending. In

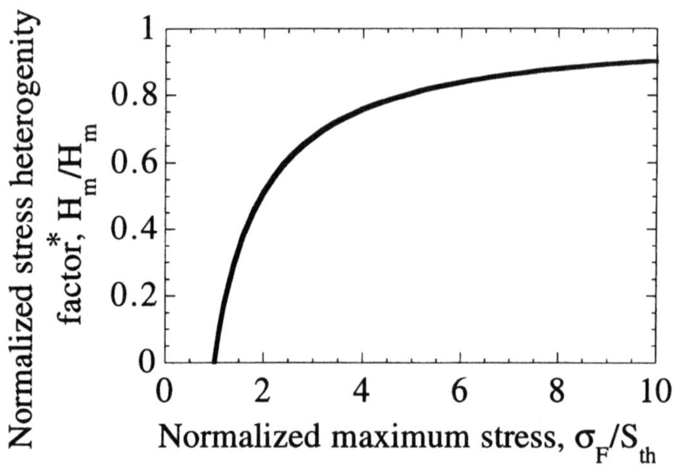

Figure 1. Normalized stress heterogeneity factor versus dimensionless maximum stress in rotating bending ($m = 26$).

the case of a complex structure, the stress heterogeneity factor and effective volume may be numerically calculated as will be shown later.

5. Analysis of fatigue tests on SG cast iron

In this section, a series of experiments performed at LMT-Cachan and Renault on specimens made of ferritic SG cast iron are analyzed in details. These experiments have been carried out at different stress levels. The specimens are tested under cyclic tension with two different load ratios ($R = -1, R = 0.1$). Each curve of a standard S-N plot can be associated with a constant failure probability.

When the fatigue limits are known, the identification can be performed in two different stages. The first step consists of the identification of the flaw size distribution. A minimization scheme is used to determine the minimum error between all the available experimental data on fatigue limits [10]. By assuming the maximum flaw size be bounded by a_M, the flaw size distribution f_0 can, for instance, be fitted either by eq. (10) or by a beta distribution

$$f_0(a) = \frac{a^{\alpha-1}\,(a_M - a)^{\beta-1}}{B_{\alpha\beta}\; a_M^{\alpha+\beta-1}}; \text{ when } 0 < a < a_M;\ \alpha > 0,\ \beta > 0 \quad (18)$$

where α and β are the parameters of the beta function, and $B_{\alpha\beta}$ is the Euler function of the first kind. The parameters to identify are α and β, the volume ratio $\frac{V}{V_0}$ and the threshold stress S_{th}. The first step of the identification is applied to the experimental results for the load ratio $R = 0.1$. It yields: $\alpha = 5.7$, $\beta = 26$, $\frac{V}{V_0} = 1$, and $S_{th} = 105$ MPa. The second step of the identification concerns the crack growth law (parameters C, n and μ). The parameter μ is used in $g(R) = (1-R)/(1-\mu R)$ [4]. In tension, the identification is performed by studying a constant cumulative failure probability (e.g., 50%). The following values are obtained: $n = 2.0$, and $C^* = 5.9 \times 10^{-5}$. In Fig. 2 predictions of the number of cycles to failure are compared with the experimental observations. It is shown that the identified laws are in good agreement with the experimental results.

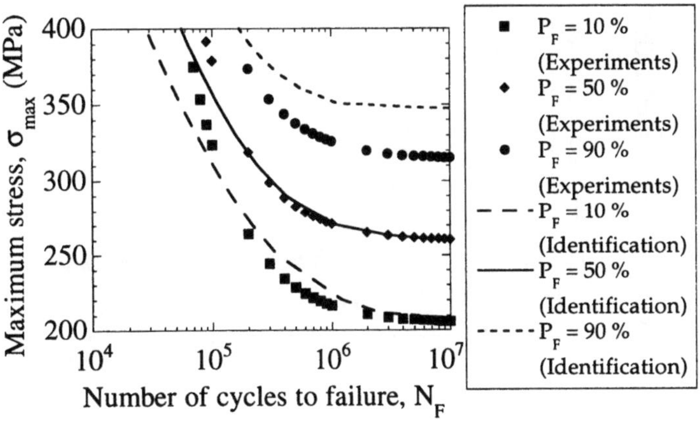

Figure 2. Identified failure probabilities compared with experiments ($R = 0.1$).

6. Validation of the model

The previous results are validated by comparing them with other experimental results obtained independently. The two stages of the numerical identification are compared separately.

6.1. FLAW DISTRIBUTION

Experimental investigations are performed to quantify the identified flaw distribution, and also to get more information about the size of the largest defect a_M. Systematic microscopic observations on 50 fracture surfaces were performed by using a Scanning Electron Microscope. Flaws with a diameter less than 80 μm were not considered to avoid confusion with graphite nodules (with maximum size on the order of 60 μm in diameter). In Fig. 3, the experimental flaw distribution is given and compared with the identified distribution. The identified distribution is in good agreement with the experimental one. The experimentally measured value of a_M is 400 μm. This result shows that we are dealing with short cracks, and that the threshold stress intensity factor can be considered as a constant for flaws of this size as shown in Ref. [11]. Equation (10) could have been used but there was not a good agreement between the identification of the flaw size distribution by analyzing fatigue limits and the measurements shown in Fig. 3.

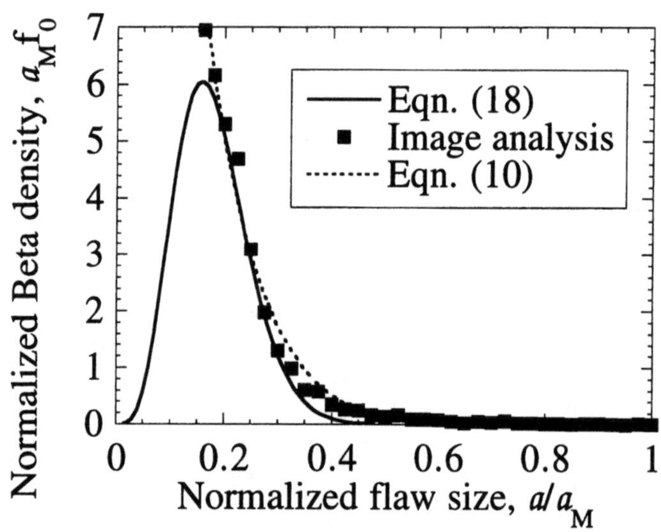

Figure 3. Experimental and identified flaw distribution.

6.2. PROPAGATION LAW

Figure 4 shows the crack growth rate as a function of the stress intensity range. The solid curve is the identified one. The other curves represent experimental results obtained on specimens made of SG cast iron. The solid squares concern an artificial short crack of initial length $a_0 = 240$ μm [12].

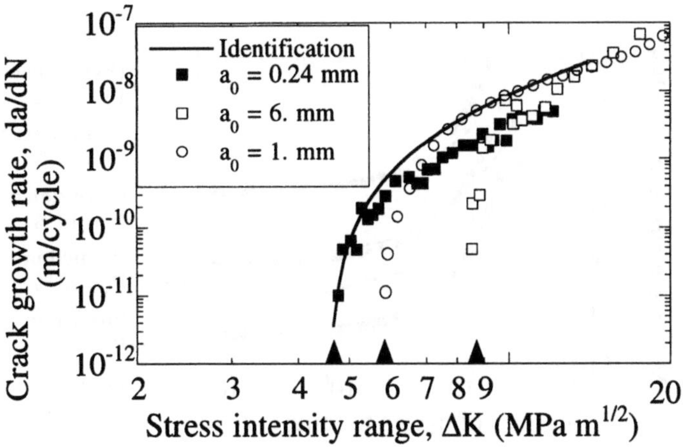

Figure 4. Crack growth rate as a function of stress intensity range.

The open circles correspond to an artificial crack of initial size $a_0 = 1$ mm. The open squares concern an artificial long crack of initial length $a_0 = 6$ mm [12]. The identified curve is in good agreement with the experimental one for short cracks especially near the threshold regime. The distinction with the curve for long cracks is mainly described by threshold differences (Fig. 4).

6.3. PREDICTION OF EXPERIMENTAL RESULTS IN TENSION

The comparison between the threshold values calculated by using eq. (7) for the load ratio $R = -1$ and $R = 0.1$ allows to identify the value of parameter μ ($\mu = 0.59$). Figure 5 shows the comparison between the experimental results for $R = -1$ and the predicted results using the parameters identified previously ($R = 0.1$) and μ.

The predictions are in reasonable agreement with the experimental data. This result shows that function g accounts for the influence of load ratio for different cumulative failure probabilities.

6.4. PREDICTION OF EXPERIMENTAL RESULTS IN ROTATING BENDING

Experiments were carried out on 30 specimens in rotating bending. These experimental results are used to assess the predictive capacity of the model in the case of complex loads. Figure 6 shows the experimental results with the 10% and 50% failure probabilities predicted by the model identified from experiments in tension.

The prediction of the 50% failure probability is good. This result shows

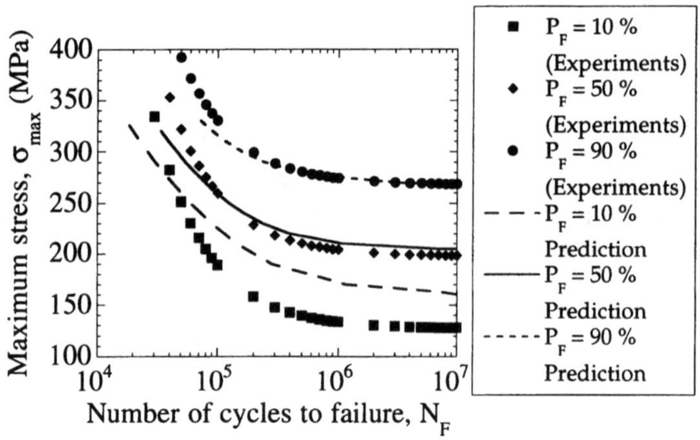

Figure 5. Predicted failure probabilities compared with experiments $(R = -1)$.

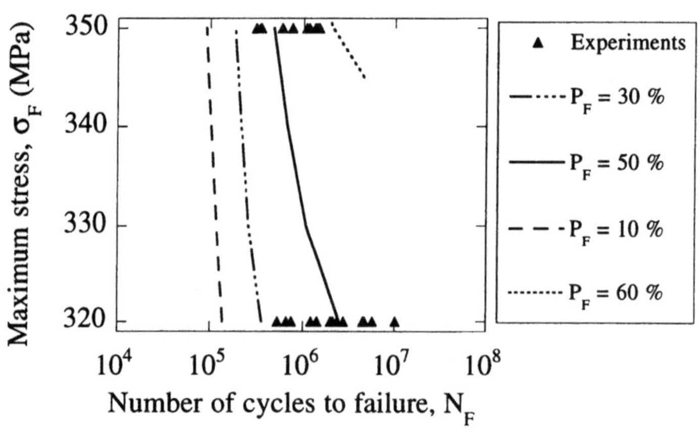

Figure 6. Predicted failure probabilities in rotating bending with $\frac{V}{V_0} = 1.1$ $(R = -1)$.

that the difference between the stress levels for the same failure probability in tension and in rotating bending are due to the stress field heterogeneity.

7. Application to a structure

The identification and validation of the present model allows to make an extension of the model to real structures. A post-processing program (ASTAR) has been developed to compute the failure probability and stress heterogeneity factors of a cast structure.

From the results of an elastic computation of a structure through a Finite Element analysis, ASTAR evaluates the equivalent stress σ_{max} at each integration point. A critical flaw size a_{c0} is associated to this stress level and to a given number of cycles N_{F}. The failure probability at the integration point is then calculated by numerical integration of eq. (8). The integration over the total volume of a finite element gives the failure probability of the element. The failure probability of the structure can be calculated according to eq. (9). This procedure has been applied to a suspension arm designed by Renault car company. The mesh consists of 3712 triangular shell elements. The industrial FE package ABAQUS [13] is used to perform the elastic analysis of the structure. Figure 7 shows the change of the effective volume as a function of the maximum applied stress. Figure 8 gives the contours of the failure probabilities P_{F0} for a maximum

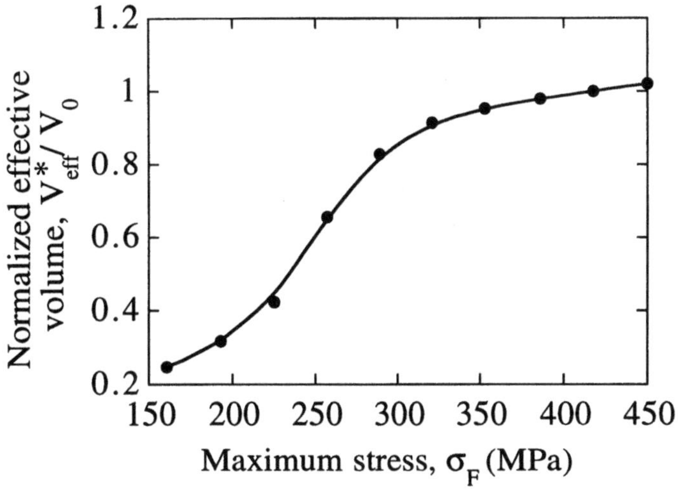

Figure 7. Effective volume versus maximum applied stress ($R = 0.1$).

stress level of the order of 300 MPa and a number of cycles $N_{\mathrm{F}} = 10^7$.

8. Conclusions

A reliability analysis taking account of flaw size distributions has been developed for components subjected to cyclic loading conditions. An expression of the cumulative failure probability is derived in the framework of the weakest link theory and by assuming that the flaws do not interact.

Experimental data on SG cast iron in tension and in rotating bending are analyzed within this framework. The predictions of the whole set of

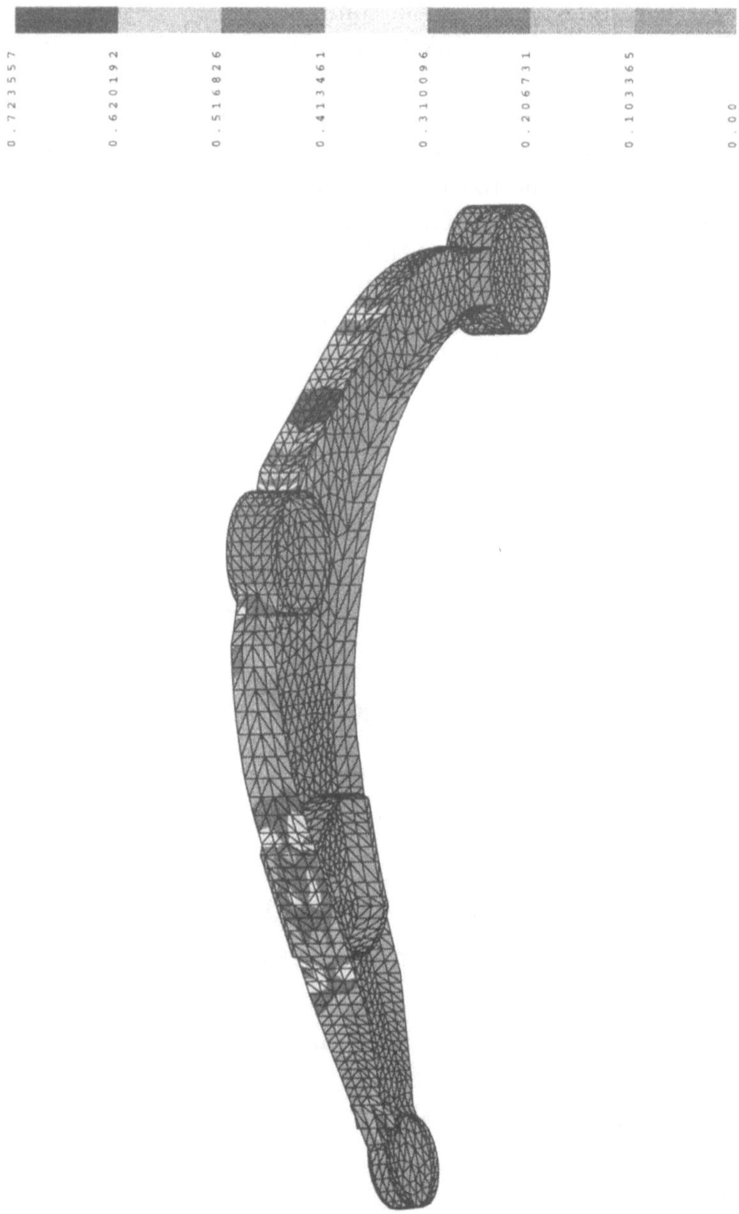

Figure 8. Contours of the failure probability P_{F0} for a maximum stress level of the order of 300 MPa and a number of cycles $N_F = 10^7$ ($R = 0.1$).

data is in good agreement with the experimental number of cycles to failure. This last result shows that the expression of cumulative failure probability proposed herein is able to model fatigue data obtained on SG cast iron and that the model can describe the influence of the load ratio. It shows that the model can also take into consideration the influence of stress field heterogeneity. Typical applications of this approach concern the reliability analysis of cast components. The post-processing approach can predict the reliability of the whole component under cyclic loading conditions. In other words, it enables to predict the number of cycles to macrocrack initiation (local failure), or the probability of reaching a certain number of cycles without failure at any point of the component.

9. Acknowledgments

HYA, RB and FH gratefully acknowledge the financial support of Renault through contract CNRS/109 (H5–24–12) with LMT-Cachan.

References

1. Weibull, W., (1939), *A Statistical Theory of the Strength of Materials*, Roy. Swed. Inst. Eng. Res., **151**.
2. Batdorf, S. B. and Crose, J. G., (1974), A Statistical Theory for the Fracture of Brittle Structures Subjected to Polyaxial Stress States, *J. Appl. Mech.* **41**, 459-465.
3. Lamon, J. and Evans, A. G., (1983), Statistical Analysis of Bending Strengths for Brittle Solids: a Multiaxial Fracture Problem, *J. Am. Ceram. Soc.* **66**, 177-182.
4. Pellas, J., Baudin, G. and Robert, M., (1977), Mesure et calcul du seuil de fissuration après surcharge, *Recherche aérospatiale* **3**, 191-201.
5. Clément, P., Angeli, J. P. and Pineau, A., (1984), Short Crack Behavior in Nodular Cast Iron, *Fatigue Eng. Mat. Struct.* **7**, 251-265.
6. Hild, F. and Roux, S., (1991), Fatigue Initiation in Heterogeneous Brittle Materials, *Mech. Res. Comm.* **18**, 409-414.
7. Hild, F., Kadouch, O., Lambelin, J.-P. and Marquis, D., (1996), Analysis of the Failure of Ceramics Due to Subcritical Crack Growth, *ASME J. Eng. Mat. Tech.* **118**, 343-348.
8. Freudenthal, A. M., (1968), Statistical Approach to Brittle Fracture, in H. Liebowitz (ed.), *Fracture* **2**, Academic Press, New-York (USA), 591-619.
9. Hild, F., Billardon, R. and Marquis, D., (1992), Hétérogénéité des contraintes et rupture des matériaux fragiles, *C. R. Acad. Sci. Paris* **t. 315**, 1293-1298.
10. Béranger, A.-S., Billardon, R., Hild, F. and Agha, H. Y., (1996), Effect of Initial Flaws on Fatigue Crack Initiation of S.G. Cast Iron, in G. Lütjering and H. Nowack (eds.), *Proceedings Fatigue '96*, Pergamon, London (UK), 1269-1274.
11. Clément, P. and Pineau, A., (1984), Amorçage et propagation de petites fissures dans la fonte à Graphite Spheroïdal, *Proceedings Journées internationales de printemps de la SFM*, 203-218.
12. Clément, P., (1984), *Propagation par fatigue de petits défauts dans une fonte G.S.*, CNAM dissertation, Paris.
13. Hibbitt, H. D., Karlsson, B. I. and Sorensen, P., (1995), *Abaqus, version 5.5.*

STUDY OF CRACK DEVELOPMENT AS THE BASIS FOR RHEOLOGY OF CEMENTITIOUS MATERIALS

P. STROEVEN and M. STROEVEN
Faculty of Civil Engineering and Geosciences, Delft University of Technology, Stevinweg 4, 2628 CN Delft, The Netherlands

Abstract

Concrete is a man-made material containing a particulate filler (roughly) designed on the basis of a sieve curve. For a river aggregate, the particles are approximately spherical and smooth-textured. The particle-matrix interface is usually the weakest chain link in the mechanical system; damage evolution starts at particle-matrix interfaces, even in the so called virgin state. For direct tension, these interface cracks will be on average perpendicular to the loading direction. For compression, they will run parallel to the loading direction. A single fracture surface is formed in tension, and a series of fracture surfaces in compression. They are the result of crack concentration within a process zone, in which the engineering crack meanders around a dividing plane. This allows us to design structural models for crack development on different resolution levels. Stereological notions are employed for that purpose.

1. Introduction

Concrete contains an aggregate with a wide range of particle sizes, the size distribution function of which is governed by the sieve curve. The building code generally allows application of a maximum grain size of (about) 30 mm or more for mass structures. The finer sand fractions have particles in the sub-millimeter range. The aggregate constitutes a dense random particulate packing (Fig. 1), thereby providing the material with a load-bearing skeleton. An increase in the range of particle sizes will have a positive influence on concrete 'quality', since smaller particles will fill up the holes left by the larger ones (Jiang *et al*, 1994; Stroeven, 1997). The particulate skeleton is glued together by a cementitious binder. The binder additionally fills the open spaces to guarantee durability. In this case also,the particle size range of the binder is a major parameter in improving 'quality (Fidjestøl, *et al*,

R. Wang (ed.), IUTAM Symposium on Rheology of Bodies with Defects, 205–222.

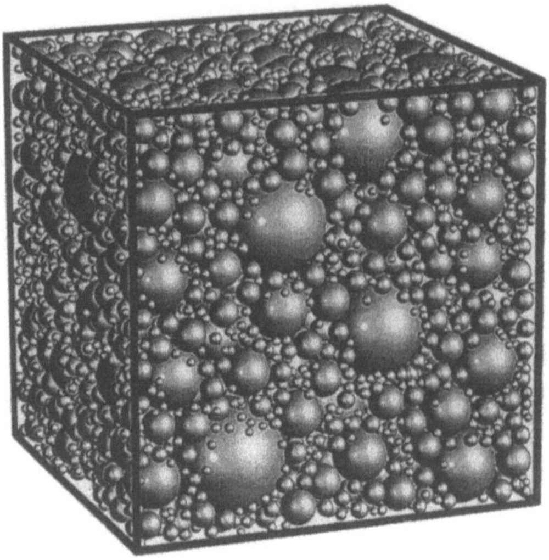

Figure 1*:* Computer-simulated state of dense random packing of aggregate.

1994). In normal weight concrete, the particulate skeleton transmits the major part of the compressive loadings. Since the tensile strength is less than the compressive strength, a steel reinforcement is applied for structural applications. Bond between aggregate and the cementitious matrix forms the weakest chain link in the micromechanical system, particularly when river aggregate is employed. This is due to the inherent discontinuity in the density of cement particles near the interface (Stroeven *et al*, 1997b). The phenomenon is also reflected in low microhardness values in the interphase layer (Lyubimova, 1962).

Rheological behaviour of hardened concrete is intimately related to the development, extension and coalescence of cracks on the various levels of the microstructure. The virgin state of cementitious materials is normally characterized by a dense structure of small cracks, predominantly along the interfacial transition zones (ITZ) between particles and matrix (Fig. 2). This will be the result of high stresses caused by shrinkage and differential settlements during hardening (Hsu, 1963), and the relative weakness of the ITZ. High residual stresses will also be scattered throughout the material volume in places where debonding failed to take place. The mechanical system of the material can therefore be conceived as a load-bearing particulate structure and a dense 3-D network of (potential) cracks; the structures are coupled. Under increasing loading the integrity of the material will gradually break down. This process of structural loosening manifests itself in cracking on the various levels of the microstructure, ultimately leading to the development of engineering cracks.

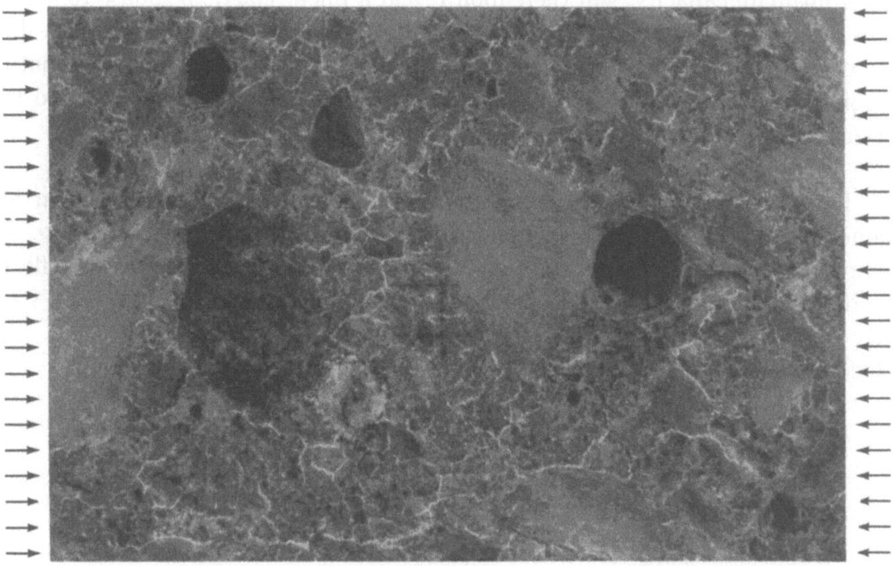

Figure 2: Cracking visualized by filtered particle method in a section of a concrete specimen previously subjected to about 75% of ultimate compressive loading; image shows interface cracking around small sand grains in a blown-up part of field.

In a global way, total crack extension is called "damage", so that crack formation leads to damage evolution. The damage evolution process is characteristic of the material structure under the particular loading conditions.

Structural models for cementitious sytems allowing estimation of 'strength' on the basis of damage (evolution) parameters should be based on these principles. This is fundamentally different for non-structural approaches like lattice and truss systems. Since this is a very complicated problem, commonly the analytical approach is based on ordered crack configurations (Nied *et al*, 1978). In most cases the model is two-dimensional. The present paper discusses only the outlines of a new type of approach to solving these problems.

2. Damage and Resolution

Under low stresses, concrete is considered to be an elastic material. A concrete specimen subjected to at least five compressive stress cycles with a maximum of (about)

50% of ultimate capacity, will thereupon reveal a linear elastic response to stresses below discontinuity, because stress concentrations from the virgin state have been released by microcraking during these load cycles. The 'elastic limit', 'discontinuity point' or 'crack initiation strength' in concrete technology is characterized by a certain 'critical' state of particle-matrix debonding (Stroeven, 1973; 1986).

For modelling of *damage in bulk*, the supposedly spherical river gravel particles (diameter d) are assumed to be debonded over a similar angular extension, 2α. The starting of crack coalescence defines on a global level the elastic limit. It is reflected by a discontinuity in mechanical behaviour, which is commonly detected by strain gauge readings or acoustic emission measurements (Dalhuisen *et al*, 1997). The surface area of a bond crack at the onset of coalescence is given by (Fig. 3)

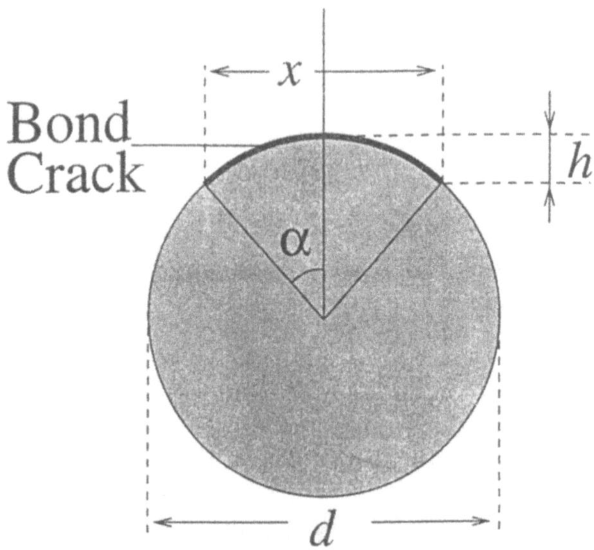

Figure 3: Debonding along the particle-matrix interface in concrete.

$$S = \pi(h^2 + \frac{1}{4}x^2) \tag{1}$$

where h and x are the height and span of the crack, respectively.
Substitution of $h = d(1 - \cos\alpha)/2$ and $x = d\sin\alpha$ yields for the average crack surface area $\overline{S} = \pi\overline{d^2}(1 - \cos\alpha)/2$, with $\overline{d^2}$ being the second moment of the particle size distribution function (psd). The upper and lower boundaries of the sieve curves in the building codes can generally be approximated by a straight line and a second order parabola in a semi-logarithmic plot. Transformation readily leads to the

approximate psd's (Stroeven, 1982)

$$f(d)_u = 2.5\frac{d_0^{2.5}}{d^{3.5}} \quad \text{and} \quad f(d)_l = 3\frac{d_0^3}{d^4} \tag{2}$$

where the subscripts u and l refer respectively to the upper and lower boundary curves corresponding to the sieve curve area in building codes, and d_o is the smallest particle size considered in the model. The first, second and third moments of these psd's are presented in Table 1.

	$\overline{d}$	$\overline{d^2}$	$\overline{d^3}$	$\overline{d_c}$	$\overline{d_c^2}$
$f(d)_{upper} = \frac{5}{2}d_o^{2.5}/d^{3.5}$	$\frac{5}{3}d_o$	$5d_o^2$	$5d_o^{2.5}d_m^{0.5}$	$3d_o$	$3d_o^{1.5}d_m^{0.5}$
$f(d)_{lower} = 3d_o^3/d^4$	$\frac{3}{2}d_o$	$3d_o^2$	$3d_o^3 \ln\frac{d_m}{d_o}$	$2d_o$	$2d_o^2 \ln\frac{d_m}{d_o}$

TABLE 1: Moments of the psd's corresponding to the boundaries of the sieve curves prescribed by the building code. Herein d_o and d_m are, respectively, the smallest and largest particles in the model; d_c is the size of a particle intersecting a plane.

It should be noted that $\overline{d_c}$ is the average size of the grains intersecting a (fracture) plane. A comparison with $\overline{d}$ demonstrates that a section, crack or fracture plane (dividing surface) yields *biased* information on the psd in bulk, hence on the bulk composition of the aggregate. The total amount of damage is $\overline{S}.N_V = S_{Vc}$, where the particle density, N_V, is given by $N_V = 6V_V/\pi\overline{d^3}$. Here V_V is the volume fraction of particles. As an example, substitution for the upper bound yields

$$\overline{S}N_V = 3(1-\cos\alpha)\frac{\overline{d^2}}{\overline{d^3}}V_V = \frac{3(1-\cos\alpha)V_V}{(d_m d_o)^{0.5}} \tag{3}$$

Table 2 presents estimates for 'damage' as a function of resolution, determined by eq (3) assuming $\alpha = 45^0$, and shows the influence of resolution or magnification ($M = d_m/d_o$). An approach with a sensitivity level of 1mm (=lower boundary for detecting crack trace length) would yield a specific crack surface area of about 0.2 mm^2/mm^3, which is quite close to experimental findings (Stroeven, 1990; 1992). Fig. 2 shows a section image of a concrete specimen subjected to about three quarters of ultimate compressive stresses taken from this investigation.

Two extreme cases for the onset of crack coalescence can be modelled. In the first, the bond cracks will develop to a constant angular extension - as used for the development of eq (3) - so that slightly out-of-plane cracks have to coalesce in order to create the meso crack. The same hold for meso cracks which join to form the macrocrack. In the second, the particle–matrix interface cracks will extend

$M = d_m/d_o$	V_V	d_o [mm]	d_m [mm]	S_{Vc} [mm^{-1}]
2	0.1	16-32	32	0.004
4	0.2	8-32	32	0.011
8	0.3	4-32	32	0.023
16	0.4	2-32	32	0.044
32	0.5	1-32	32	0.078
64	0.6	0.5-32	32	0.132
128	0.7	0.25-32	32	0.217

TABLE 2: Damage as a function of resolution

to a common dividing suface. This implies the *angular bond crack extension to be a variable.* Since the curvature of the dividing surface will be considerably less than that of the largest particles in the mix, for modelling purposes the dividing surface can be assumed planar. This is illustrated by the computer-simulated image of a fracture surface of Fig. 4, which is modelled according to this concept. It will form the basis for modelling damage evolution in the so called fracture process zone. Although the width of this zone will be governed by the *largest* particles in the mixture, d_m, the average distance of the coalescing microcracks to the dividing plane can be expressed in terms of average size, $\overline{d_c}$, of the particles intersecting with the fracture plane, which is governed by the *minimum* grain size, as shown by Table 1. Though relatively small, the friction resistance of the fracture surface is due to this phenomenon.

In practice, the bond cracks which will constitute part of the future fracture surface will not necessarily all be parallel to the dividing plane. Instead a system with a *partially planar* orientation distribution has to be taken as a realistic manifestation of damage states found in experimental investigations (Stroeven, 1990; 1992). The cracks with a uniform random orientation will be due to influences like shrinkage, whereas the planar portion arises from the unidirectional loading. The size distribution of the circles in the dividing plane delineating the tip of the 2-D bond cracks is governed by an integral equation of Abel's type (Stroeven, 1973). It is assumed that this size distribution function is also valid for the partially ordered case.

The relevant parameter to characterize damage in the model is the total crack surface area per unit of the dividing surface, S_A. The increase in surface area of the dividing surface is due to the particle indentations. Individual contributions are

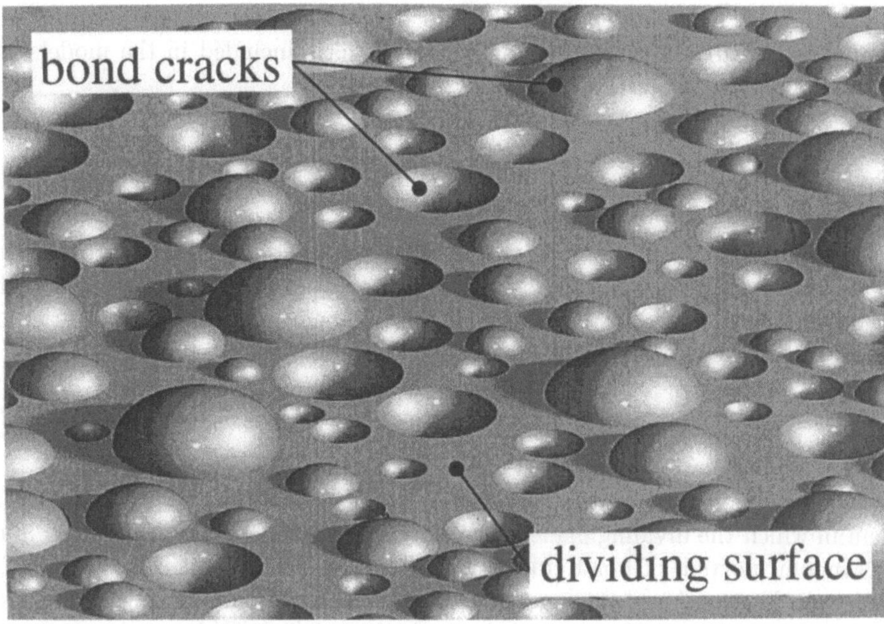

Figure 4: Computer-simulated fracture surface in concrete based on a dense random packing of the aggregate.

given by eq (1). Using the moments of the psd's presented in Table 1, we determine S_A for the two different sieve curves given by eq (2). Hence, for crack parameters in a section we have

		$f(d)_l = 3d_0^3/d^4$	$f(d)_u = \frac{5}{2}d_0^{2.5}/d^{3.5}$
$\overline{x}$	$= J_2\overline{d^2}/\overline{d}$	$= \frac{\pi}{2}d_0$	$= \frac{3\pi}{4}d_0$
$\overline{x^2}$	$= J_3\overline{d^3}/\overline{d}$	$= \frac{4}{3}d_0^2 \ln \frac{d_m}{d_0}$	$= 2d_0^{1.5}d_m^{0.5}$
$\overline{h}$	$= \overline{d_c}/4$	$= \frac{1}{2}d_0$	$= \frac{3}{4}d_0$
$\overline{h^2}$	$= (\overline{d_c^2} - \overline{x^2})/4$	$= \frac{1}{6}d_0^2 \ln \frac{d_m}{d_0}$	$= \frac{1}{4}d_0^{1.5}d_m^{0.5}$
$\overline{S}$	$= \pi(\overline{h^2} + \frac{1}{4}\overline{x^2})$	$= \frac{\pi}{2}d_0^2 \ln \frac{d_m}{d_0}$	$= \frac{3}{4}\pi d_0^{1.5}d_m^{0.5}$
N_A	$= (6V_V\overline{d})/(\pi\overline{d^3})$	$= \frac{3}{\pi}\frac{V_V}{d_0^2}(\ln \frac{d_m}{d_0})^{-1}$	$= \frac{2}{\pi}V_V(d_0^{1.5}d_m^{0.5})^{-1}$
$\overline{S}N_A$	$= S_A$	$= \frac{3}{2}V_V$	$= \frac{3}{2}V_V$

in which $J_2 = \int_o^{\pi/2} \sin^2\theta d\theta$ and $J_3 = \int_o^{\pi/2} \sin^3\theta d\theta$ (Kendall *et al*, 1963). Note again that d_o and d_m are the smallest and largest particles included in the model, and that magnification $M = d_m/d_o$.

3. Roughness of Fracture Surfaces

The average value, S_A, of the bond crack surface area per unit of the corresponding area of the dividing plane is also the major parameter defining the roughness of the fracture surface. The planar roughness index, R_S, being the ratio of total fracture surface area to the corresponding area of the dividing plane, is given by

$$R_S = A_{Am} + S_A = 1 + S_A - V_V \tag{4}$$

with A_{Am} represents the areal fraction of the matrix in the divided surface.
Equation (4) holds only for a 2-D (=planar) portion of cracks in a partially planar system in which the dividing plane is the orientation plane of the 2-D portion. The 3-D 'random' portion of bond cracks in a partially planar system has also to be considered. The total surface area of the 3-D portion of bond cracks per unit area of the circular planes enclosed by their crack tip is 3/2 ($=\overline{S/A}$), as demonstrated earlier for the 2-D portion. Further, the total area of these circular planes enclosed by the crack tips is twice their projected area on the dividing plane ($=\overline{A/A'}$). Here a prime indicates a projected area. Hence, $\overline{S/A'} = 3$. This situation is sketched in Fig. 5. Hence, eq (4) can be modified to encompass both the 2-D and 3-D portions,

$$R_S = 1 - V_V + 3V_{V3} + \frac{3}{2}V_{V2} = 1 + 2V_V(1 - \frac{3}{4}\omega) \quad (\omega = \frac{V_{V2}}{V_V}) \tag{5}$$

where the indices 2 and 3 refer to the 2-D and 3-D portions, respectively. The linear roughness index, R_L can be approximated by $R_L = 1 + \pi(R_S - 1)/4$. Substitution of eq (5), assuming $\omega = \frac{1}{3}$ and $\omega = 0$, respectively, gives

$$R_L = 1 + \frac{\pi}{4}.\frac{3}{2}V_V \qquad \text{and} \qquad R_L = 1 + \frac{\pi}{4}.2V_V \tag{6}$$

The sieve curves lead to near self-similarity situations as to the texture of the fracture surface at different resolution levels. The resolution-dependency can therefore be expressed by means of the fractal equation (Stroeven, 1991; 1996)

$$\log R_L = (D_l - 1)\log M + C \tag{7}$$

D_l is the fractal profile dimension. Because D_l is obtained from the slope of the curve defined by eq (7), it is not necessary to determine the value of the constant C. Solutions of eq (7) using eqs (6) and (2) have been determined, yielding fractal dimensions around 1.08 to 1.13, which fall close to the available experimental data

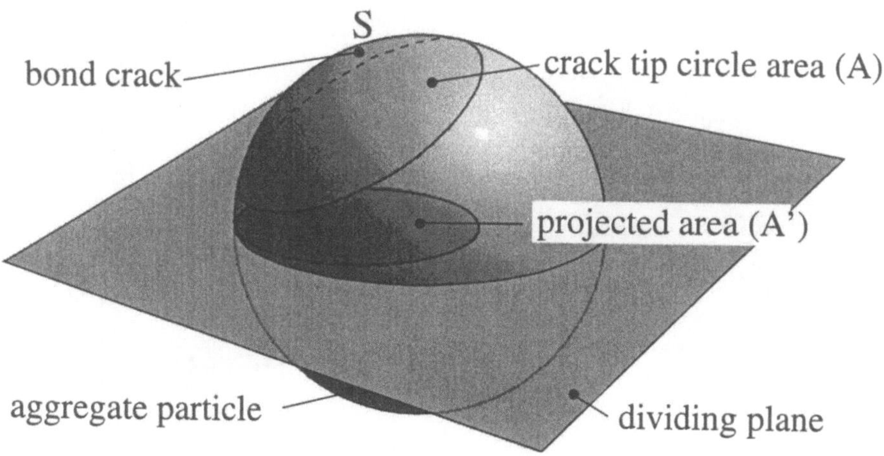

Figure 5: Geometry of a bond crack in a 3-D system with respect to the dividing fracture surface. Note that $\overline{S/A} = S_A$=3/2, and $\overline{A/A'} = A'_A$=2.

(El-Saouma *et al*, 1990). In general, slightly curved lines either with positive or with negative curvature will be obtained, revealing the non-ideal fractal properties of the fracture surface. High density concretes reveal an increased brittleness. As a result, cleavage of particles will be more dominant, reducing fractal dimension. This has been experimentally confirmed (Rawicki *et al*, 1992).

4. **Crack Coalescence**

Crack coalescence will be governed by a function of the distance to the nearest neighbour. In a global approach to crack coalescence in the dividing plane, the relevant parameter will therefore be the average nearest neighbour distance between cracks, $\overline{\Delta}_2$. This is given by

$$\overline{\Delta}_2 = \frac{0.5}{\sqrt{N_A}} \tag{8}$$

where N_A can be taken from the list of crack parameters. Upon substitution in eq (8) it is found for the respective sieve curves that

$$\overline{\Delta}_2 = 0.5 d_0 \sqrt{\frac{\pi}{3V_V} \ln M} \tag{9}$$

$$= 0.5 d_0 \sqrt{\frac{\pi}{2V_V} M^{0.5}} \tag{10}$$

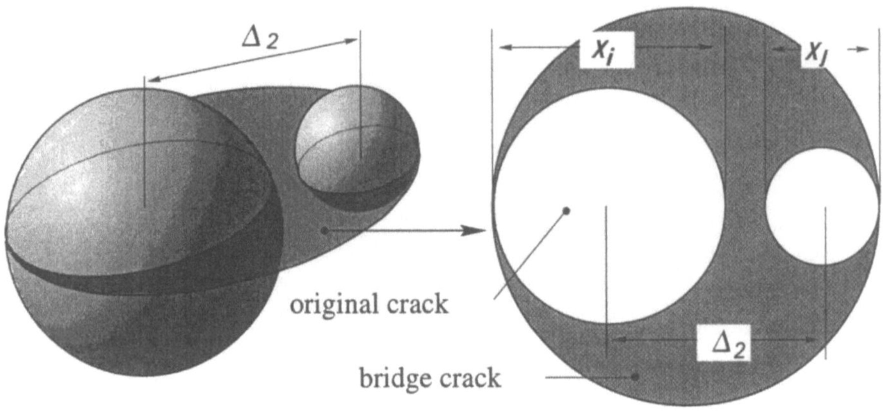

Figure 6: Coalescence of bond cracks in fracture process zone of concrete specimen.

It can easily be checked that $\overline{x}$ and $\overline{\Delta}_2$ have about the same value (on the various levels of the microstructure), indicating a high probability for coalescence of large numbers of neighbouring cracks. The evolution under relatively low loadings of damage on microstructural level is additionally stimulated by the high residual stresses, mentioned earlier.

If it is assumed that all nearest neighbour pairs of cracks would join in the dividing plane over the full loading range of the material body, then a lower bound estimate for increase in 'damage evolution' *in the dividing plane* would be (Fig. 6)

$$\frac{1}{4}\pi\overline{(2x)^2}\frac{N_A}{2}/\frac{1}{4}\pi\overline{(x)^2}N_A = 2 \tag{11}$$

Here x is the diameter of the crack in the dividing plane and N_A is the number of cracks per unit area of the dividing plane. Of course, $\frac{\pi}{4}\overline{x^2}N_A = A_A = V_V$, in which A_A is the areal fraction of the dividing surface being cracked. As a consequence, average crack size would be doubled. Roughly speaking, this process leads to a new situation in which the smaller distances are eliminated, so that a more uniformly spaced crack pattern is obtained. In direct compression tests of concretes such a doubling of damage evolution was recorded from discontinuity to ultimate at a measuring sensitivity of about 0.5 mm for crack traces in the section plane (Stroeven, 1990; 1992). This may demonstrate that for such a sensitive approach the damage development at ultimate is not very advanced. In fact, ultimate is a structurally insignificant development stage in the gradual degradation process of the material body. In direct tension, experiments have demonstrated that this development stage is even less advanced, i.e. specific crack surface area showed a

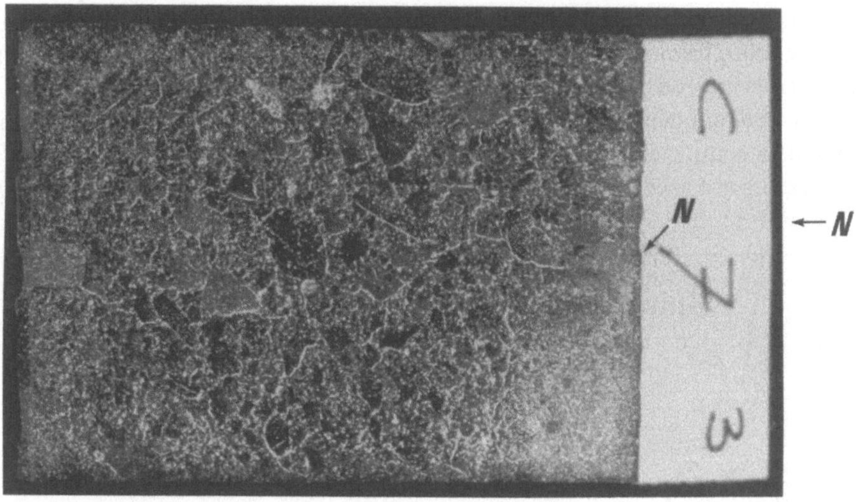

Figure 7: Central section of two-sided notched fine-grained concrete specimen previously subjected to direct tension. Specimen has been yielding leaving a load-carrying capacity of about 75% of ultimate (N=notch).

less pronounced increase from discontinuity to ultimate than in the case of direct compression. Moreover, two-sided notched sections of prismatic fine-grained concrete specimens, previously subjected to direct tensile stresses, which had declined to about 75% of ultimate during yielding, did not reveal an obviously delineated macrocrack (Fig. 7). Crack concentration increased only slightly toward the notched section. A fracture process zone with clear boundaries was proven a fiction: this concept is a convenient but structurally unrealistic mathematical model.

In a regular pattern of cracks, (average) crack spacing, s, would be twice as large as the average nearest neighbour spacing in a random point system (on which the present arguments are based). Hence, s could amount to twice the average crack size, $\overline{x}$ on a high level of the microstructure (say, the mesolevel). Full coalescence with the nearest neighbour would in this case lead to the following estimates for damage evolution:
in the starting situation before two bond cracks will join:

$$S_A = \tfrac{3}{2}.\tfrac{\pi}{4}(\overline{x})^2 N_A = \tfrac{3}{8}\pi(\overline{x})^2 N_A$$

and after joining:

$$S_A = (\tfrac{7\pi}{4}(\overline{x})^2 + 2.\tfrac{3}{2}\tfrac{\pi}{4}(\overline{x})^2)\tfrac{1}{2}N_A = \tfrac{5}{4}\pi(\overline{x})^2 N_A.$$

Spatial damage around the dividing plane would thus increase by a factor 10/3.

The nearest neighbour distance distribution of the aggregate particles is significantly skewed to the left, as demonstrated by Fig. 8. Small distances occur in reality more frequently than estimated by the theory based on a 'random' particulate structure consisting of finite size particles (a similar result as produced by a so called 'random generator'). Clustering is a natural phenomenon which is underestimated in computer generated systems of particulate matter in which such random generators are used. This implies a large proportion of the particles in the aggregate

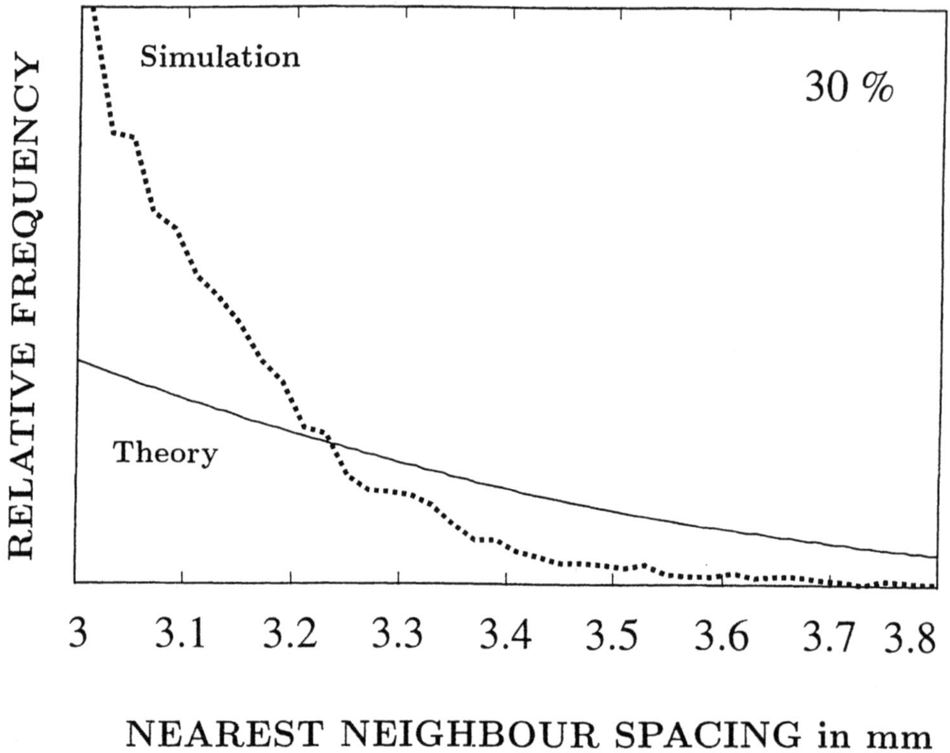

Figure 8: Distribution function of the 3-D nearest neighbour spacing.

to be spaced very close to the nearest neighbour. The effect on crack coalescence is that instead of bilateral crack coalescence, clusters of cracks will join, as demonstrated by the computer simulated crack pattern in the dividing surface of Fig. 9. To model this process, the spatial dispersion of the particles should be combined with the actual size of the particles to yield realistic information on individual tip-to-tip spacing of cracks. This is impossible to accomplish in a global approach. The computer simulation system for material structure, which will be introduced in this paper, will be used for this very purpose.

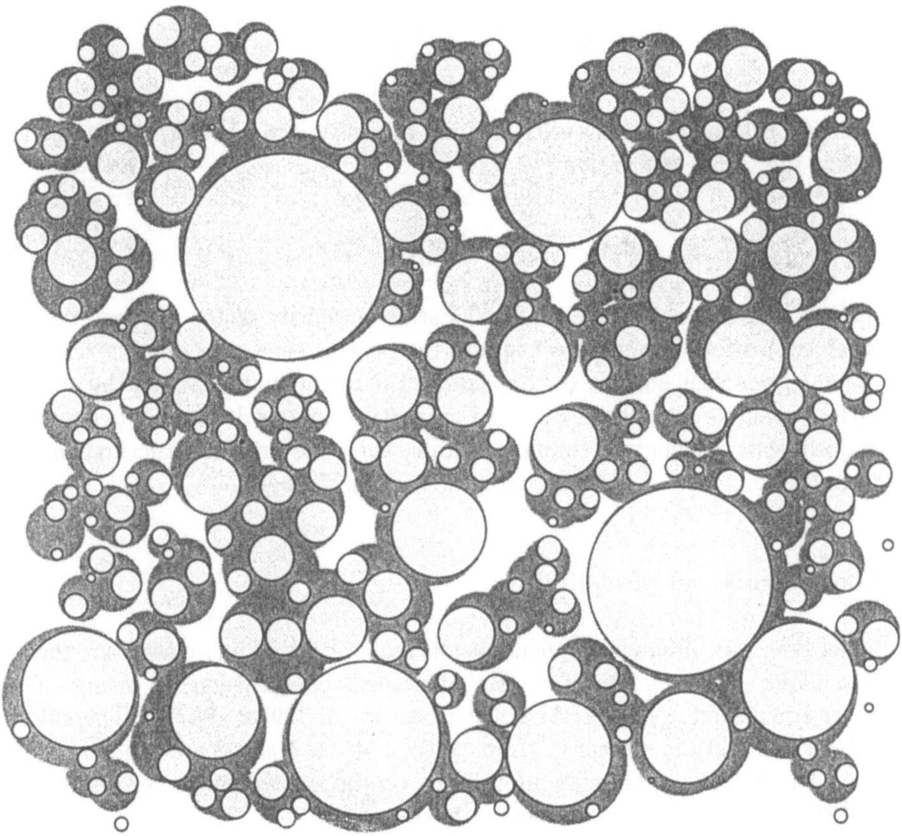

Figure 9: Distribution of cracks in the dividing surface (in light grey), and the crack development due to coalescence of nearest neighbours (in dark grey).

5. Global Structural Model

A structural model for the 'crack initiation strength', or cracking strength could incorporate this crack interaction phenomenon in a global way. The computer-simulation approach will be used in the near future to elaborate more local features. The model focuses on the highest microstructural level, so the leading parameters are $\overline{x}$ for size and s for spacing. Koiter's solution (Koiter, 1959) for the stress intensity factor of an array of collinear cracks was transformed by Nied & Arin (1978) into a multiple flaw fracture mechanics model for a planar array of interacting circular cracks, which is highly relevant for the present case. The critical stress, σ_{cr} , is given by

$$\sigma_{cr} = \frac{\pi}{2} \frac{K_{cr}}{\sqrt{s \tan \frac{\pi}{2} \frac{\overline{x}}{s}}} \tag{12}$$

in which K_{cr} is a the material's critical stress intensity factor. Since, $s \simeq 2\overline{x}$, eq (12) reduces to the simple expression

$$\sigma_{cr} = C(\frac{V_V}{d_0^2})^{\frac{1}{4}} \tag{13}$$

where C is proportional to the material's stress intensity factor. In fact this is a 'classical' expression, because crack size in the present concept is directly related to (minimum) grain size, so that $\sqrt{\overline{x}}$ appears in the denominator. When the cracking strength of concrete can indeed be based on particle-matrix debonding as the weakest chain link in the micromechanical system, then the present structural formula indicates the influence of particle characteristics on strength.

6. Local Structural Modelling

To model the crack interaction phenomenon realistically in and closely around the dividing plane, we intend to study the structural phenomenon by means of the computer simulation system SPACE (Stroeven *et al*, 1997a; 1997b). Therefore, a short description of the system is given. Also, Figs. 1, 4 and 8 are produced along this way. Further, Fig. 7 presents information on the nearest neighbour distribution of bond cracks which is also obtained with this system.

To describe the non-homogeneous granular nature of the internal structure, the model assumes the material to be composed of a finite number of discrete elements. Each element represents by its shape and size a characteristic phase in the material. Aggregate can be interpreted in this connection as pores or sand and gravel grains, or - on a lower level of the microstructure - as binder particles (cement, fly ash, silica fume). These elements are dispersed in a presumably homogeneous and uniform matrix and are contained in a volume that corresponds to the shape of the specimen. For example, concrete is on a mesolevel considered to be composed of aggregate elements distributed within a cuboidal or cylindrical container. The associated particle size distribution functions can easily be derived from sieve curves. The model simulates the material by describing the shape of each element and its position and orientation within the container

The computer-simulation model has to reproduce particulate structures realistically, and avoid the problem of 'particle' overlap. A commonly used method to distribute particles randomly in a limited volume is to place particles one after the other (sequentially) at random locations, taking care not to overlap other particles or the boundary (Zaitsev *et al*, 1981). Especially at higher densities, several trials may be necessary for each particle to find a free (non-overlapping) space. Indeed the number of unsuccessful attempts increases dramatically with fractional densities exceeding 0.3 (Pelikán *et al*, 1994). Hence, near-neighbours will be more remote than in reality. Moreover, it has been established by Widom that the numerical procedure of sequentially adding spheres at random non-overlapping locations in 3-D space significantly influences the limiting density, i.e. fractional density of mono-sized spheres will be in the range 0.35 to 0.38 (Widom, 1966). In sequential rejection processes, the spatial positions of successfully placed particles are not influenced by the particles added later. Due to this lack of mutual influence a special order of selection is required, especially when the particles in the distribution vary considerably in size or shape. A process in which all particles are affected equally by their environment in finding an appropriate location seems therefore far more realistic.

A single aggregate element will undergo influences from all sources in finding its destination (position, orientation) inside the container during the production/generation process. It is the internal influence of 'particle' interference (i.e. chemical, physical or mechanical interaction), and the external influences such as the container walls, gravity forces, and 'particle'-matrix friction that determine the final distribution of the particles.

To overcome the fundamental shortcomings of the random generator process, we simulate the actual production characteristics of cementitious materials. At the first production stage, design information like the shape and size distribution of each ingredient, and the volumetric density, is used to obtain an actual set of discrete particles. In the next stage the particles and the matrix are mixed, then moulded into the container, and finally agitated. In this stage most of the influences that control the particle distribution are active. Particle overlap is avoided because two particles that collide during the mixing and moulding process will not be able to penetrate each other.

In summary, the following steps are provided by the algorithms:

1. initially, a structured or random 3-D dilute distribution of elements with predefined shape and size distribution is generated within the boundaries of a cuboidal container. Random velocity and rotation vectors are assigned to each element;

2. the second stage is an iterative procedure where the displacement (and the rotation) of each element is given by a Newtonian model for particle motion. The model relates the changes in displacement and velocity of an element to a set of point forces acting on it. By temporarily freezing all other elements, the trajectory of each element is checked for contacts with the boundary of the cube or with

other elements. The object is displaced for one time step or up to its first contact. Velocities of the objects immediately after contact are calculated by the contact model and are assigned to the objects, whereupon the next contact is searched within the remaining time. The number of contact searches within a single time step is limited to avoid excessive calculations;

3. the iteration stops when certain convergence conditions are reached.

As illustrated by Fig. 8, application of the SPACE system will allow us to perform a structural investigation of damage evolution in the fracture process zone, provided a micromechanical concept is adopted. We can assess the significance of the micromechanics by comparing outcomes with quantitative image analysis data on damage evolution.

7. Conclusions

Stereological modelling of cracking in bulk and in the fracture plane is achieved for concretes containing river aggregate. The approach will also offer a good approximation to describe damage characteristics in concretes in which crushed aggregate is used, provided particle-matrix debonding is the weakestlink in the micromechanical chain. Three-dimensional measures for damage, such as crack density or specific crack surface area, have been related to characteristics of the load-bearing particulate skeleton. The same holds for damage evolution inside the fracture process zone, leading ultimately to the formation of the fracture surface. Roughness of the fracture surface could also be linked up with characteristics of the particulate phase. This allow us to demonstrate the resolution-dependence of damage in experimental observations and in modelling approaches. The fractal concept was used for this purpose. It could be demonstrated that ultimate stress in tension or compression involves only a relatively small amount of crack coalescence. The major part of crack coalescence necessary for 'controlled' development of the fracture surface will occur during the yielding stage. By using a global fracture mechanical model developed for ceramic materials, we could relate the cracking strength of cementitous materials to characteristics of the particulate phase. The development of a local structural approach to 'strength' by way of the realistic 3-D structural computer simulation program SPACE is foreseen for the near future.

8. References

Dalhuisen, D.H., Stroeven, P. and Moczko, A.T. (1997) Global characterization of damage evolution in cementitious composites in direct compression, *Proc. Brittle Matrix Composites 5*, Warsaw Oct. 13-15 1997 (to be published)

El-Saouma, V.E., Barton, C.C. (1990) Gamaleldin, N.A., Fractal characterization of fracture surfaces in concrete, *Engr. Frac. Mech.*, 35-1, 47-53.

Fidjestøl, P. and Frearson, J. (1994) High performance concrete using blended and triple blended cements, *High Performance Concrete* (Eds. V.M. Malhotra), ACI, Detroit, 135-157.

Hsu, T.C.(1963) Mathematical analysis of shrinkage stresses in a model of hardened concrete, *J. Amer. Concr. Inst.*, 60, 3, 371-390.

Jiang, W.,Roy, D.M. (1994) Strengthening mechanisms of high-performance concrete, *High Performance Concrete, Proc ACI Int Conf Singapore* (Ed V.M. Malhotra), ACI, Detroit, 753-767.

Kendall, M.G. and Moran, P.A.P. (1963) Geometrical Probability, Griffin % Co., London.

Koiter, W.T. (1959) An infinite row of collinear cracks in an infinite elastic sheet, *Ingenieur Arch*, 28, 168-172.

Lyubimova, T.Y., Pinus, E.R. (1962) Crystallization structure in the contact zone between aggregate and cement in concrete, *Colloid J. USSR*, 24, 5, 491-498.

Nied, H.A., Arin, K. (1978) Multiple flaw fracture mechanics model for ceramics, *Fracture Mechanics of Ceramics, Flaws and Testing 3* (Eds. R.C. Bradt, D.P.H. Hasselman, F.F. Lange), Plenum Press, New York, 67-83.

Pelikán, K., Saxl, I. and Ponížil, P. (1994) Germ-Grain model of short fibre composites, *Acta Stereol*, 14/1, 75–82.

Rawicki, Z., Wojnar, L. (1992) On quantitative fractographic analysis of high density concretes, *Acta Stereol.*, 11/2, 185-189.

Stroeven, P. (1973) Some aspects of the micromechanics of concrete, PhD Thesis, Delft University of Technology, Delft.

Stroeven, P. (1982) Structural modelling of plain and fibre reinforced concrete, *J. Comp.*, April, 129-139.

Stroeven, P. (1986) Stereological approaches to cementitious composites, *Science on Form* (Eds. S. Ishizaka, Y. Kato, R. Takaki, J. Toriwaki), KTK Scientific Publ, Tokyo, 175-182.

Stroeven, P. (1990) Some observations on microcracking in concrete subjected to various loading regimes, *Engr. Fract. Mech.*, 35, 4/5, 775-782.

Stroeven, P. (1991) Fractals and fractography in concrete technology, *Brittle Matrix Composites 3* (Eds. A.M. Brandt, I.H. Marshall), Elsevier Appl. Sc., London, 1-10.

Stroeven, P. (1992) Structural loosening approach to the concept of the fracture process zone in cementitious materials, *Phase Interaction in Composite Materials* (eds. S.A. Paipetis, G.C. Papanicolaou), Omega Sc., Wallingford, UK, 287-296.

Stroeven, P. (1996) Stereological estimation of fractal number of fracture planes in concrete, *Disordered Materials and Interfaces* (Eds. H.Z. Cummins, D.J. Durian, D.L. Johnson, H.E. Stanley), MRS 407, Pittsburgh, 343-348.

Stroeven, M., Stroeven, P. (1997a) Computer-simulation approach to the ITZ; effect of fine particle additions, *Brittle Matrix Composites 5* (Eds. A.M. Brandt, V.C. Li, I.H. Marshall), Woodheead Publ. Co., Cambridge/Warsaw, 310-319.

Stroeven, M., Stroeven, P. (1997b) Simulation of hydration and the formation of microstructure, *Computational Plasticity* (Eds. D.R.J. Owen, E. Oñate, E. Hinton), CIMNE, Barcelona, pp. 981-987.

Stroeven, P., Stroeven, M. (1997c) Aspects of a computer-simulation approach to the design of high strength concrete, *Proc. Engr. Found. Conf. on High Strength Concrete*, Kona, Hawai'i, July 13-18 (to be published)

Widom, B. (1966) Random sequential addition of hard spheres to a volume, *J. of Chem. Phys.*, 44, 3888–3894.

Zaitsev, J.B., Wittmann, F.H. (1981) Simulation of crack propagation and failure of concrete, *Mat. and Struct.*, 14, 357-365.

RATE SENSITIVE DAMAGE BEHAVIOR OF MORTAR IN COMPRESSION *

LI ZHAOXIA HUANG YAOPING
College of Civil Engineering, Southeast University
Nanjing, 210018, P.R. China

ABSTRACT This study investigates the knowledge on rate sensitive behavior in cracking process or damage behavior for brittle materials by testing concrete constituent, mortar at different rates of straining. The material response is compared based on tests of different strain rates. Cracking process of specimens are observed and recorded synchronously with loading and deformation response for the mortar at different rates of straining. For the mortar in a range of strain rates from 0.2 - 20,000,000 microstrains/sec, the compressive strength increases and its nonlinearity decreases with increasing strain rate. As the strain rate increases, the failure shows comparatively more violence and faster crack propagation, with an increasing number of cracks and fragments. The strain rate sensitive damage behavior indicates the reason why compressive strength, nonlinearity, viscoplasticity and other mechanical properties of mortar and concrete are sensitive to the rate of straining.

1. Introduction

There have been a lot of experimental study on the phenomenon of rate sensitivity in material strength for several decades, in which the strain rate sensitive behavior of concrete and its constituents has been measured in terms of the strength, modulus of elasticity, Poisson's ratio and nonlinearity of stress-strain curves in compression or

---

* Project 19472021 supported by National Nature Science Foundation of China

R. Wang (ed.), IUTAM Symposium on Rheology of Bodies with Defects, 223–234.

tension[1,2].

Now, it is more important to explain why the materials show their strain rate sensitivity and formulate a constitutive theory for evaluating the quasi-static and dynamic behavior of brittle materials over a range of strain rates. Since cracking appears to be the major process responsible for the nonlinear deformation, strength and other material properties in concrete and its constituents, the study of cracking process would be helpful for explaining the strain rate sensitivity of the material. The effect of strain rate on damage or cracking process in the material is to be studied here.

There have been many studies in which cracking or submicrocracking were observed in concrete, cement paste and mortar in compression[3] . However, much less attention has been paid to studying the strain rate sensitive behavior of cracking process and failure mode. In the work by Suaris and Shah[4], it was concluded that cracking appears to be the major process responsible for the observed strain rate effect in concrete and fiber reinforced concrete tested in their study. However, there was not sufficient work to show the difference of cracking and failure for the material tested at different rates of straining.

The purpose of this study is to improve the knowledge on strain rate sensitive behavior in compressive strength, cracking process and failure mode for brittle materials by testing concrete constituent, mortar, at different rates of straining. The material response is compared based on different strain rates. The cracks on the surface of specimens are observed with two Long Distance Microscopes (LDM) and recorded synchronously with loading and deformation response for the mortar at different rates of straining.

2. Experimental procedure

2.1. MATERIALS AND SPECIMENS

The experiments in this study are made on specimens prepared from mortar mixed by 425 (for Mortar R) or 525 (for Mortar F) cement and the standard sand for plastic mortar strength test. The grain size of the standard sand is 0.4 mm. Two water-cement ratios (w/c) of 0.4, 0.3 were used for Mortar R and Mortar F respectively, and the two types of mortar had the same sand-cement ratio (s/c) 2.0. For Mortar R, a water reducer at a dosage rate of 0.6 percent by mass of the cement was also used.

Cylinder test specimens, $\phi 2.5 \times 5.0$ cm, were prepared using steel moulds. The mortars were mixed according to the Standard for Mechanical Mixing of Hydraulic Cement Paste and Mortar of Plastic Consistency. The specimens were cast in a vertical position. During the first 24hr, the specimens remained in the moulds and were then removed from the moulds and stored in lime-saturated water at a controlled room temperature of 20 ℃ for 28 days.

2.2. EXPERIMENTAL SETUP

2.2.1. Loading

Specimens were loaded in uniaxial compression by using a rigid testing system with an automatically controlled loading setup. The rigid testing system consists of the normal loading setup and a three-point bending beam with great rigidity and the elastic relation of deformation and loading. By controlling the rate of loading f, it allows the specimens to be loaded at controlled rates of straining (see Fig. 1), and by which strain softening response can be monitored successfully.

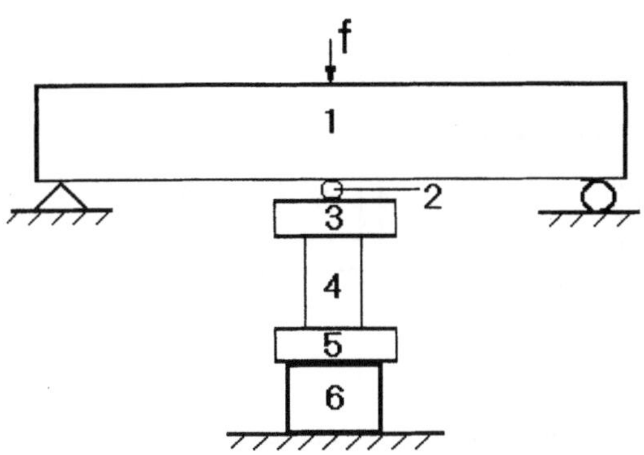

1. Beam, 2. Ball, 3. Gasket, 4. Specimen, 5. Gasket, 6. Load gauge

Figure 1. A schematic diagram of loading setup

The two displacement gauges provided the average longitudinal strain for the entire height of the specimen. The tests were conducted at nine different (average) strain rates ranging from 0.2 microstrains/sec (2.0×10^{-7} / sec) to over 20,000,000 microstrains/sec (20 / sec). The strain rate was calculated from the linear portion of the strain- time curves in the tests. The load -displacement response and deformation- time curves during the test were recorded by a bridge amplifier meter and a personal computer. The stress-strain curves and strain-time curves were then obtained from the load-displacement response and stored in the computer.

2.2.2. Crack Observation and Recording

Surface cracks on the test specimens were observed by two Long Distance Microscopes (LDM) placed opposite to each other as shown in Fig. 2. The resolving power of LDM is 2 μm, and the magnification is 77 times at a distance of 65 cm from the testing specimen. Two CCD cameras of horizontal 600 TV lines were connected to the two LDMs, respectively. During testing, crack patterns on the surface of specimens could be seen in a monitor and recorded by a video tape recorder. At the same time, the deformation and load in the specimen were recorded by a bridge amplifier meter and a PC computer. In this way, the cracking process observed on the surface of specimens were recorded synchronously with the load-deformation response.

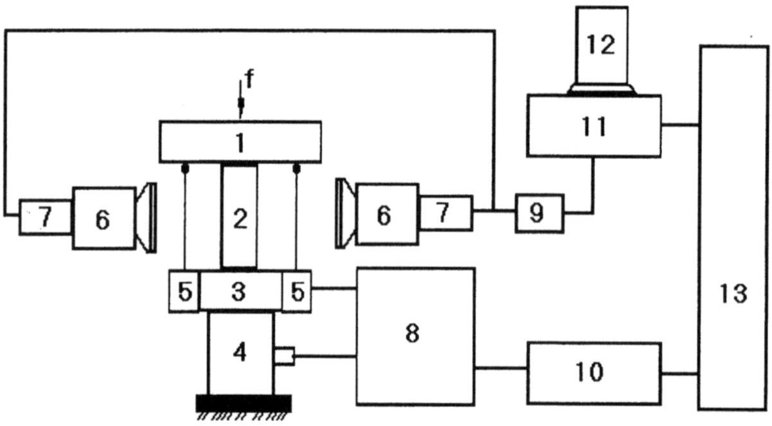

1. Gasket 5. Displacement gauge 9. Channel selector
2. Specimen 6. Long Distance Microscope 10. A/D converter
3. Gasket 7. CCD Camera 11. Video cassette recorder
4. Load gauge 8. Multi-channel bridge amplifier 12. Monitor 13. Computer

Figure 2. A layout diagram of the testing system

3. Results and Discussion

3.1. RATE SENSITIVITY ON COMPRESSIVE STRENGTH, NONLINEARITY OF STRESS-STRAIN CURVES

The average values of strength (peak stress) for Mortar R and F tested respectively by a set of specimens are shown in Fig. 3, in which the curves are plotted against the strain rates on a semi log scale. Fig. 4 gives stress-strain curves for the specimens loaded at varying rates of straining in which each curve is given for a typical specimen of Mortar F loaded at one of the strain rates ranging from 0.2 microstrains/sec (2.0×10^{-7} / sec) to 20,000,000 microstrains/sec (20 / sec). It can be seen that, over the range of strain rates considered in this study, the strength of mortar increases with an increasing strain rate, and the nonlinearity of stress-strain curves in the hardening and softening regimes decreases with the increase of strain rate. At a high strain rate, the stress-strain curve is seen as almost linear. However, the response of deformation for the specimen at a low strain rate shows nonlinearity clearly.

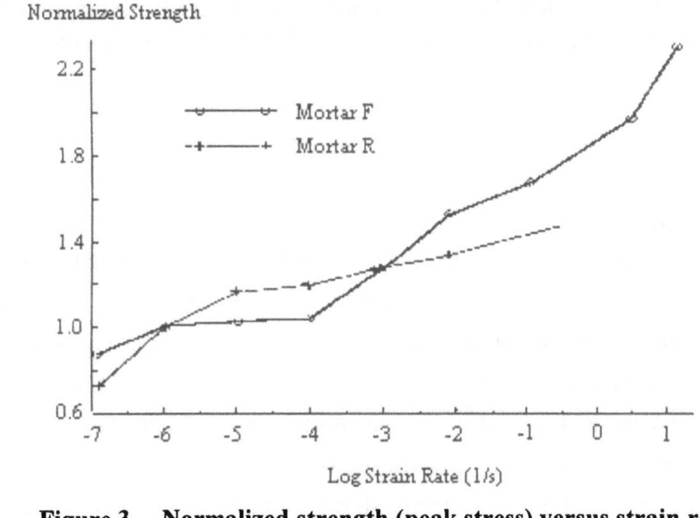

Figure 3. Normalized strength (peak stress) versus strain rates for two types of mortar

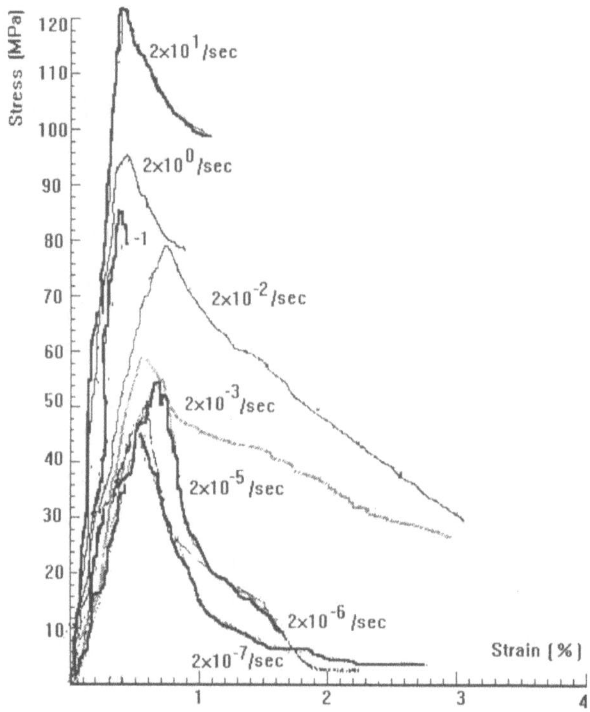

Figure 4 Stress-strain curves for the specimens of Mortar F loaded at varying rates of straining ranging from 0.2 to 20,000,000 microstrains/sec

3.2. RATE SENSITIVITY ON DAMAGE

3.2.1. Failure mode

It is observed from the test that, as the strain rate increased, specimens failed with an increasing number of cracks and fragments. At a low strain rate, the specimens may fail without sudden rupture and the failure occurs when multiple cracks become connected and run through the specimen. At a high strain rate, specimens failed more abruptly, with several fragments and a louder cracking noise. This behavior suggests that the failure mode of mortar specimens is sensitive to the strain rate. As the strain rate increased, failure mode of the specimens shows comparatively more violence and rapidity in time of crack propagation.

3.2.2. Cracking process

In order to identify the mechanisms responsible for failure and cracking of specimens at different strain rates, cracking process synchronous with the load-deformation response was studied, in which images recorded by the video tape recorder are transformed to digital images by the use of an image processing system composed of a computer and image processor. The images at a low and a high strain rate are shown in Figs. 5 and Figs. 6. Fig. 5(a) shows a stress-strain curve with special points marked 1,2,....,8 for Mortar F loaded at a strain rate of 2.0 microstrains/sec. The images of cracking at the marked points of Fig. 5(a) are shown in Fig. 5(b) which display images on developed surface of the specimen observed by two LDMs placed opposite to each other. It can be seen that, on the specimen loaded at a low strain rate surface cracks were generated at a low stress state prior to the peak stress, and propagated with increasing strain until failure occurred. Fig. 6(a) gives the stress-strain curve for Mortar F loaded at a higher strain rate of 2000 microstrains/sec. Each graph in Fig. 6(b) shows the cracking state (on one face only) at the marked points in Fig. 6(a) . It can be seen that, at a higher strain rate, the surface cracks were generated near the peak point of the stress-strain curve and the cracking process occurred mainly within the softening regime.

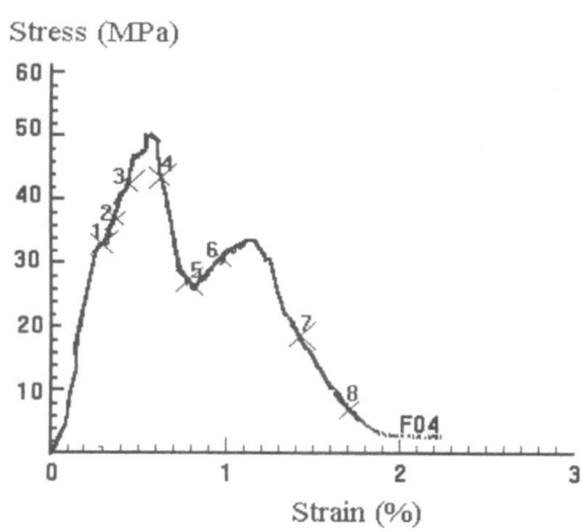

Figure 5(a) A stress-strain curve with special points marked for Mortar F loaded at a strain rate of 2.0 microstrains/sec

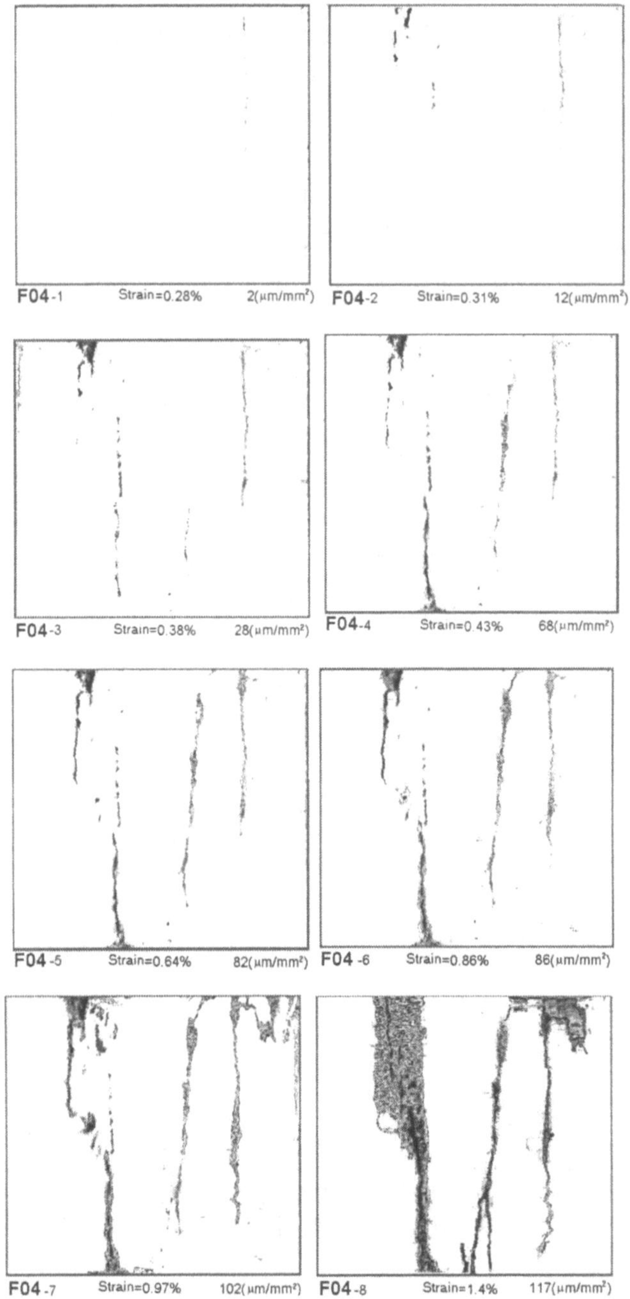

Figure 5(b) The cracking states at marked points for Mortar F loaded at a strain rate of 2.0 microstrains/sec

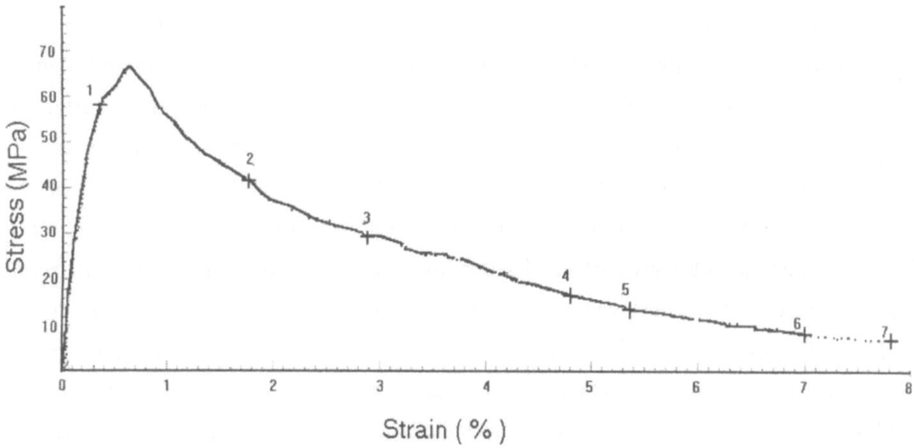

Figure 6(a) A stress-strain curve with special points marked for Mortar F at a strain rate of 2000 microstrains/sec

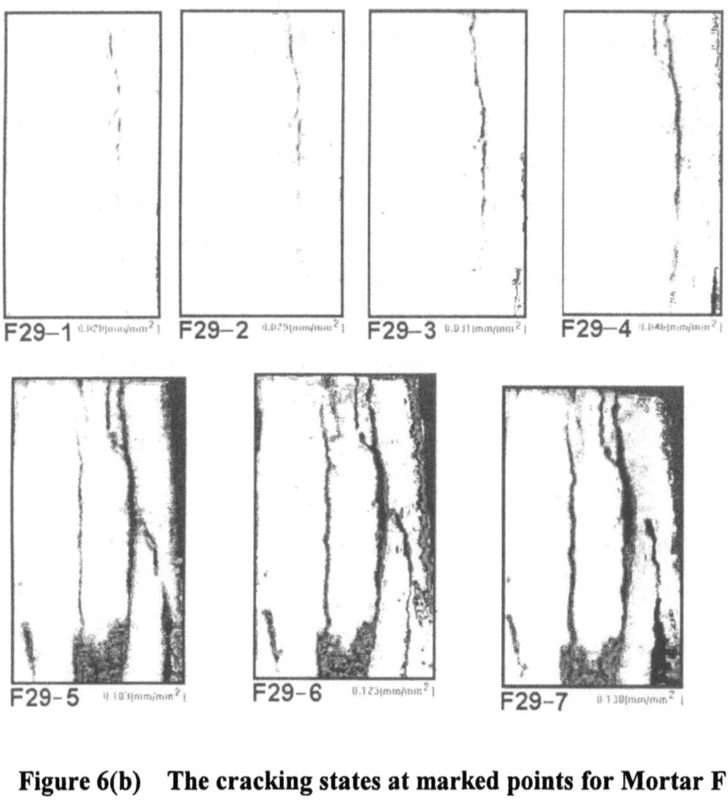

Figure 6(b) The cracking states at marked points for Mortar F at a strain rate of 2000 microstrains/sec

The digital images with the cracking information are analyzed statistically. In order to get quantitative results of cracking, values of total area (that is, length multiplied by the width of cracks on the surface of the specimen) of cracks per area on the surface of the specimen are calculated statistically by checking first the trunk of cracks and its branches and computing then the total area of the trunk and branches[5]. For the specimens loaded at varying rates of straining (the stress-strain curves are shown in Fig. 4), the statistical total area of cracks per area on the surface of the specimen was obtained respectively. These values versus strain at different rates are showed in Fig 7.

It is evident that there exists a significant difference between the cracking process of mortar specimens loaded at low and high strain rates. At a low strain rate, surface cracking starts earlier and at a lower stress state, and the cracking process occurs over a longer duration. Crack propagation in specimens at a low strain rate takes place within both the hardening and softening regimes, while crack propagation in specimens at a high strain rate occurs mainly within the softening regime. These observation on the cracking process indicate why specimens loaded at low strain rates have lower strengths. The cracking process appears to be responsible for the strain rate effect on strength.

Fig. 8 shows the maximum values, before fracture, of cracked area densities for the specimens loading at different rate of strain. It is seen that, as the strain rate increased, specimens failed with a decreasing value of cracked area. Considering the above observation on the failure mode of specimens loaded at different strain rate, it is clear that, there exists different types of cracking process for the specimens loaded at lower and higher rate of strain. At a low strain rate, cracks generated early and past a longer duration of propagation; At a high strain rate, cracks have not enough time to propagate and the specimen fails due to too many yet tiny cracks, with very little of load supporting area.

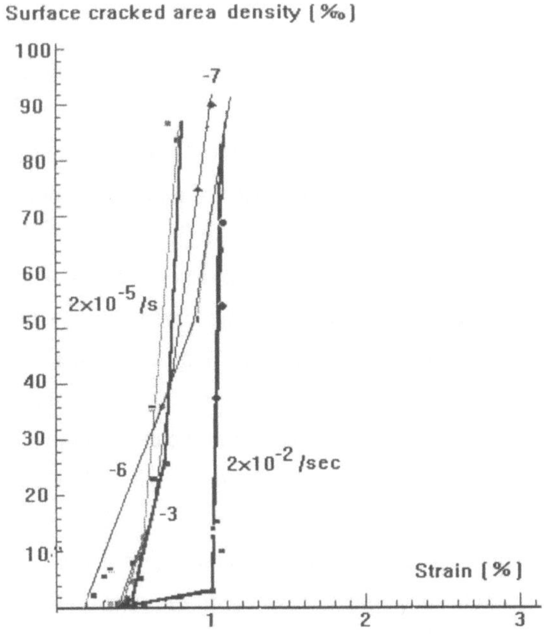

Figure 7 The statistical values of cracked area density versus strain at varying rates of straining ranging from 0.2 to 20,000,000 microstrains/sec

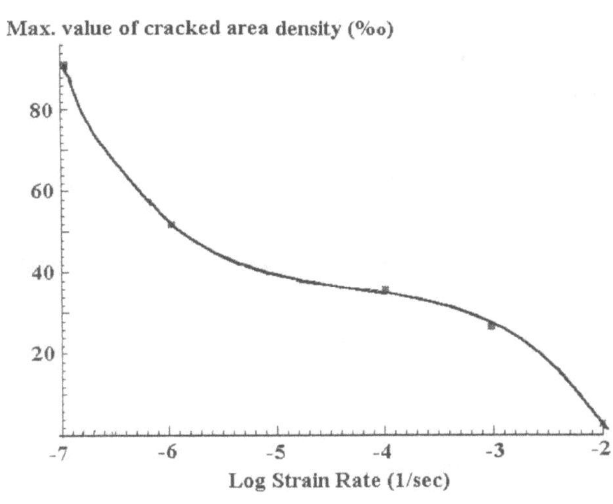

Figure 8 The maximum values of cracked area density for the specimens loading at different rate of strain

4. Conclusions

From the experimental results and the discussion presented above, the following conclusions can be drawn:

★ For the mortar in a range of strain rates from 0.2-20,000,000 microstrains/sec, the compressive strength increases and its nonlinearity of material response decreases with an increasing strain rate.

★ The failure mode and crack process of mortar specimens is sensitive to the strain rate. As the strain rate increased, the failure shows comparatively more violence and rapidity in cracking propagation, with an increasing number of cracks and fragments;

★ At a lower strain rate, cracking on the surface of specimens begins earlier and at a lower stress state than that at a higher strain rate, and the cracking process occurs over a longer duration. As the strain rate increases, specimens fail with a decreasing value of cracked area.

★ The strain rate effect on failure mode and cracking process indicates the reason why the compressive strength of mortar is sensitive to the strain rate. In order to further clarify the strain rate effect on damage and cracking, additional studies are needed in the micro- and meso-scale .

REFERENCES

1. Watstein, D. (1953) Effect of straining rate on the compressive strength and elastic properties of concrete, ACI Journal, Proceedings 77 (5), 729-744
2. Harsh, S., Shen, Z. and Darwin D. (1990) Strain-rate sensitive behavior of cement paste and Mortar in compression, ACI Materials Journal 87 (5), 508-516
3. Attiogbe, E. K. and Darwin, D. (1987) Submicrocracking in cement paste and mortar, ACI Materials Journal 84 (6) 491-500
4. Suaris, W. and Shah, S (1983) Properties of concrete subjected to impact , Journal of Struc. Eng., ASCE, 109 (7) 1727-1741
5. Wang Sunyu, Li Zhaoxia et al (1996) A method for detecting surface meso-crack by use of message from scanning images, Shanghai Mechanics 17, 149-152 (in Chinese)

COUPLED EFFECT OF CREEP AND STRESS RELAXATION OF SOFT CLAY

Junshi Li Jun Sun

East China Airport Design and Construction Corporation, 200335, Shanghai
Department of Geotechnical Engineering, Tongji University, 200092, Shanghai

Abstract

The rheological behavior of soil mass is usually not governed simply by a single process of creep and stress relaxation. Here, a Mesri stress-strain-time relation function (Mesri., G. etc. 1981) is applied to describe the coupled effect of creep and stress relaxation of soft clay. Numerical results are compared with those obtained from data acquired from coupled creep-stress relaxation tests of clay samples. The results of the study are satisfactory. Some relevant factors influencing the coupled effect are also discussed.

1. Introduction

Creep and stress relaxation represent two different aspects of the rheological characteristics of materials. Although it is recognized that creep process will induce stress relaxation, and stress relaxation will also induce creep, yet they are considered to be two separate and independent processes. Analyses in physical rheology(Liu Xiong 1994) indicated that creep and stress relaxation are two idealized mechanical concepts of long-term mechanical properties; in essence they are governed by the same physical mechanism. For example, the macroscopic difference between creep and stress relaxation lies in the following: in the process of creep there exist external energy sources to supply energy to the stressed system, in which the strength of its internal structure weakens, the structure relaxes, and in consequence deformation increases with time. The internal friction in the same stage of material aging will consume partially or even all of the creep deformation energy. While the process of stress relaxation there is no consumption of external energy; the decrease in internal stress is due to the weakening of material structure; the internal friction accompanying by the weakening process will consume energy at the cost of the initially accumulated energy in the material. In soil clay, the structure includes defects such as fissures, cavities and weak connecting zones. During creep, micro-fissures close, micro-cracks build up, and then macrocracks appear.

R. Wang (ed.), IUTAM Symposium on Rheology of Bodies with Defects, 235–240.

After the soil mass has been excavated, instantaneous deformation will be induced in the residue. This kind of deformation consists of both elastic and plastic components. Under different boundary conditions there will be stress relaxation in some parts of the soil mass where the deformation is restrained. At the same time a definite part of the loading will be transferred to the adjacent regions, and cause creep in the soil mass; the development in creep will in turn induce stress relaxation. In this context, the creep and stress relaxation are mutually affected and coupled(Liu Xiong 1994).

The unstable process of earth slope failure include the dual action of soil creep and stress relaxation. When the strength of soil is completely exhausted, a potential sliding surface occurs in the soil slope. On the potential sliding surface, a certain part of it may slide down due to creep while on another part of it there exists stress relaxation. It is a sophisticated problem to determine on which part there is stress relaxation and on which part there is creep. As a compromise we deal with creep and stress relaxation as an unified process(Zhu Changqi & Guo Jianyang 1990).

In this paper, a Mesri function for describing the stress-strain-time nonlinear relation of clay is adopted, we use coupled creep-stress relaxation tests(Xiong Junmin & Li Zuoqin 1993) to verify the feasibility of the stress-strain-time function on the one hand and to discuss the influences of some factors on the coupled effects on the other hand.

2. Coupled creep-stress relaxation tests of clay

Coupled creep-stress relaxation tests have been carried out to determine the long-term strength of clay. In this kind of tests the stresses and strains are functions of time. The advantages of the coupled tests consist in their simplicity of operation which allows us to save test time and to apply them widely in practical engineering.

The principle of the tests is as follows. Initial stress is imposed on a specimen by a dynamometer, and the deformation is kept constant. Then, the dynamometer is relaxed, causing a decrease in stress. Creep is induced in the specimen by the variation in stress, while the increment in deformation of the specimen will be equal to the deformation recovered in the dynamometer. Meanwhile stress relaxation takes place in the specimen due to the variation in deformation. Therefor the test process incorporates a coupled process of creep and stress relaxation.

Fig.1 shows one result of experiments for Dark yellow clay.

3. Mechanical model of the coupled tests

Let the total deformation of the "dynamometer-specimen" system be ΔL, in which the deformation of the dynamometer is ΔL_R and that of the specimen is ΔL_s, then we have

$$\Delta L = \Delta L_R + \Delta L_s = \text{constant} \tag{1}$$

Let the elastic modulus of the dynamometer be E_R, the cross-sectional area of the specimen A_s, the height H_s, the axial strain ε_s and the axial stress σ, then

$$\Delta L_R = A_s\sigma / E_R, \Delta L_s = \varepsilon_s H_s \tag{2}$$

Substituting (2) into (1), we have

$$\Delta L = A_s\sigma / E_R + \varepsilon_s H_s = \text{constant} \tag{3}$$

Differentiating the above equation with respect to time gives

$$A_s \dot{\sigma} / E_R + \dot{\varepsilon}_s H_s = 0, or A_s \dot{D} / E_R + \dot{\varepsilon}_s H_s = 0 \tag{4}$$

where $D = \sigma_1 - \sigma_3$, is the difference in principal stresses.

4. Constitutive equation with respect to stress-strain-time function in a coupled problem

Mesri function(5) is a general expression to describe the stress-strain-time relation of clay which includes the time effect of soil mass in the whole extent of stress levels.

$$\varepsilon = \frac{2}{(E_u / S_u)_1} \frac{\overline{D_1}}{1-(R_f)_1 \overline{D_1}} (\frac{t}{t_1})^\lambda \tag{5}$$

where $\overline{D_1}$ is stress level, $(E_u / S_u)_1$ and $(R_f)_1$ are respectively the ratio of shear strength and the ratio of failure at time t_1, λ is a parameter accounting for the time effect.

Differentiating (5) with respect to time we obtain

$$\dot{\varepsilon} = \frac{2}{(E_u / S_u)_1} \frac{\overline{D_1}}{1-(R_f)_1 \overline{D_1}} (\frac{t}{t_1})^\lambda \left[\frac{\lambda}{t} + \frac{\dot{\overline{D_1}}}{\overline{D_1}(1-(R_f)_1 \overline{D_1})} \right] \tag{6}$$

The above equation can describe the time dependent response of soil mass in the arbitrary extent of differences in principal stress levels.

Substituting expression (4) into expression (6), we obtain the equation of coupled effect of creep and stress relaxation:

$$-\frac{A_s}{E_R H_s} \dot{D} = \frac{2}{(E_u / S_u)_1} \frac{\overline{D_1}}{1-(R_f)_1 \overline{D_1}} \left[\frac{\lambda}{t} + \frac{\dot{\overline{D_1}}}{\overline{D_1}(1-(R_f)_1 \overline{D_1})} \right] (\frac{t}{t_1})^\lambda \tag{7}$$

Substituting $\overline{D_1} = \dfrac{D}{2S_u}$ into the above equation, we obtain

$$-\frac{A_s}{E_R H_s}\dot{D}=\frac{2}{(E_u/S_u)_1}\frac{D}{2S_u-(R_f)_1 D}\left[\frac{\lambda}{t}+\frac{2S_u\dot{D}}{D(2S_u-(R_f)_1 D)}\right](\frac{t}{t_1})^{\lambda} \quad (8)$$

This is the constitutive equation for the coupled effect of creep and stress relaxation, in which D is the difference in principal stresses $(\sigma_1-\sigma_3)$, $2S_u$ is the triaxial shear strength of soil mass $(\sigma_1-\sigma_3)_f$.

5. Numerical solution of the constitutive equation in the coupled problem

An improved Euler algorithm applied to the numerical difference is employed to solve the initial value problem of the coupled constitutive equation (8). The initial value problem of equation (8) can be written as

$$\dot{D}=f(t,D), D(t_0)=D_0 \quad (9)$$

where D_0 is the initial D at time t_0.

Let the time step length be Δt, time stations are taken as

$$t_n=t_0+n\Delta t, n=1,2,3,\ldots\ldots \quad (10)$$

and $D_n=D(t_n)$.

The improved Euler formulae for the initial value problem in expression (9) are

$$\begin{aligned} D_p &= D_n+\Delta t\cdot f(t_n,D_n) \\ D_c &= D_n+\Delta t\cdot f(t_{n+1},D_p) \\ D_{n+1} &= (D_p+D_c)/2 \end{aligned} \quad (11)$$

The following conclusions can be drawn from the computations: (1) Fig.1 indicate that, the coupled soil creep-relaxation process simulated by Mesri function is in good agreement with the results of test and suitable to describe the comprehensive and complicated rheologic behaviour of soil. (2) As shown by Fig.2, λ is an important factor affecting the coupled response, the larger it is, the smaller will be the stress at a given time station; and with the increase in coupled time duration, the extent of decrease tends to increase. (3) Fig.3 illustrates the effect of elastic coefficient of dynamometer E_R on the coupled responses when they vary from 10^2 to 10^8 N/m. As compared with other factors, the effect of E_R on stress change is not large; when E_R varies in the extent of $10^2 \sim 10^5$ N/m, the effect is greater. (4) In Fig.4, the effect of the heights of soil samples H_s on the response is shown when they change from 0.02m to 0.2m. It can be seen that, with the increase in H_s, the extent of decrease in stress increases at first and then gradually attenuates.

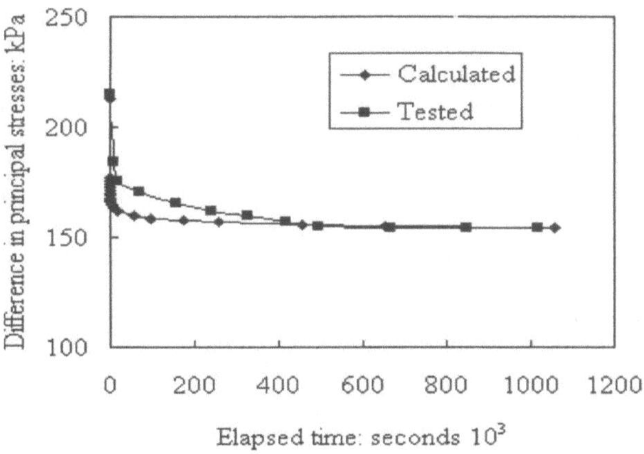

Fig.1 Stress-time curves of coupled tests for Dark yellow clay

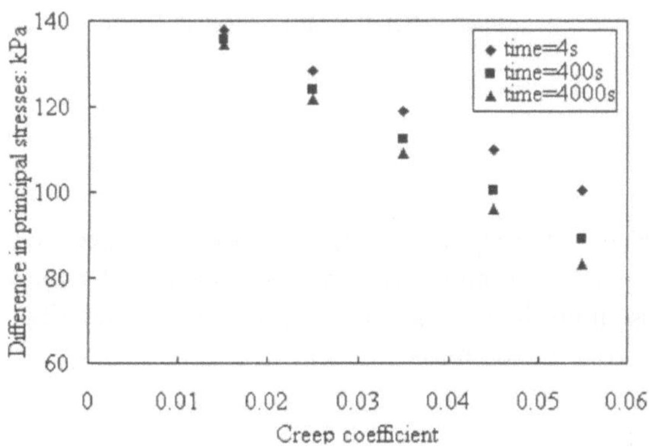

Fig.2 Effect of λ on the coupled response

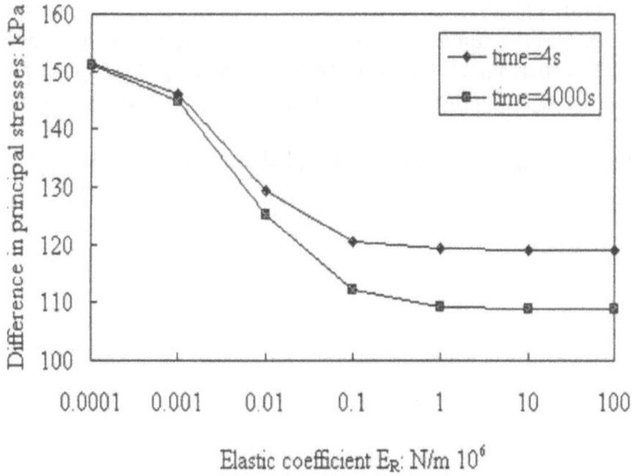

Fig.3 Effect of E_R on the coupled response

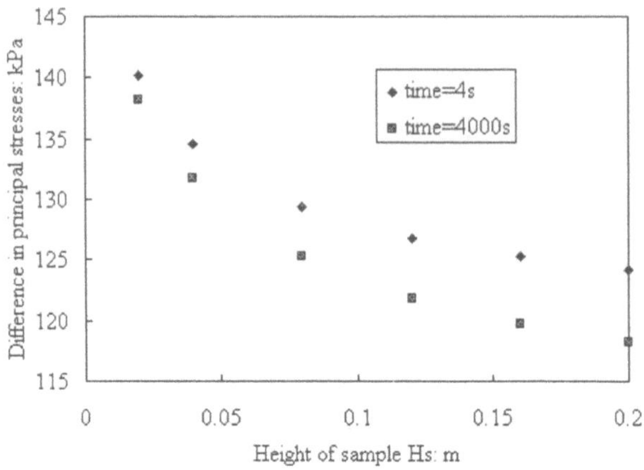

Fig.4 Effect of H_s on the coupled response

6. Conclusions

The coupled process of creep and stress relaxation simulated by the Mesri function for soil mass is in good agreement with the data obtained by the coupled tests. So the proposed stress-strain-time relation is suitable to describe the comprehensive rheological behaviour of soil mass.

References

[1] Mesri, G., Febres-Cordero, E., Shields, D. R., & Castro, A. (1981) Shear stress-strain-time behaviour of clays, *Geotechnique*, **31**(4):537-552.

[2] Liu Xiong(1994) *An Introduction to Rheology of Rocks*, Publishers of Geology, Beijing(In Chinese)

[3] Zhu Changqi and Guo Jianyang(1990) A further understanding of the rheological characteristics of clay and the new method of determining the long-term strength of clay, *Geomechanics*, **11**(2): 15-22 (In Chinese).

[4] Xiong Junmin and Li Zuoqin(1993) Tests on coupled creep-relaxation of clay, *Geomechanics*, **14**(4): 17-24 (In Chinese).

PREDICTION OF ABRUPT FAILURE OF CRACKED ROCKMASS

Ping CAO Changliang PAN Li LUO
Central South University of Technology
Changsha, Hunan, 410083, China

Abstract: *In practice, there are geo-engineering constructions which fail abruptly without additional excavation, load or geometry change, this kind of abrupt failure of rockmass can cause serious disaster. For cracked rockmass under a condition of quite low level of field stress, it is considered in the paper that a delayed abrupt failure may be induced by a long term of crack propagation with a subcritical crack growing process, this weakening process of rock causes the final crack fracture when the stress intensity factor at the tip of crack reaches the fracture toughness of the rock. Based on this principle a method to predict the abrupt failure of cracked rockmass is proposed.*

1. Introduction

In practice, there are geo-engineering constructions which fail abruptly without additional excavation, load or geometry change. Its relation to the excavation and geometry change appears to have a certain delay period of time after the excavation, so that it is characterized with abrupt collapse, and causes geological disaster. Abrupt failure in hard rockmass can take place even in a condition with quite low level of field stress in geo-engineering.

Rockmass is a typical damaged material, in which the macro-defects such as joints and micro-cracks usually play an important or key role in its deformation and failure. Fracture as a typical brittle failure can occur in the case of quite low level of field stress. It is known that a fracture failure in rock, with high rate of crack extension, can be determined if the stress intensity factor at the tip of crack equals to the fracture toughness. On the other hand, for rock, as most brittle materials, there exists a period of subcritical crack growth before fracture failure, within the process crack extends with slow rate. Subcritical crack growing of rock can be considered as an initial source of inducing fracture in rock, it indicates that a mechanism of abrupt failure in hard rock mass is just a final result of rock mass being weakened by subcritical crack growth.

2. Rate of Subcritical Crack Growth

The brittle failure of rock is mainly due to fracture along crack, if stress intensity factor at the tip of a crack is equal to fracture toughness of rock, a fracture failure, characterized with fast

Supported by the National Natural Science Foundation of China

R. Wang (ed.), IUTAM Symposium on Rheology of Bodies with Defects, 241–248.

crack growing, will occur. Besides the fracture with fast crack growth, in rock material there exists a period of subcritical crack growth, which is a process of crack propagation with slow rate. A lot of fracture testing results show that the subcritical crack growth in rock can carry on even in the case when load is kept unchanged.

The crack propagation during the process of subcritical crack growth increases the crack length gradually and indirectly increases the stress intensity factor of the crack, it leads to crack fracture at last when the length of the crack is long enough to make stress intensity factor equal to the fracture toughness of the rock. The determination of the time interval from the beginning of subcritical crack growth to the final failure is very important. It can be used to predict the period of time in which a geo-engineering construction will abruptly collapse after excavation.

It is known that there exists a threshold value of stress intensity factor K_{ISCC}, which is less than the fracture toughness of the rock, if the stress intensity factor of a crack is equal to or greater than the threshold value, a subcritical crack growth will take place. According to Charles, R. J., the subcritical crack growth is considered to be caused by the exchange of the ions at the tip of the crack and by the stress concentration at the crack tip, this is a typical interaction of mechanics and chemistry. Charles had set up an expression of subcritical crack growth rate which is a function of the stress intensity factor of a crack and the environmental temperature, etc.(Charles, R. J. 1958). In the case of constant environmental temperature during the period of subcritical crack growth, the expression will be simplified as (Atkinson, B. K. 1987):

$$V = AK_I^n \qquad (1)$$
$$\text{if} \quad K_{IC} > K_I \geq K_{ISCC}$$

Where V is the rate of the subcritical crack growth, K_I is the stress intensity factor of the crack, A and n are constant values which can be evaluated from subcritical crack growing tests.

3. Duration of Subcritical Crack Growth and Abrupt Failure Prediction

From the above analysis it is known that the abrupt failure in hard rock mass can be considered as a phenomenon of fracture failure after a period of crack propagation with low growing rate. The time span of subcritical crack growing process is important in predicting the abrupt failure. While an excavation in rock mass is made, it will induce changes of stress distribution around the excavated space. For such a problem, in the rock with cracks when the stress intensity factors at the tip of some cracks equal to or greater than K_{ISCC}, these cracks will undergo subcritical crack growth. The initial state of the crack growth is related to the stress condition just after the excavation, and the end state of the subcritical crack growth is relevant to the critical state of fracture if the stress intensity factor equals to the fracture toughness of the rock.

The duration of subcritical growing process can be defined by the time interval between initial subcritical crack growth and critical crack growth. This duration will help us to predict the time in which an abrupt failure will take place after excavation in hard rock mass. For example, in the following a model of infinite plane with a central crack is used to analyze the duration. Assuming the half-length of the central crack in the infinite plane is a_i (Fig. 1) at the initial state of subcritical propagation, the relative stress intensity factor at the tip of the crack is found to de:

$$K_{Ii} = \sigma\sqrt{\pi a_i} \quad (2)$$

Where σ is the field stress acted in the plane. Suppose the stress increases to the critical value σ_{IC} of fracture while the crack length remains unchanged, so the critical stress satisfies:

$$\sigma_{IC}\sqrt{\pi a_i} = K_{IC} \quad (3)$$

Where K_{IC} is the fracture toughness of the rock. On the other hand, if the stress σ remains

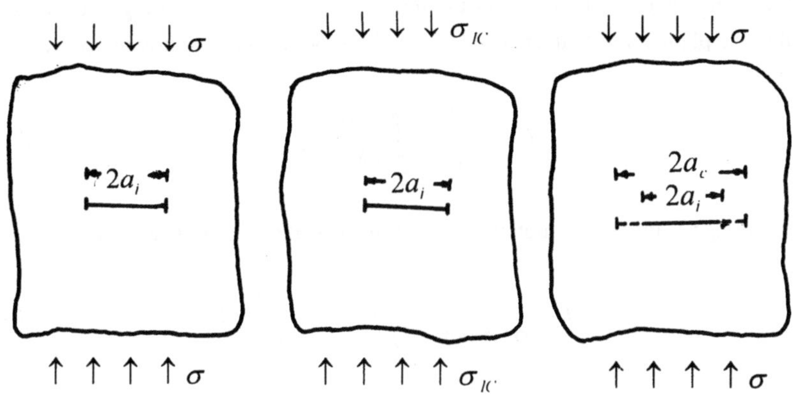

(a) $K_{Ii} = \sigma\sqrt{\pi a_i}$ (b) $\sigma_{IC}\sqrt{\pi a_i} = K_{IC}$ (c) $\sigma\sqrt{\pi a_c} = K_{IC}$

Fig. 1 Infinite plane with a central crack

constant but the crack length increases to the critical length $2a_c$ of fracture, there is the following critical formula:

$$\sigma\sqrt{\pi a_c} = K_{IC} \quad (4)$$

For an arbitrary case, assuming that the stress intensity factor K_{Ii} at the tip of a crack satisfies $K_{ISCC} \leq K_{Ii} < K_{IC}$, the crack, which is under the condition of subcritical crack propagation, will extend with a slow rate, say $2V$, so the growing rate can be defined with the half-length of the crack a and time t:

$$da = Vdt$$

It is evident that the duration of subcritical crack growing process can be calculated by integrating the above expression, that is:

$$t = \int_{a_i}^{a_c} \frac{da}{V} \tag{5}$$

For conventional plane problem with a central crack of length $2a$, there is $K_I = \sigma\sqrt{\pi a}$, so Eq.(5) can be rewritten as:

$$t = \int_{K_{Ii}}^{K_{IC}} \frac{2K_I dK_I}{\pi\sigma^2 V} - \int_{\sigma_{Ii}}^{\sigma_{IC}} \frac{2a d\sigma}{\sigma V} \tag{6}$$

Where K_{Ii} is given by (2). Substituting (1) into (6) and supposing that the field stress σ acted in the plane remains unchanged within the subcritical crack growing process, we have:

$$t = \frac{2}{A\pi\sigma^2} \int_{K_{Ii}}^{K_{IC}} \frac{dK_I}{(K_I)^{n-1}} \tag{7}$$

Integrating (7) gives the duration t of the subcritical crack growing process:

$$t = \frac{2}{\sigma^2 \pi A(n-2)} [\frac{1}{K_{Ii}^{n-2}} - \frac{1}{K_{IC}^{n-2}}] \tag{8}$$

Rewriting (8) as:

$$t = \frac{2}{\sigma^2 \pi A(n-2) K_{Ii}^{n-2}} [1 - (\frac{K_{Ii}}{K_{IC}})^{n-2}] \tag{9}$$

A lot of testing results show that the constant $(n-2)$ is usually quite a large positive value, so formula (9) can then be approximately expressed as the following by considering $(K_{Ii} / K_{IC}) < 1$:

$$t = \frac{2}{\sigma^2 \pi A(n-2) K_{Ii}^{n-2}} \tag{10}$$

In fact, K_{Ii} in formula (10) can not be easily determined because the initial crack length $2a_i$ is difficult to be given beforehand. Hereby a simplified method is to be used to calculate the time span t. From formulas (2) and (3) there exists a relationship between K_{Ii} and K_{IC}:

$$K_{Ii} = \sigma K_{IC} / \sigma_{IC} \tag{11}$$

The value of σ_{IC} can be obtained with conventional rock fracture testing. Substituting (11) into (10) gives the duration of subcritical crack growing process:

$$t = \frac{2\sigma_{IC}^{n-2}}{\pi A(n-2)\sigma^n K_{IC}^{n-2}} \tag{12}$$

From formula (12) the time span of the subcritical crack growing process can be determined by considering the parameters $A, n, \sigma, \sigma_{IC}$ and K_{IC}, among them A, n, σ_{IC} and K_{IC} are constants of rock properties, the only variation is field stress σ. If the stress intensity factor at the tip of the crack satisfies $K_{ISCC} \leq K_{Ii} < K_{IC}$, as well as the parameters of rock sample and the stress σ are known, the time of the abrupt failure after excavation in rock mass can then be predicted by evaluating the duration t.

In practice, the rockmass consists of macro-joints with random distribution (Fig.2), the interaction of joints will change the stress intensity factor of crack. The stress intensity factor of a central crack in Fig.2(b) can then be written as:

$$K_{Ii} = \eta\sigma\sqrt{\pi a_i} \tag{13}$$

Where η is the influence coefficient and $\eta > 1$, so there is:

$$\eta\sigma_{IC}^{*}\sqrt{\pi a_i} = K_{IC} \tag{14}$$

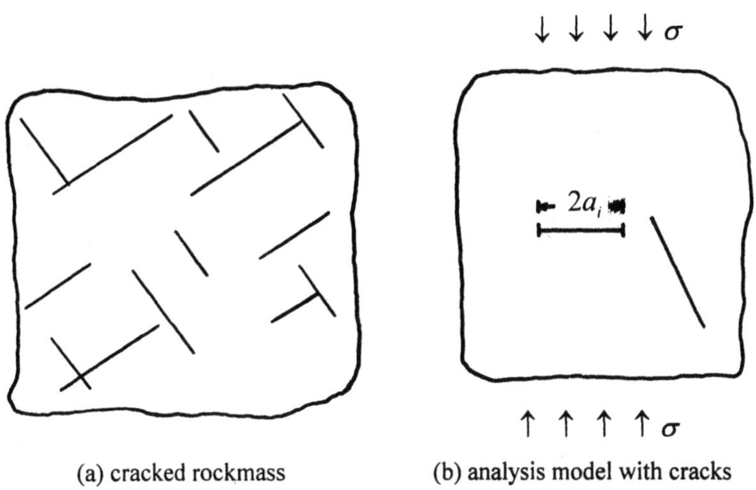

(a) cracked rockmass (b) analysis model with cracks

Fig. 2 Plane model of cracked rockmass

σ_{IC}^{*} here indicates the critical stress of fracture for rock or rockmass with random distribution defects. From formulas (13) and (14) there is:

$$K_{Ii} = \sigma K_{IC} / \sigma_{IC}^{*}$$

Now the relative duration expression will be:

$$t = \frac{2}{A\pi\eta^2\sigma^2}\int_{K_{Ii}}^{K_{IC}}\frac{dK_I}{(K_I)^{n-1}} \tag{15}$$

The relevant time duration in the case can similarly be derived by assuming that the coefficient η remains unchanged during the process of subcritical crack growth:

$$t = \frac{2(\sigma^*_{IC})^{n-2}}{\pi A(n-2)\sigma^n\eta^2 K_{IC}^{n-2}} \tag{16}$$

It is evident that the time duration of the subcritical crack growth in rockmass with joints is much less than that in plane with a relative central crack.

4. Case

For the rock specimens from Dongchuan Copper Mine of China, the above mentioned values of K_{IC}, A and n had been indirectly evaluated from the results of double torsion tests (Cao, P. 1993). The critical stress σ^*_{IC} of the rock samples is obtained with conventional tension tests. The relevant parameters obtained are shown in the following table.

Table 1. Parameters used in calculating the duration

$K_{IC}(MN/m^{3/2})$	$\sigma^*_{IC}(MPa)$	$A(m/s)$	K_{ISCC}/K_{IC}	n
1.63	3.98	5.90×10^{-15}	0.85	34.08

In the ultimate case $\eta = 1$, we can get the relative expression of time duration for the copper mine by substituting the parameters into formula (12):

$$t = 2.27\times10^{25}/\sigma^{34.08} \tag{17}$$

$$\text{if} \quad K_{ISCC} \le K_{Ii} < K_{IC} \quad \text{or} \quad \sigma^*_{ISCC} \le \sigma < \sigma^*_{IC}$$

Where duration unit is second and stress unit is MPa, σ^*_{ISCC} is the threshold stress related to the threshold value of stress intensity factor K_{ISCC} for rock or rockmass with defects. Considering the relative values in the above table and formulas (13), (14), the relationship between σ^*_{ISCC} and σ^*_{IC} for specimens of the mine can be calculated as:

$$\sigma^*_{ISCC} = \sigma^*_{IC}K_{ISCC}/K_{IC} = 3.38\,MPa$$

From formula (17) the calculated duration of the whole subcritical crack growing process is about 100 days for typical specimen of Dongchuan copper mine by assuming that the value of stress σ is just the threshold stress σ^*_{ISCC}. It indicates that the subcritical crack growth is quite a long process. In this case the relationship between stress σ and duration t of rock sample from Dongchuan copper mine is shown in Fig. 3.

It is not difficult to see from Fig.3 that if the stress σ tends to σ^*_{IC}, the relative duration t will tend to zero. For an arbitrary value of stress σ ($\sigma^*_{ISCC} \leq \sigma < \sigma^*_{IC}$), in most cases the duration of the subcritical crack growth is so long that the delayed f racture is characterized as abrupt collapse.

Above mentioned principle had been employed to predict the caving law of orebody of Dongchuan copper mine which was suggested to be mined with block caving method. Block caving is a mass productive mining method, the caving process of orebody is done gradually by a series of delayed abrupt failure of orebody. From formula (16) the relative critical stress σ^*_{IC} can be determined with the following expression if the time duration t is given, say t_0:

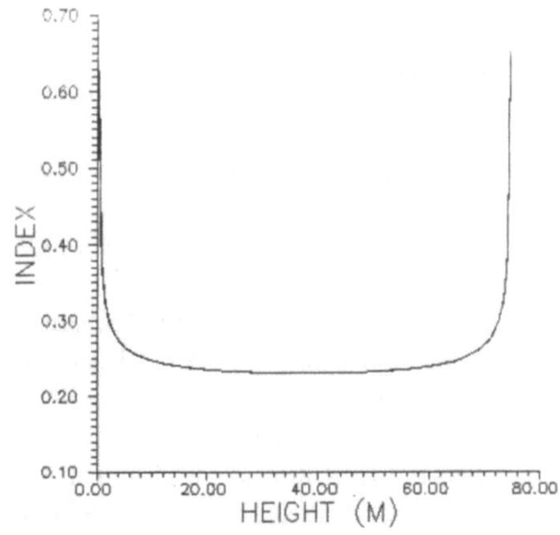

Fig. 3 A relationship between σ and t

$$\sigma^*_{IC} = [\pi A(n-2)t^0 / 2]^{1/(n-2)} K_{IC}(\sigma^n \eta^2)^{1/(n-2)} \tag{18}$$

Based on the experience of caving law of the orebody, it is known that the maximum time interval between any two times of orebody caving , that is any two adjacent abrupt failures of orebody, is usually within 72 hours. Taking the 72 hours as a given time duration t^0 of a step of abrupt failure, the critical stress σ^*_{IC} can then be determined by formula (18). With the critical stress so obtained, the caving laws of the orebody had been predicted after evaluating the stress distribution in the orebody using conventional numerical analysis methods. Then the caving rate and the caved ore weight of orebody as well as the zone of caved orebody had been predicted for the mine (Cao, P. 1995).

5. Conclusion

The abrupt failure of rockmass can cause a serious disaster for human, effective prediction and prevention of the disaster is important for rock engineering. For hard and brittle rock under a condition of quite low level of field stress, an delayed abrupt failure may be induced by a long term of crack propagation within the subcritical crack growing process. The weakening progress of rock causes the final crack fracture when the stress intensity factor at the tip of crack reaches the fracture toughness of the rock. The results show us that the time duration of the subcritical crack growing process is a function of the field stress acted at the crack if the environment temperature remains constant and the stress intensity factor of the crack is greater than or equal to the threshold value enabling subcritical crack growth.

In practical engineering, the stress condition in rockmass is usually in compression and/or shear rather than only in tension, the principle and method mentioned earlier can be utilized to analyze the possibility of abrupt failure and then to predict its duration by employing available numerical analysis method.

References

Charles, R. J. J., (1958) Static Fatigue of glass I, Journal of Applied Physics, Vol.29, 1549-1553

Charles, R. J. J., (1958) Static Fatigue of glass II, Journal of Applied Physics, Vol.29, 1554-1560

Atkinson, B. K., (1987) Fracture Mechanics of Rock. London; Academic Press, London

Cao, P., Wang, H. Sun, Z., (1993) Testing Study of Subcritical Propagation Rate Using Double Torsion Method, Proceeding of the Fifth Symposium on Fracture and Strength of Rock and Concrete. Press of National University of Defense Technology, Changsha, 190-194 (in Chinese with English Abstract)

Cao, P., Pan, C., Sun, Z., Luo, L., (1995) Prediction of Block Caving Rate Using Subcritical Crack Growing Velocity, Transaction of Nonferrous Metals Society of China, Vol.5, No.4,14-17

ON THE STUDY OF CREEP RUPTURE OF STRUCTURE*

REN WANG
Department of Mechanics and Engineering Sciences
Peking University, Beijing, 100871, P. R..China

Abstract The study of creep rupture and the life expectancy of structures has long been a subject of interest. The paper makes a brief survey of the continuum damage mechanics approach to this subject which began in the early fifties and has achieved quite satisfactory progress especially for metallic structures. In recent years due to the need for prolonging the lifetime of existing structures and the application of new materials, especially polymer materials, it is still under vigorous development. This paper begins from some early basic work, introduces the development of constitutive relations from uni-axial stress state to multi-axial stress state, and methods for calculating the creep life together with estimating it by upper and lower bounds. In an attempt to assess the creep rupture process of fiber reinforced composites, an analytic calculation of creep rupture lifetime of a bi-material 3-bar truss under vertical and horizontal loads is presented as an example. Finally, subjects for future research are mentioned.

1. Introduction

Creep rupture of a structure occurs after the creep deformation reaches a certain stage in which the damage in the interior of the material accumulates to such an extent that the structure can no longer carry the given load and fails, or the deformation is considered as excessive. It is somewhat like the load carrying capacity in plasticity theory, but the structure now fails after a prolonged time. The calculation and estimation of the failure time, i.e. the lifetime of the structure under such a loading, is a problem of great interest. It is noted that not only metals working under high temperature reveal creep deformation, but even rock and earth media have creep deformation under certain conditions[1]. For polymers and polymer composites, creep deformation is their basic character; there have been a number of studies of creep rupture in polymer composites in high speed aircraft covering[2], electronic packaging[3], underground piping[4] etc.

To study the lifetime of a body one must rely on experiment. Many tests under uni-axial stress condition have long been carried out together with the microscopic examination of

* Supported by National Natural Science Foundation of China

R. Wang (ed.), IUTAM Symposium on Rheology of Bodies with Defects, 249–266.

the ruptured specimen and the study of the mode of failure. It is well known that creep deformation can be divided into three stages: in the primary creep stage, the material deforms continuously under constant stress, however, the strain rate decreases gradually due to the adjustment of internal deformation mechanisms. For metallic materials, especially under high stress, this period may be rather brief, but it could be quite long for polymers, Findley[5] has reported his 26-year creep tests on poly-(vinylchloride) (PVC) and polyethylene (PE), the creep strain rate is still decreasing gradually as the deformation continues. The secondary creep stage arises as the strain rate decreases to a minimum and remains constant with the deformation growing steadily. The total deformation at this stage is sometime used as a criterion to set the allowable lifetime of the structure. For polymers, the later part of the primary creep stage may sometimes be taken approximately as a steady state creep. Under low stress, the strain rate may become zero and the deformation stop growing; the corresponding stress is known as the endurance strength (often referred to as non-failure under cyclic or random loading). For larger stress, after a certain amount of deformation, the strain rate increases with an accelerating rate, leading to final rupture. This is the tertiary creep stage. For some semi-brittle materials, such as ceramics and some polymers, brittle failure occurs and the tertiary creep stage is negligible.

For high temperature applications in power plant piping, the secondary creep stage is sometimes neglected. Wilshire[5] suggested that the idea of steady state strain rate should be discarded, saying that it is actually the decreasing strain rate overtaken by an increasing strain rate due to damage development that leads to final rupture, for creep in brittle material one may consider that there is only a primary stage. The θ-projection concept and its modification have no secondary creep stage; they fit the experimental data of high temperature CrMoV alloy steel very well with 4 or 3 parameters. On this basis, Wilshire[5] also discusses the extrapolation of short time creep data to the long time creep curve, emphasizing in so doing that the suitable stress-temperature regime according to the relevant deformation mechanism should be taken care of.. However, following traditional concepts, we shall use the three-stage division in the following and be interested especially in the tertiary creep stage.

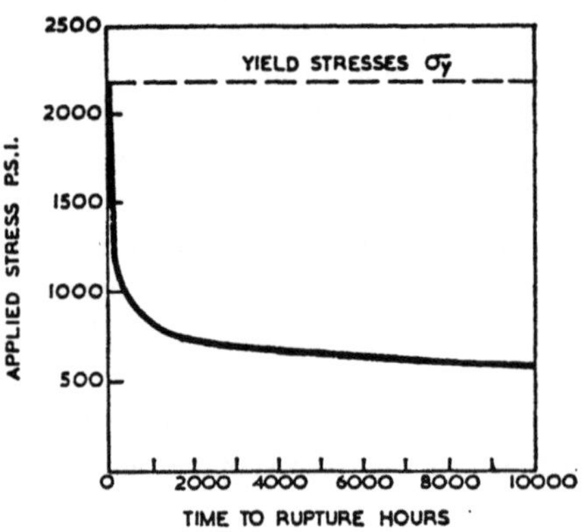

Fig.1a. Applied uni-axial stress vs. time to rupture

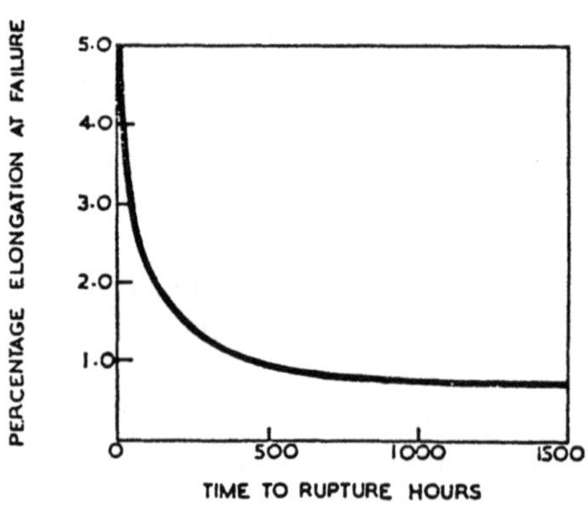

Fig.1b. Failure strain vs. time to rupture[7]

It may be of interest to look at the above figures taken from Hayhurst[7] for some uni-axial test results of Al-Mg-Si alloy steel at 210 °C. Fig.1a shows that under lower stress, the lifetime increases; in Fig.1b, the failure strain also decreases with increasing lifetime. This should mean that under lower stress both the strain and strain rate are small and the practical creep failure strain may usually be small.

A number of tests have been carried out under multi-axial stress condition: thin-walled tubes under different combination of stress conditions, tensile loading of notched plates, bending of beams etc. Together with the research into the deformation mechanisms, various criteria of rupture have been proposed: maximum principal stress, effective shear stress, etc. Like the yield condition in plasticity theory, an isochronous rupture surface, which represents equal lifetimes for different stress conditions, is used to express the failure criterion. For polymer materials, since its deformation mechanism is more complicated and the physical aging effect is more pronounced, results on multi-axial stress tests are rather scarce.

In theoretical investigations, it has long been recognized that the internal structure of a material deteriorates under the secular action of external load. Researchers have tried to define the damage and sought for the rule which links its accumulation to structural failure. According to Golub[8], the attempts began in the twenties known in the literature as the linear cumulative damage rule. Figure 2 gives the plot of damage ω against the rupture time t_R; they are straight lines for different stresses. The total damage is determined from a simple summation, and fracture occurs when the total damage reaches 1.

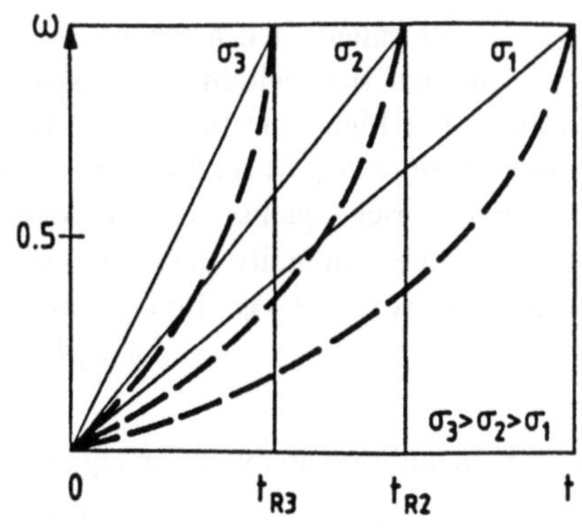

Fig.2. Cumulative damage vs. time

The continuum damage mechanics approach toward creep rupture analysis was first proposed by Kachanov[9]. He related the deterioration of internal structure to the degree of breakage of continuity, called ψ, in the material. When the material deteriorates from $\psi = 1$ to $\psi = 0$, it fails. It is now common to use the damage parameter $\omega = 1 - \psi$; the damage accumulates from 0 to 1 for final failure. The accumulation rule is a nonlinear one as shown in dashed lines in Fig.2. We shall dwell on its recent development in the following. Section 2 will give the progress in constitutive relations and some lifetime assessment methods for metallic material, section 3 will concern polymer and polymeric composites. In an attempt to assess the creep rupture process of fiber reinforced composites, we give an analytic calculation of creep rupture lifetime of a bi-material 3-bar truss under vertical and horizontal loads as an example in section 5. Finally we discuss future research topics.

2. Constitutive Relations for Creep Rupture and Lifetime Assessment Methods for Metallic Materials

2.1. UNI-AXIAL CONSTITUTIVE RELATIONS

The earliest macro-mechanical analysis on creep rupture is due to N. J. Hoff[10] who analyzed a rod subjected to constant tensile load. He used a power type viscous flow law

$$\dot{\varepsilon} = B\sigma^n \tag{1}$$

where the dot stands for time derivative, and B and n are material constants dependent on temperature. He made a finite deformation analysis of a uniform rod under tension. Considering incompressibility, so that the cross sectional area decreases and approaches zero as the rod elongates, one obtains the rupture time t_R. It is inversely proportional to the initial strain rate

$$t_R = 1/n\dot{\varepsilon}_0 \tag{2}$$

Kachanov[9] called such a scheme as *viscous failure,* saying that the deformation is due to internal flow within the crystals in the material; it is suitable for low stress condition. For higher stress, and considering the tertiary creep stage, he suggested the idea of *semi-brittle failure*; this is due to the formation and propagation of cracks at the grain boundaries against a background of growing creep deformation. Thus he introduced the continuity parameter ψ which is related to the now commonly used parameter ω by $\omega = 1 - \psi$. Eq.(1) may now be written as

$$\dot{\varepsilon} = B[\sigma/(1-\omega)]^n \tag{3}$$

There is in addition an evolution equation of ω:

$$\dot{\omega} = A[\sigma/(1-\omega)]^\nu \tag{4}$$

where A, ν are material constants dependent on temperature; A can also vary with time to reflect changes in internal structure due to aging or radiation. The damage evolution equation in a generalized form was also given by Golub[8] as

$$\frac{d\omega}{dt} = C\left(\frac{\sigma}{1-\omega}\right)^m \cdot \frac{\omega^\beta}{(1-\omega)^q} \tag{5}$$

where $C>0$, $m\geq 1$, $q\geq 0$, $\beta\geq 0$ are parameters determined by experiment. Eq.(4) is the case when $\beta = q = 0$. With $q = 0$,it will be the equation given by Rabotnov[11],and with $\beta = 0$, it will be the equation given by Lemaitre and Plumtree[12]. They all follow the form

$$\dot{\omega} = f(\omega)g(\sigma) \tag{6}$$

This is the form for the damage rate equation which Cocks and Leckie[13] discussed from the viewpoint of damage micro-mechanisms.

To include the primary creep stage, a factor t^m would be attached to the r.h.s. of Eqs. (3) and (4) respectively[14]. For constant stress they can be integrated for ω from 0 to 1 to get the rupture time

$$t_R = \{(1+m)/[A\sigma^{\nu}(1+\nu)]\}^{1/(1+m)} \tag{7}$$

Eqs.(3) and (4) are 1-dimensional constitutive relations for creep rupture.

The θ projection concept discards the secondary creep stage[6] and expresses the creep strain by

$$\varepsilon^c = \theta_1(1-\sinh\theta_2 t)+\theta_3(\sinh\theta_4 t-1) \tag{8}$$

The first term on the r.h.s represents the primary creep strain, and the second term the tertiary creep strain. From the viewpoint of deformation mechanism, it is the adjustment of internal structure without damage in the primary stage, while the initiation of damage, developing very slowly at the beginning, together with its buildup, form a continuous process in the later stages.

2.2. MULTI-AXIAL CONSTITUTIVE RELATIONS

To extend these equations to multi-dimensional stress state, researchers first changed σ to the maximum tensile stress. After many investigations with the micro-mechanisms of destruction, it was noted that the first stress invariant and the second deviatoric stress invariant, i.e. the effective shear stress, all contribute to the evolution of damage for different materials, it is now accepted to use a dimensionless function ϕ (σ_{ij}/σ_0) so that

$$\dot{\varepsilon}/\dot{\varepsilon}_0 = [\phi^n \partial\phi / \partial(\sigma/\sigma_0)]/(1-\omega)^n \tag{9}$$

where ϕ is a homogeneous convex function of σ_{ij}/σ_0 of the first order. Under uni-axial stress σ_{ij} reduces to σ, eq.(9) goes back to eq.(3). The general form of ϕ can be written as [7]

$$\phi = \alpha\sigma_I/\sigma_0 + \beta\sigma_e/\sigma_0 + \gamma\sigma_m/\sigma_0 \tag{10}$$

It is a linear combination of σ_I, the maximum principal stress; σ_e, the effective shear stress; and σ_m the mean normal stress; α, β, and γ are weighing parameters with $\alpha + \beta + \gamma = 1$.

The parameters are determined from tests, e.g. [7] gives $\alpha = 0$, $\beta = 0.09$, $\gamma = 0.91$ for Aluminium; $\alpha = 0.848$, $\beta = 0.064$, $\gamma = 0.088$ for pure copper; $\alpha = 0$ $\beta = 0.25$ $\gamma = 0.75$ for low alloy steel.

The more commonly used one is independent of σ_m, i.e. $\gamma = 0$; eq. (10) reduces to[15]

$$\phi = \alpha\sigma_I/\sigma_0 + (1-\alpha)\sigma_e/\sigma_0 \tag{11}$$

Figure 3 shows an isochronous rupture surfaces under plane stress for aluminium and pure copper[15]. Recently, Trivandey and Delobelle[16] made an extensive experimental study of austenitic stainless steel, and concluded that damage is both time and strain dependent. However the scalar or tensorial nature of damage is not clearly proved. They suggested adding a small contribution of the second principal stress in ϕ.

The damage evolution equation will now be written as:

$$\dot{\omega} = A\left[\phi(\sigma_{ij}/\sigma_0)/(1-\omega)\right]^{\nu} \qquad (12)$$

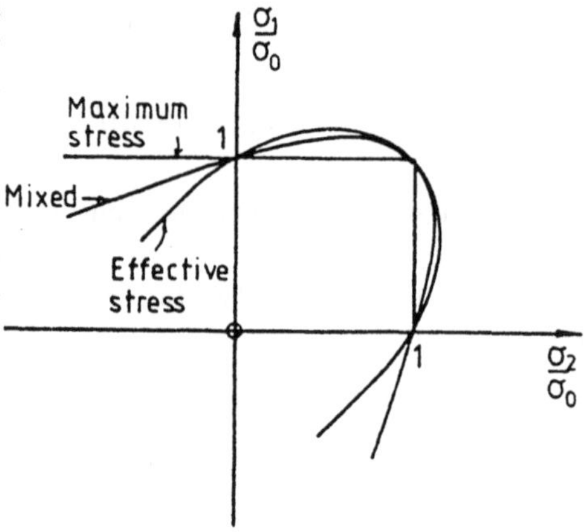

Fig.3. Isochronous rupture surface in plane stress

Eq. (9) and (12) are 3-dimensional constitutive relations for creep rupture. Eq.(9) represents a normality rule similar to the theory of plasticity with the strain rate vector normal to the isochronous rupture surface.

Researchers have also considered strain hardening, the creep strain rate can be expressed as a function of stress and the history of creep strain:

$$\dot{\varepsilon}_{ij}^{c} = \tfrac{1}{2} f(\sigma_e, q) S_{ij}/\sigma_e \, ; \quad \dot{q} = (2/3\dot{\varepsilon}_{ij}^{c}\dot{\varepsilon}_{ij}^{c})^{1/2} \qquad (13)$$

where the S_{ij} are the deviatoric stress, q is the creep hardening variable. Because q cannot adequately describe the recovery of material hardening after the stress change, Murakami and Ohno[17] introduced a creep hardening hyper-sphere in the creep strain space:

$$g = 2/3(\varepsilon_{ij}^{c} - \alpha_{ij})(\varepsilon_{ij}^{c} - \alpha_{ij}) - \rho^2 \leq 0 \qquad (14)$$

where α_{ij} and ρ are the center and radius of the hyper-sphere with their own evolution rules respectively. When $g = 0$, the values obtained after a given deformation history is then used to get the creep hardening variable q :

$$q = [\rho + (\varepsilon_{ij}^{c} - \alpha_{ij})S_{ij}/\sigma_e]/2\lambda_0 \qquad (15)$$

where λ_0 is a material constant specifying the rate of development of ρ.

2.3. LIFETIME ASSESSMENT METHODS

Using the above constitutive relations and the damage evolution equation together with the equilibrium and continuity equations, one forms an initial value problem starting with elastic solution as the initial condition, see e.g. [18]. Kachanov[9] has given several interesting examples of brittle failure: tubes under internal pressure, pure bending of a beam, a beam under concentrated load, a stretched plate with a hole. He has investigated the propagation of the failure front; in his case it is the surface of $\psi =$

0. The examples show that the failure front initially propagates rather slowly, then the rate of formation of the failed region increases, and in the last stage failure bears an "avalanche" character. This shows that for a comparatively large portion of the lifetime the structure may still resist the external load. This seems to be a common feature in most of the other problems solved by numerical methods also. In [19] Kachanov has given other solutions on the brittle fracture of a thin interlayer, both in plane strain case and in the axisymmetric case under a central concentrated tensile load. There are a few other examples, such as a thick wall cylinder, a rotating disk etc. solved analytically (see e.g. [18]). They may be used as a check for numerical methods. However, one can see that the solution for any practical structure will be very complicated and will rely on numerical and other approximate methods. An iteration method is described in [18], but it is the finite element method that is most frequently used, especially in the crack propagation problems in creeping materials[20,21,22]. Recently Needleman[23] in his review of the computational modeling of material failure has also discussed some of these problems.

Approximate methods that give lower and upper bounds are also used in estimating the lifetime of structure. They may be useful in the early design stage.

LOWER BOUND---Reference stress method. Leckie and Hayhurst[15] introduced a reference stress which is defined as an uni-axial stress σ_D that causes a creep energy dissipation rate equal to the average dissipation rate of the structure; from

$$V\sigma_0\dot{\varepsilon}_0\left(\frac{\sigma_D}{\sigma_0}\right)^{n+1} = \int_V \dot{D}_s dV \tag{16}$$

where V is the volume of the deformed material of the structure, and $\dot{D}_s$ is the energy dissipation rate, we obtain the reference stress

$$\sigma_D/\sigma_0 = \left\{\frac{1}{V}\int_V (\dot{D}_s/\sigma_0\dot{\varepsilon}_0)dV\right\}^{1/(n+1)} \tag{17}$$

Many examples show that σ_D is nearly independent of material constants. After being calculated from an exact solution for a certain value of n, it may be used for all other values of n. A reference rupture stress is then introduced which is the stress for a uni-axial specimen to have the same rupture time as the structure. For kinematically determinate structures, it can be used to get a lower bound for the rupture time. It is not proved for other cases. For aluminium alloy and steel which rupture according to effective shear stress, i.e. $\alpha = 0$ in eq.(11), the reference rupture stress σ_R has the following form:

$$\sigma_R/\sigma_0 = \left\{\int_V (\dot{D}_s/\sigma_0\dot{\varepsilon}_0)^{(n+1+\nu)/(n+1)} dV \Big/ \int_V (\dot{D}_s/\sigma_0\dot{\varepsilon}_0)dV\right\}^{1/2} \tag{18}$$

the corresponding rupture time is [15 appendix]:

$$t_R \geq 1/A(1+\nu)\sigma_R^\nu \tag{19}$$

For structures like plates and shells, which collapse when the entire cross section reaches yield stress,

$$\sigma_R = (P / P^L)\sigma_y \tag{20}$$

where P is the external load, P^L is the plastic limit load, σ_y is the yield stress.

UPPER BOUND. Ponter[24] gave some upper bounds for creep rupture life of structures subjected to variable load and temperature. He based his derivation on ϕ being a convex function. For constant load he gives:

$$t_R \leq 1 / A(1+\nu)\sigma_D^{\nu} \tag{21}$$

He went on to give upper bounds for variable loads and temperature history. Cocks and Leckie[13] have extended his results by considering the damage rate equation that can be written in the form as eq.(6). Boyle and Spencer[18,Ch.9] have also discussed the estimate of failure time from energy principles and given two examples. They show that their simple estimates provide an indication of lifetime which can be adequate for design purpose.

Liaw et al[25] have considered the influence of primary creep in the estimation of lifetime. They conclude that for steam pipes under high temperature, the inclusion of primary creep decreases the creep crack growth life. Thus the life prediction analysis without including primary creep can lead to a non-conservative estimate of remaining life of a steam pipe. They developed a life prediction method using time dependent fracture mechanics which is not considered in this paper.

3. Constitutive Relations for Creep Rupture and Lifetime Estimation for Polymers and Polymer Composites

3.1. CONSTITUTIVE RELATIONS

Although the phenomenal appearances and even the forms of constitutive relations may look quite similar for metals and polymers, the micro-mechanisms of deformation and rupture are quite distinct. For metals, the micro-mechanism occurs at the atomic scale; for polymers it is on the molecular scale, due to its basic micro-structure being extensively intertwined, cross linking long molecular chains. Disentanglement, rotation of these chains and slipping between them give rise to deformation. The basic damage mechanisms in polymers are crazing and shear cracking on the molecular scale. Even for the crystalline phase in semi-crystalline polymers, dislocation motion on a molecular (rather than atomic) scale may be a mechanism of deformation in this phase. The deformation is usually larger, and the dependence on temperature and strain rate cover a larger range. Due to its increasing field of application, the study of constitutive relations for polymers is currently intensive, as can be seen from some of the recent review papers[26-29]. Some updated studies will be given below.

The works on constitutive relations may be grouped into two kinds of approaches:

1. **Thermodynamic and statistical approach.** Schapery[30] adopted the frame-work of thermodynamics of unequilibrium processes, introduced internal state variables and derived general constitutive equations for nonlinear viscoelastic composites from a general potential function. However, there is little discussion of the physical mechanism for the nonlinear time dependent behavior. Altenbach et al[31] also discussed constitutive relations from the assumption of the existence of a potential. Chen[32] considered a new relaxation mechanism of polymer deformation, and treated the entanglements as a kind of fuzzy constraint due to the combined effects of cohesive forces and steric hindrance. The anelastic dissipation was treated as a thermally activated process, promoted by the thermodynamic force, for the conformation reorganization of segments through the coordinated inner rotation and local slippage of links. The static retardation dissipation was treated as a thermal-mechanical activated anchoring-disanchoring-reanchoring process of the chain, and the scission of the primary bonds as the chemical-mechanical activated process. She set up a visco-elastic-plastic constitutive relation which covers several existing ones as its limiting cases. Chen et al[33] further derived a unified three-dimensional constitutive relation suitable for describing the relaxation and transition of polymeric materials under small or large deformation. It reproduces yielding at small deformation, the softening effect within the moderate deformation as well as the hardening effect at large deformation. Vujosevic and Krajcinovic[34] considered a statistical model for the creep deformation and failure of thermoset-resin. The model is a two-dimensional triangular central force lattice of perfect geometry. Neighboring nodes are connected by elastic links of identical stiffness, identical strength, identical length and identical activation energy. Within the framework of the reaction rate theory, the bonds (links) of the lattice are viewed as "coupled oscillators in a state of thermal vibration". The bond rupture was treated as a random process activated by spatially and temporally random thermal fluctuations. An estimate of time to creep rupture was obtained. They suggest that it provides a foundation for the formulation of a rational design model for the creep failure of resin specimens.

2. **Simple rheological modelling.** Most papers have used the differential representation to avoid stress history dependence and to cope with nonlinear behavior. This again can be divided into two types, depending on whether the rheological elements are connected in series or in parallel.

In Series, this means that the strain is the sum of an elastic strain and a viscous or visco-plastic strain,

$$\varepsilon = \varepsilon^{e} + \varepsilon^{v} \ (\text{or}\ \varepsilon^{vp}) \tag{22}$$

Zhang and Moore[29] have reported an extensive series of uni-axial compression tests on high-density Polyethylene (HDPE) used for buried pipes. They used true stress and true strain to cover finite deformation. The model consists of an elastic spring in series with six viscous dashpots to fit the experimental curves, so that:

$$\varepsilon=\varepsilon^{e}+\varepsilon^{v}=\frac{\sigma_n}{E_0}+\sum_{i=1}^{6}\frac{\sigma_n}{E_i}\left\{1-\exp\left(-\frac{t}{\tau_i}\right)\right\} \tag{23}$$

It can describe the primary and secondary creep stages over a range of stress and a time range between 0.1 sec to 10^4 sec. With 9 dashpots in series it can cover a time range up to years. Findley[5] also used a series representation,

$$\varepsilon=\varepsilon_0+\varepsilon_t t^m \quad \text{(Findley)} \tag{24}$$

where ε_0 and ε_t are functions of stress. Li and Dasgupta[3] used ε_0 = a sinh(σ/b) and ε_t = c sinh (σ/d) with a, b, c, d as material constants.

In parallel, this means the total stress is the sum of a elastic stress and a viscous or visco-plastic stress, so that:

$$\sigma=\sigma^{e}+\sigma^{v}\ (\text{or}\ \sigma^{vp}) \tag{25}$$

The rheological model used by Teoh et al[35] is shown in Fig.4; the right hand part is the model in parallel, and it is further connected in series with a elastic spring on the left. The applied stress σ_{ap}, is equal to the sum of elastic stress σ_{re} and effective stresses σ_{ef}. The viscous element shown obeys a Eyring fluid flow law:

$$\dot{\varepsilon}_a=K\sinh B\sigma_{ef} \tag{26}$$

where ε_e is the elastic strain and ε_a is the time dependent anelastic strain, K, B are material constants. They have fitted it with many kinds of polymers.

Raghagan et al[27] and Bardenhagen et al[28] have used similar models, but the viscous dashpots obey different viscous flow laws. They also formally extend the relations to three-dimensions and discussed the anisotropic case.

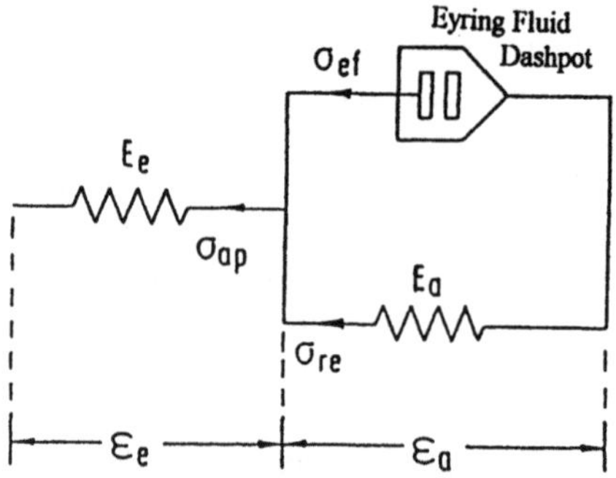

Fig.4. 3-element rheological model

3.2. FAILURE CRITERIA

The constitutive relations above, except that of Vujosevic et al[34], do not include damage effects, thus do not represent the tertiary creep stage. To estimate lifetime, we must attach failure criteria. One kind of failure criterion is the deformation limit. When the deformation reaches that limit the material is said to fail. Another is by energy fracture criterion. The one commonly used[27,35] is that the total elastic energy stored in the resistive springs equals to the critical energy for fracture. Brueller's[36] criterion is for the stored deviatoric strain energy to reach a critical value. Mai et al[37,38,49] have presented an Essential Work of Fracture as a failure criterion; they separated a damaged body into a fracture process zone and the remainder. The total work is then separated into the essential fracture work imported into the fracture process zone and the nonessential fracture work absorbed by the outer region. When a new damage (to be specific a crack) surface is created, an energy rate G is released which will be the energy input into the fracture process zone if the crack propagation is autonomy. So when the generalized energy release rate reaches a critical value say G_c, the crack grows and leads to material failure. The criterion was supported by tests on sheets and under plane strain.

There are many other works on crack and void growth in creeping material, they are reviewed in the paper by Yang[39] in this Symposium.

3.3. CALCULATION OF RUPTURE TIME

Teoh et al[35] give a detailed calculation of rupture time with the rheological model of Fig.4. Note, that ε_e and ε_a are the elastic and anelastic strain respectively, E_e is the elastic modulus along the main fiber, and E_a is the time dependent anelastic modulus transverse to the main fiber. The criterion of critical elastic stored energy is used for its rupture, viz.

$$W_e = \int_0^{t_R} (\dot{W} - \dot{D})dt = R \tag{27}$$

where $\dot{W}$ is the input power, $\dot{D}$ is the dissipated power, W_e is the elastic stored energy, R is the critical value. According to Fig.4, we get

$$\int_0^{\varepsilon_e^*} \sigma_{ap} d\varepsilon_e + \int_0^{\varepsilon_a^*} \sigma_{re} d\varepsilon_a = R \tag{28}$$

where ε_e^* is the elastic strain at rupture, ε_a^* is the anelastic strain at rupture. When σ_{ap} is constant, Eq. (28) can be integrated to get:

$$\frac{\sigma_{ap}^2}{E_e} + \frac{1}{2}\left(\varepsilon_a^*\right)^2 E_a = R \tag{29}$$

from which we get the anelastic strain at rupture to be

$$\varepsilon_a^* = \left[\frac{2}{E_a}\left(R - \frac{\sigma_{ap}^2}{E_e}\right)\right]^{1/2} \tag{30}$$

Since the applied stress does not change, $\dot{\sigma}_{ef} = -\dot{\sigma}_{re}$, and since $\varepsilon_a = \sigma_{re}/E_a$, we may substitute these into eq.(26) and integrate,

$$\int_{\sigma_{ap}}^{\sigma_{ef}} \frac{d\sigma_{ef}}{\sinh B\sigma_{ef}} = -\int_0^t E_a K dt \tag{31}$$

we get:

$$\sigma_{ef} = \frac{2}{B \tanh^{-1}\left[\tanh(B\sigma_{ap}/2)\exp(-E_a BKt)\right]}$$

and

$$\varepsilon_a = \frac{\sigma_{ap} - \sigma_{ef}}{E_a} = \frac{\sigma_{ap}}{E_a} - \frac{2}{BE_a}\tanh^{-1}\left[\tanh\left(B\sigma_{ap}/2\right)\exp(-E_a BKt)\right]$$

When $t = t_R$, $\varepsilon_a = \varepsilon_a^*$; from which we get the rupture time as:

$$t_R = (1/E_a BK)\ln\left[\tanh(B\sigma_{ap}/2)/\tanh(BH/2)\right] \tag{32}$$

where

$$H = \sigma_{ap} - \left[2E_a(R - \sigma_{ap}^2/E_e)\right]^{1/2}$$

We see, when $\sigma_{ap} = (E_e R)^{1/2}$, $H = \sigma_{ap}$, $t_R = 0$ instantaneous failure,

when $\sigma_{ap} = \left[R/(1/E_e + 1/2E_a)\right]^{1/2}$, $t_R = \infty$ endurance strength.

4. An Example: Creep Rupture Analysis of a Bi-material Three-bar Truss

Figure 5 shows the three bar truss, in which bars 1 and 3 are elastic creeping material, bar 2 is a purely elastic material. Their constitutive relations are respectively:

$$\dot{\varepsilon}_1 = \dot{\varepsilon}_1^e + \dot{\varepsilon}_1^c$$

$$\dot{\varepsilon}_1^e = \frac{1}{E_1}\dot{\sigma}_1(t) \tag{33)a}$$

$$\dot{\varepsilon}_1^c = B\left[\sigma_1(t)/(1-\omega_1(t))\right]^n \tag{33)b}$$

$$\dot{\omega}_1 = A\left[\sigma_1(t)/(1-\omega_1(t))\right]^\nu \tag{33)c}$$

$$\dot{\varepsilon}_2 = \frac{1}{E_2}\dot{\sigma}_2(t) \tag{33)d}$$

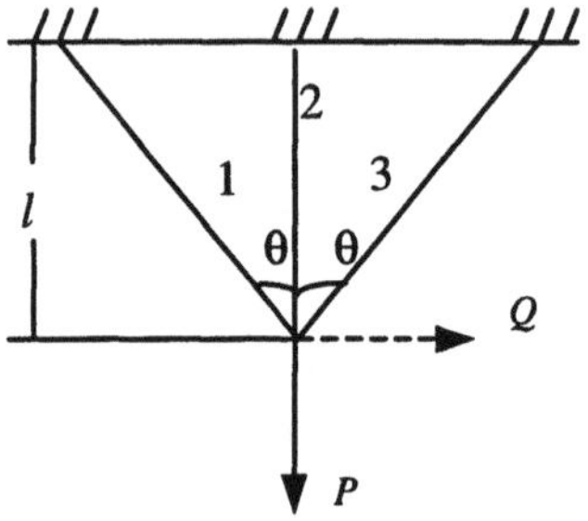

Fig.5. Bi-material three-bar truss

Bar 3 is the same as bar 1.

Case 1. Loaded vertically by P

This may be looked upon as the case when the purely elastic fiber is embedded in a visco-elastic matrix subjected to a pull along the fiber. From symmetry, $\sigma_1 = -\sigma_3$, equilibrium gives:

$$2\sigma_1(t)\cos\theta\, S_1 + \sigma_2(t)S_2 = P \tag{34}$$

where S_1, S_2 are cross sectional areas of bars 1 and 2 respectively. Compatibility yields:

$$\varepsilon_2 = \varepsilon_1/\cos^2\theta \tag{35}$$

At t = 0, the elastic response is:

$$\sigma_2 = (E_2/E_1)\sigma_1/\cos^2\theta$$

Thus, the initial conditions at the beginning of creep deformation are:

$$\sigma_1 = PE_1\cos^2\theta[2E_1\cos^3\theta\, S_1 + E_2S_2]^{-1} \equiv MP \tag{36)a}$$

$$\sigma_2 = PE_2[2E_1\cos^3\theta\, S_1 + E_2S_2]^{-1} \equiv NP \tag{36)b}$$

For t > 0, substituting the constitutive equations (33) into eq.(35), we get:

$$\frac{\dot{\sigma}_1(t)}{E^*}+B\left(\frac{\sigma_1(t)}{1-\omega_1(t)}\right)^n=0 \tag{37}$$

where $E^*=E_1E_2S_2/(E_2S_2+2E_1S_1\cos^3\theta)$.

Integrating eq. (33)c from $\omega_1=0$, we get

$$\omega_1(t)=1-\left\{1-A(1+v)\int_0^t[\sigma_1(\tau)]^v\,d\tau\right\}^{1/(1+v)} \tag{38}$$

Substituting back to eq. (37), we get an integral-differential equation:

$$\frac{\dot{\sigma}_1(t)}{E^*}+B\frac{[\sigma_1(t)]^n}{\left\{1-A(1+v)\int_0^t[\sigma_1(\tau)]^v\,d\tau\right\}^{n/(1+v)}}=0 \tag{39}$$

Under initial conditions (36), Goncalves Filho[40] gets closed form expressions for $\sigma_1(t)$, $\sigma_2(t)$, $\omega_1(t)$ and t_R after complicated calculations. Taking $\omega=\omega_{CR}$ as failure condition, we may divide the solution according to structural parameters into two different situations:

(a) $$\frac{k_1}{k_2}\equiv\frac{2S_1E_1\cos^3\theta}{S_2E_2}=\frac{E_1B}{A}-1 \tag{40)a}$$

(b) $$\frac{k_1}{k_2}\neq\frac{E_1B}{A}-1 \tag{40)b}$$

with eq. (40)a as the critical situation. Here we have already neglected the primary creep stage, further let $v=n-1$, expressions for t_R are

(a) $$t_R=\left(\frac{1}{\sigma_1(0)}\right)^{n-1}\frac{1}{A}\ln\frac{1}{1-\omega_{cr}} \tag{41)a}$$

(b) $$t_R=\left(\frac{1}{\sigma_1(0)}\right)^{n-1}\left(\frac{1}{nA-(n-1)E^*B}\right)\left[1-(1-\omega_{cR})^{[nA-(n-1)E^*B]/A}\right] \tag{41)b}$$

For k_1/k_2 larger or less than $(E_1B/A)-1$ (concerning the properties of the elastic creeping material only) their responses are different. When k_1/k_2 is larger, it corresponds to the elastic-creeping bars carrying more load, and damage will grow at an increasing rate; when it is equal, the damages grow at a constant rate; and when it is smaller, at a decreasing rate. Figures 6 and 7[40] (with $\omega_{CR}=0.9$) show some of the numerical results of situation (b) for Titanium Aluminium alloy as the elastic creeping material, with $k_1/k_2=18$ (above the critical situation of 13), $n=6.8$, $v=5.8$. The curves show the beneficial effect of the stress redistribution process on the elastic creeping bar, the lifetime of the 3-bar configuration is shown to be 230% of that without the elastic bar.

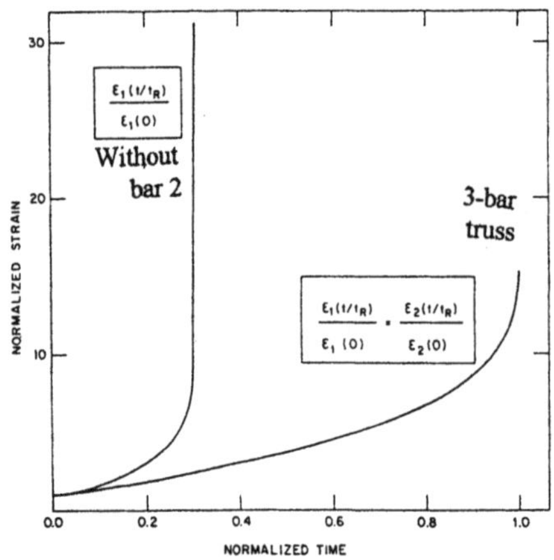

Fig. 6. Strain vs. normalized time [40]

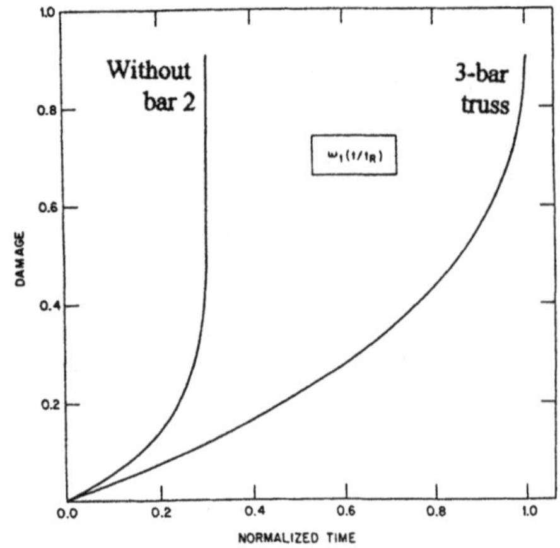

Fig. 7. Damage vs. normalized time [40]

These results are those of [40], corresponding to θ = 0. For the present case, the magnitude of θ reflects the role of the matrix; with increasing θ, k_1/k_2 gets smaller and tends to be less than the critical value $(E_1B/A) - 1$, damage will grow in a decreasing rate, and the lifetime will be prolonged and depend more on the elastic bar.

Case 2. Loaded horizontally by Q

This corresponds to the fibers being loaded transversely. Bar 1 is now under tension, bar 3 under compression while bar 2 is free from load under small deformation. From equilibrium we have

$$\sigma_1(t) = -\sigma_3(t) = Q/(2S_1 \sin\theta), \quad \sigma_2 = 0 \tag{42}$$

Compatibility gives

$$\varepsilon_1(t) = -\varepsilon_3(t) \tag{43}$$

At t = 0, elastic solution gives

$$\sigma_1 = Q/(2S_1 \sin\theta) = \sigma_0 \tag{44}$$

therefore

$$\dot{\varepsilon}_1^c = B\{\sigma_0/[1-\omega_1(t)]\}^n \tag{45}$$

$$\dot{\omega}_1 = A\{\sigma_0/[1-\omega_1(t)]\}^\nu \tag{46}$$

After integration we have

$$1-\omega_1 = \left[1 - A(1+\nu)\sigma_0^\nu t\right]^{1/(1+\nu)}$$

or

$$A(1+\nu)\sigma_0^\nu t = 1 - [1-\omega_1(t)]^{1+\nu} \tag{47}$$

When $\omega = \omega_{CR}$, $t = t_R$,

$$t_R = \frac{1-(1-\omega_{cr})^{1+\nu}}{A(1+\nu)(Q/2S_1 \sin\theta)^{\nu}} \tag{48}$$

Fig.8 shows the damage vs. normalized time. It should be a lower bound, since when finite deformation is considered, the elastic bar 2 will contribute and increase the life. Chun and Daniel[41] have tested and analyzed the transverse creep behavior of a uni-directional SiC/Al composite. They observed that during creep, a gradual stress transfer takes place between matrix and fibers, followed by stress redistribution and stress relaxation in the matrix, resulting in higher creep resistance. No lifetime was considered.

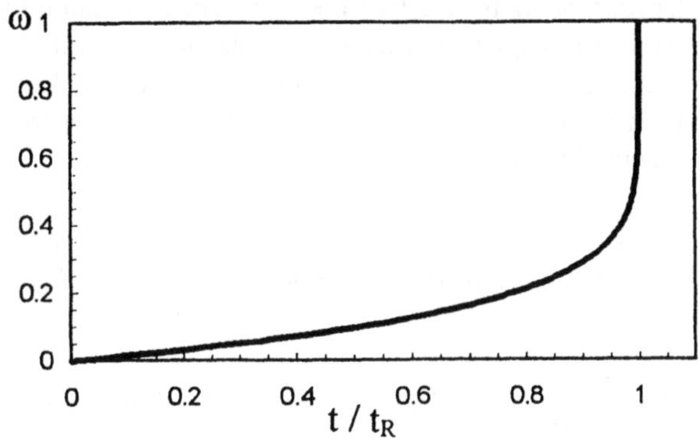

Fig.8. Damage vs. Normalized Time

5. Concluding Remarks

We have given a rough survey on the present status of creep rupture theory, but have really touched very little on structure, There are many conferences and symposia on application to structures and the problems arising therein; readers can refer to their proceedings. For example, IUTAM holds "Creep in Structures" every ten years, University College, Swansea holds "Creep and Fracture of Engineering Materials and Structures" every three years, and there are many others. Those presented above are also far from complete. I shall make some final remarks as follows.

1. Aside from the theoretical works on creep rupture, for practicing engineers, there is the problem of deciding the remaining life of an existing structure. On the one hand, there are the questions of where to measure, and how to improve the precision of *in-situ* measuring techniques[42]. On the other hand, there is the interpretation of the measured values, which again relies on theoretical work; there are many such problems cited in the symposia organized e.g. at Swansea [43].

2. Although the study on creep rupture has made quite a big progress in recent years, there are many subjects that still need further investigations, such as how to extend

the analysis to the 3-dimensional case, and how to include anisotropy induced by finite deformation.Non-stationary loading, severe environmental effects especially for polymers, and radiation effects etc. are important. There are problems of modelling that will reduce the number of parameters and suitable for computation[16,23]. For computer software see e.g. Chattopadhyay and Ghosh[44].

3. Crack growth in creeping material is one of the fundamental problems in deciding lifetime; it is surveyed in [39] by Yang. I wish to mention one supplementary work which is on the effect of crack tip shape on lifetime. Yokobori Jr. and Yokobori[45] compared uncracked specimens to those with rounded and sharp ended notches. They used three terms: t_f being the fracture life, t_i the crack initiation life, and t_v the crack growth life so that $t_f = t_i + t_v$ .For uncracked specimens, $t_i = 0.98\ t_f$, while for cracked specimens, t_i is negligible. For specimen with a rounded hole, t_i lies between these extremes. This means that for specimens with no apparent defects, or rounded holes, t_i should be used to estimate t_f.

4. For polymer materials, the study of the effect of physical aging is still important; some important literatures in recent years have been surveyed in [26], others will be found in [46].

5. For fiber reinforced composite, the response under compressive loading is an active field of research, see e.g.[47].

6. When the mechanical behavior is understood, an optimization study may be carried out, this has been recently surveyed by Zyczkowski[48].

Acknowledgement: The careful editing and correction in English made by Professor G. Gladwell to the manuscript, and comments and typing of equations by Dr. X.H.Chen are gratefully acknowledged.

References:

1. Kelly A., Cook, A. H. and Greenwood, G.W. (1978) *Creep of Engineering Materials and of the Earth,* Royal Soc. London.
2. Brinson, L .C. and Gates, T.S. (1995) Effects of physical aging on long term creep of polymers and polymer matrix composites, *Int. J. Solids Structures*, **32**, 827-846.
3. Li, J. and Dasgupta, A. (1993) Failure mechanism models for creep and creep rupture, *IEEE Trans. on Reliability*, **42**, 339-353.
4. Otani, H. Phoenix S. L. and Petrina, P. (1991) Matrix effects on lifetime statistics for carbon fibre-epoxy micro-composites in creep rupture, *J. Mater. Sci.* **26**, 1955-1970.
5. Findley, W.N. (1987) 26-year creep and recovery of Poly(Vinyl-Chloride) and Polyethylene, *Polymer Engrg. Sci.*, **27**, 582-585.

6. Wilshire, B. (1991) Microscopic models and macroscopic constitutive laws for high temperature creep and creep fracture of metallic and ceramic materials, in Cocks, A. C. F. and Ponter, A. R. S. (eds.) *Mechanics of Creep Brittle Materials-2*, Elsevier Sci. Publ., New York, pp. 112-123.

7. Hayhurst, D. R. (1972) Creep rupture under multi-axial states of stress, *J. Mech. Phys. Solids*, **20**, 381-390.

8. Golub, V.P. Non-linear models of creep damage accumulation, in Cocks, A. C. F. and Ponter, A. R. S. (eds.) *Mechanics of Creep Brittle Materials-2,* Elsevier Sci. Publ., New York, pp. 254-267.

9. Kachanov, L. M. (1961) Rupture time under creep conditions, in Radok, J. R. M. (ed.) *Problems in Continuum Mechanics,* JIAM, 202-218.

10. Hoff, N. J. (1954) Approximate analysis of structures in the presence of moderately large creep deformation, *Q. J. Appl. Math.* **12**, 49-55.

11. Rabotnov, Yu. N. (1969) *Creep Problems in Structural members*, North-Holland Publ. Co. Amsterdam.

12. Lemaitre, J. and Plumtree, A., (1979) Application of damage concept to predict creep-fatigue failures, *J. of Engineering Materials and Technology, Trans. ASME,* **101**, 284-292.

13. Cocks A. C. F. and Leckie, F. A., (1987) Creep constitutive equations for damaged materials in Wu, T. Y. and HutchinsonJ. W. (eds.) *Advances in Applied Mechanics*, **.25**, 239-295.

14. Othman, A. M. and Hayhurst D. R. (1990) Multi-axial creep rupture of a model structure using a two parameter material model, *Int. J. Mech. Sci.* **32**, 35-48.

15. Leckie, F. A. and Hayhurst, D. R. (1974) Creep rupture of structures, *Proc. R. Soc. London* **A340**, 323-347.

16. Trivaudey F. and Delobelle P. (1993) Experimental study and modelization of creep damage under multi-axial loadings at high temperature, in Wilshire, B. and Evans, R. W. (eds,) *Creep and Fracture of Engineering Materials and Structures*, The institute of Materials, London, 137-147.

17. Murakami, S. and Ohno, N. (1982) A constitutive equation of creep based on the concept of a creep-hardening surface, *Int. J. Solids Structures*, **18**, 597-609.

18. Boyle, J. T. and Spence J. (1983) *Stress Analysis for Creep,* Butterworths, London.

19. Kachanov, L.M. (1984) in Dvorak, G. J. and Shield, R.T. (eds.) *Mechanics of Material Behavior*, Elsevier, Amsterdam, pp.191-199.

20. Conway, J. B.(1967) *Numerical Methods for Creep and Rupture Analysis*, Gordon and Breach Sci. Publ., New York, Ch.VIII Stress rupture analysis.

21. Singh, R. N. and Ramakrishnan, C. V. (1995) Fracture Behavior of creeping materials under biaxial loading by finite element method, *Engrg. Fracture Mech.*, **51**, 637-648.

22. Hall, F. R. and Hayhurst, D.R. (1991) Continuum damage mechanics modelling of high temperature deformation and failure in a pipe weldment, *Proc. R. Soc. London* **A433**, 383-403.

23. Needleman, A. (1994) Computational modelling of material failure, Appl. Mech. Rev. 47 no.6 part 2 S34-S42.

24. Ponter, A.R.S. (1977) Upper bounds on the creep rupture life of structures subjected to variable load and temperature, *Int. J. Mech. Sci.* 79-92.

25. Liaw, P. K., Saxena, A. and Schaefer, J. (1997) Creep crack growth behavior of steam pipe steels: Effects of inclusion content and primary creep, *Engrg. Fracture Mech.*, **57**,105-130.

26. Wang, R. (1996) A review on creep failure of polymer and polymer composite, in Abe,T. and Tsuta,T. (eds.) *Advances in Engineering Plasticity and its Applications*, Pergamon Press, 43-52.

27. Raghavan, J. and Meshii, M. (1997) Creep rupture of polymer composites, *Composites Sci. & Tech.* **57**,375-388.

28. Bardenhagen, S. G. , Stout, M. G. and Gray, G.T. (1997) Three-dimensional, finite deformation, viscoplastic loading, *Mechanics of composite materials*, **31**, 511-518.

29. Zhang C. and Moore I. D. (1997), Nonlinear mechanical response of high density polyethylene, Part I: experimental investigation and model evaluation, Part II: uniaxial constitutive modeling, *Polymeer Engrg.and Sci.*, 404-413, 414-420.

30. Altennbach, H., Altenbach, J. and Zolochevsky, A. (1995) A generalized constitutive equation for creep of polymers at multiaxial constitutive models for polymeric materials *Mechanics of Materials*, **25**, 235-253.
31. Schapery, R. A. (1994) Nonlinear viscoelastic constitutive equations for composites based on work potentials, *Appl. Mech. Rev.* **47** no.6 part 2, S269-S275.
32. Chen, X.-H. (1995) Statistical mechanics of fuzzy random polymer networks, *Science in China* Ser.**A** **v.38** (Eng. ed.), 1095-1104.
33. Chen, X.-H., Tong P. and Wang R. (1998) Non-equilibrium statistical thermodynamic theory for viscoelasticity of polymers, *J. Mech. Phys. Solids*, **46**, 139-152.
34. Vujosevic, M. and Krajcinovic D. (1997) Creep rupture of polymer: A statistical model, *Int. J. Solid Structures*, **34**, 1105-1122.
35. Teoh, S. H. Cherry, B.W. and Kausch, H. H. (1992) Creep rupture modelling of polymers, *Int. J. Damage Mech.*, **1**, 245-256.
36. Brueller, O. S. (1981) Energy related failure criteria of thermoplastics, *Polymer Engrg. and Sci.* **21,** 145-150.
37. Mai, Y. W. and Powell, P. (1991) Essential work of fracture and J-integral measurements for ductile polymers, *J. Polymer Sci.. Part B polymer physics*, **29**, 785-793.
38. Wu, J. and Mai, Y. W. (1996) The essential fracture work concept for toughness measurement of ductile polymers, *Polymer Engrg. and Sci.* **36**, 2275-2288.
39. Yang, T. Q. (1998) Rheological behavior and failure characteristics of viscoelastic solids with defects, this volume.
40. Ganclaves Filho, O. J. A. (1995) Closed form solution for the isothermal creep rupture behavior of a two bar structure under constant load, *Int. J. Solids Structures*, **32**, 3087-3104.
41. Chun, H. J. and Daniel, A.J. (1997) Transverse creep behavior of a unidirectional metal matrix composite, *Mechanics of Materials*, **25,** 37-46.
42. Cane, B.J. and Aplin P. F. (1994) Creep life assessment methods, *J. Strain Analysis*, **29**, 225-232.
43. Wilshire, B. and Evans, R. W. (1993) *Creep and Fracture of Engineering Materials and Structures*, The institute of Materials, London.
44. Chattopadhyay, L. and Ghosh, R. N. (1996) CLIP computer software for creep life prediction of engineering materials, *Engrg. Fracture Mech.*, 54, 71-73.
45. Yokobori Jr., A. T. and Yokobori, T. (1993) A new concept on high temperature creep crack initiation, growth and creep fracture life, in Wilshire, B. and Evans, R.W.(eds.) *Creep Fracture of Engineering Materials and Structures*, The institute of Materials, London, 81-97.
46. Bradshaw, R. D. and Brinson, L. C. (1997) Physical aging in polymer and polymer composites: An analysis and method for time-aging time superposition, *Polymer Engrg. and Sci.* **37**, 31-44.
47. Veazie,D. R. (1997) Compressive creep of IM7-K3B composite and the effect of physical aging on viscoelastic behavior, *Experimental Mech.* **37**, 62-68.
48. Zyczkowski, M. (1996) Optimal structural design under creep conditions, *Appl. Mech. Rev.* **49**, 433-446.
49. Mai, Y.-W., Chen, X.-H. and Wong, S.-C. (1998) Fracture Characterization of Structure-Property Relationship of Polymer Blends, book chapter for *Polymer Characterization Techniques and Their Application to Blends*, Simon, G. P. (ed.), American Chemical Society, Washington, DC, to be published.

DEVELOPMENT OF NON-UNILATERAL DAMAGE FIELD IN CREEPING PLATES

A. BODNAR, M. CHRZANOWSKI
Cracow University of Technology
ul.Warszawska 24, 31-155 Kraków, Poland

Abstract

Paper deals with damage development in plates made of materials which exhibit asymmetric behaviour with respect to its deterioration (different tension and compression failure). Material properties are assumed to be time dependent, and analysis is carried out within two time periods: in the first damage is nucleated till macrodefect appearance, and in the second one macrodefects are spreading until they span plate thickens. Ratio of these two time periods is considered as safety margin of analysed structure.

1. Introduction

There are two characteristic features which are inherent to materials exhibiting time dependent behaviour: accumulation of irreversible deformations and deterioration. Both are interrelated, though for a long time they were treated independently. As many theories of creep were pursued [1] the effects of time dependent growth of material deterioration was described initially only in the form of so called damage summation rules (Palmgren [2] and Miner [3] for fatigue, and Robinson for creep [4]). It was not earlier than mid-fifties when it became notorious that both phenomena should be encompassed within a uniform set of constitutive equations. It was mainly due to Kachanov's damage evolution law [5], and subsequently Rabotnov [6] who put both damage and strain rates into the frame of so called „kinetic equations of creep". Their works were then followed by numerous authors (Hayhurst [7], Chaboche [8], Murakami and Ohno [9], Krajcinovic and Fonseka [10] and others) and resulted in extensive description of material behaviour related to damage growth in the frame of Continuum Damage Mechanics which introduces a new state variable responsible for material deterioration.

However, the behaviour of structures made of materials which exhibit above properties received less attention. For engineering application it is of significant importance to study the influence of time dependent phenomena like deformation and deterioration upon overall behaviour of structures. With respect to structures' safety development of damage plays decisive role and has to be studied with reference to local mechanisms of material deterioration.

R. Wang (ed.), IUTAM Symposium on Rheology of Bodies with Defects, 267–276.

In the present paper the effect of compressive stresses upon the creep failure of clamped and simply supported plates is analysed. A combined theory by Hayhurst [7] and Lemaitre [11] is used here to describe damage evolution law, whereas a non-stationary creep theory coupled with damage is used to describe structure's deformation. The main interest is focused on analysis of progressive development of damage which results in three distinctive stages corresponding to nucleation of first macro-crack, its propagation throughout structure's body, and to the formation of collapse mechanism. The corresponding times limiting these periods are denoted as t_1, t_2 and t_3, and referred to as times of First Crack Appearance (FCA), Through-body Crack Appearance (TCA) and Time of Structure Collapse (TSC), respectively. However, the analysis in third period $t_2 \leq t \leq t_3$ will not be carried out here as a combine approach of Fracture Mechanics and Continuum Damage Mechanics should be applied in this period.

Time t_1 can be considered as a lower limit of structure's safe exploitation and the ratio t_2/t_1 can be therefore used as an approximation of safety margins.

The paper is in sequel to [12] and [13] but here, in contrast to these papers, the process is considered for time $t > t_1$ i.e. for period after FCA.

2. Constitutive equations

Time dependent deformations of structure's body are described by steady-state creep theory coupled with damage variable. As elastic deformations are assumed to be non-negligible, the non-stationary creep-damage theory governs the structure's behaviour through the following set of differential equations:

$$\varepsilon_{ij} = \varepsilon^{e}_{ij} + \varepsilon^{c}_{ij}, \tag{1}$$

$$\varepsilon^{e}_{ij} = D^{-1}_{ijkl}\,\sigma_{kl}, \tag{2}$$

$$\frac{\partial \varepsilon^{c}_{ij}}{\partial t} = \gamma \left(\frac{\sigma_{eff}}{1-\omega}\right)^{n} \frac{\partial \sigma_{eff}}{\partial \sigma_{ij}}, \tag{3}$$

where: ε_{ij}, ε^{e}_{ij}, ε^{c}_{ij} - total, elastic and creep strain tensors, respectively, σ_{ij} - stress tensor, D_{ijkl} - elastic constants matrix, γ and n - steady-state creep material constants, σ_{eff} - Huber-Mises effective stress, and ω is scalar damage parameter ($0 \leq \omega \leq 1$), t - time.

The evolution law for damage parameter ω used here is [12]:

$$\frac{\partial \omega}{\partial t} = A\left[\alpha \max\left\{\frac{\langle \sigma_{\max}\rangle}{1-\omega}, \frac{\langle -\sigma_{\min}\rangle}{1-h\omega}\right\} + (1-\alpha)\frac{\sigma_{eff}}{1-\omega}\right]^{m}, \tag{4}$$

where: A and m - damage evolution law constants, α - parameter which characterises failure mechanism mode ($0 \leq \alpha \leq 1$), $\sigma_{\max}$ and $\sigma_{\min}$ - maximal positive and minimal negative principal stresses, h - parameter responsible for direct influence of negative principal stress upon deterioration process ($0 \leq h \leq 1$), and $\langle\ \rangle$ denote Macauley brackets.

The case of $\alpha \neq 1$ and $h \neq 1$ Eq. (4) defines so called *non-unilateral* behaviour of materials, which respond differently for tensile and compressive stresses. For $h = 1$

material is equally sensitive to tension and compression independent of value of α, and is called *bilateral* one.

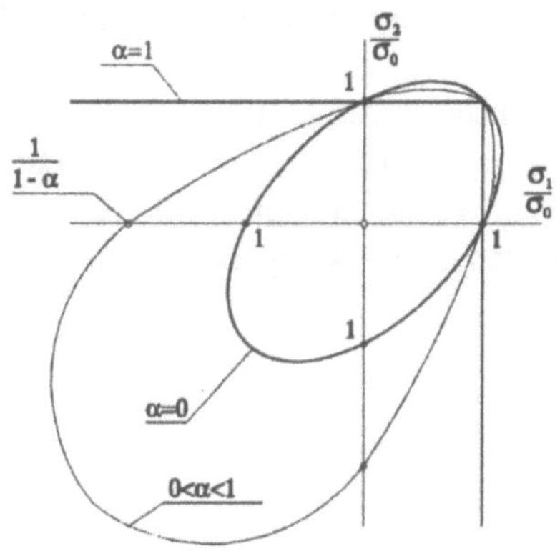

Figure 1. Isochronous failure curves for *unilateral* material.

A material described by the simplified version of Eq. (4), used by many authors

$$\frac{\partial \omega}{\partial t} = A\left[\alpha \frac{\sigma_{\max}}{1-\omega} + (1-\alpha)\frac{\sigma_{eff}}{1-\omega}\right]^m, \tag{5}$$

for which direct effect of compressive stress is neglected, will be called a *unilateral* one.

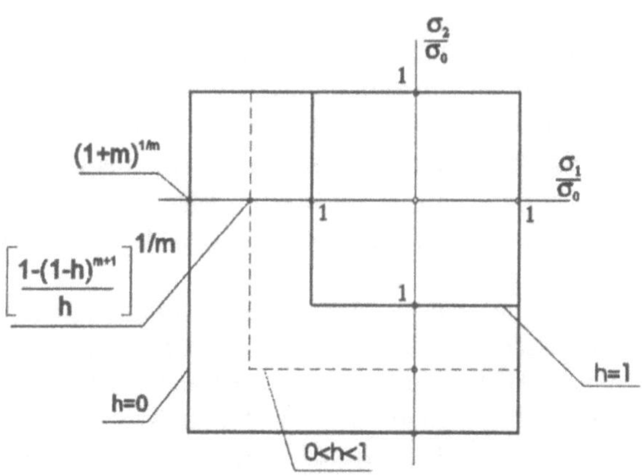

Figure 2. Isochronous failure curves for *non-unilateral* material and α = 1.

The different behaviour of above materials can be illustrated by the shape of isochronous curves, which are loci of different principal stresses combination to yield failure at the same time. In *Fig.* 1 the influence of α is shown for a *unilateral* material.

Two limiting cases $\alpha = 1$ and $\alpha = 0$ correspond to two different failure modes: inter- and trans-crystalline ones, respectively. The case of $\alpha = 1$ reduces Eq. (5) to that of classical Kachanov-Rabotnov theory of brittle creep failure [5,6], whereas the case of $\alpha = 0$ yields Huber-Mises ellipse.

The effect of h parameter is demonstrated by *Fig.* 2 which shows isochronous curves for $\alpha = 1$. Finally, combined effect of both α and h is shown in *Fig.*3: for $\alpha = 0,5$ and $m = 2$ different values of h give different curves within limit curves of *unilateral* and *bilateral* materials (shown by dashed lines).

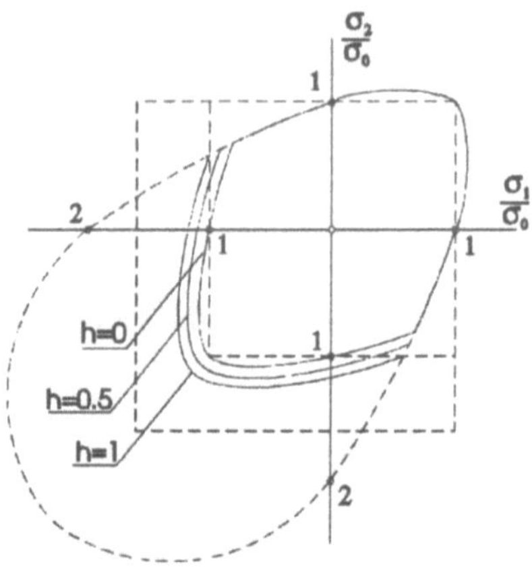

Figure 3. Isochronous failure curves for *non-unilateral* material.

3. Numerical examples and results

The rectangular clamped and simply supported plates with sides length equal to 1.0 and 2.0 m and thickness of 0.10 m under uniform pressure were analysed for different values of h equal to 0, 0.5 and 1.0. Parameter α was set to two values of 0.5 and 1.0 (in the case of α =0 the direct effect of compressive stress is cancelled, c.f. Eq. (4)). Remaining material constants are: E=0.102·10^6 MPa, ν=0.33, n =6.8, m =5.79, γ=1.38·10^{-24}(MPa)$^{-n}$ h^{-1} , A=1.08·10^{-20}(MPa)$^{-m}$ h^{-1} (Ti-6Al-2CR-2Mo alloy at temperature 675 K).

Immediate goal of the analysis carried out by means of the Finite Element Method for structure discretisation and Euler's procedure for time integration is evaluation of FCA and TCA times, and - in particular - their ratio t_2/t_1. In the computer code developed by authors, that enables analysis of thin plates as well as those of moderate thickness, the layered degenerated isoparametric eight-node Serendipity shell elements with reduced integration are employed. Ten layers and two-point Gaussian quadrature for volume integration were adopted. The time t_1 (FCA) is identified with $\omega = 1$ condition fulfilled in any layer and Gaussian point. When this condition is reached in all ten layers of a Gaussian point, the time is referred to as t_2 (TCA). For time $t > t_1$ calculation is performed for modified structure geometry i.e. excluding integration points in which $\omega = 1$.

The results were then transformed by a computer program which drew the lines connecting consecutive points in which condition $\omega = 1$ was fulfilled to provide visualisation of plates cracking process for time $t > t_1$.

Results of calculations are given in Table 1 (clamped plates) and Table 2 (simply supported plates).

TABLE 1. Safety margins in clamped plates made of *non-unilateral* and *unilateral* materials

α	*h*	*Non-unilateral* and *bilateral* material (direct compression effect)			*Unilateral* material (no compression effect [14])		
		t_1 [10^5 hrs]	t_2 [10^5 hrs]	t_2/t_1	t_1 [10^5 hrs]	t_2 [10^5 hrs]	t_2/t_1
0.5	0	0.7259	1.5399	2.1214	0.8709	1.9720	2.2644
	0.5	0.6995	1.4370	2.0543			
	1.0	0.6177	1.0344	1.6746			
1.0	0	0.4509	1.1725	2.6004	0.5220	1.3593	2.6041
	0.5	0.4334	1.1139	2.5702			
	1.0	0.3682	0.7176	1.9489			

TABLE 2. Safety margins in simply supported plates made of *non-unilateral* and *unilateral* materials

α	*h*	*Non-unilateral* and *bilateral* material (direct compression effect)			*Unilateral* material (no compression effect [14])		
		t_1 [10^5 hrs]	t_2 [10^5 hrs]	t_2/t_1	t_1 [10^5 hrs]	t_2 [10^5 hrs]	t_2/t_1
0.5	0	0.4264	0.5217	1.2235	0.5334	0.6645	1.2458
	0.5	0.4087	0.4932	1.2068			
	1.0	0.3626	0.4056	1.1186			
1.0	0	0.3647	0.4731	1.2972	0.4306	0.5603	1.3012
	0.5	0.3489	0.4496	1.2886			
	1.0	0.2949	0.3380	1.1462			

The final stages of plates cracking, corresponding to above Tables, are shown in *Figs* 4 to 7 where the location of FCA and TCA are depicted by ○ and ▲, respectively.

4. Conclusions

The results of analysis has shown considerable influence of compressive stress upon creep rupture behaviour of analysed structures and that the process for clamped and simply supported plates differ quantitatively though they exhibit similarity with respect to the influence of parameters α and h.

It is seen from Table 1 and 2 that for higher value of α both times FCA and TCA decrease. But in all cases, independent of value of h, the ratio of t_2/ t_1 is higher for higher value of α. It means that for materials which exhibit inter-crystalline failure mode rather then trans-crystalline one, the safety margin is greater. This observation is valid also for materials which are compression insensitive (*unilateral materials*).

For *non-unilateral materials* the influence of compressive damage is sound when compared with *unilateral materials.* Times of FCA and TCA are always shorter (for both, clamped and simply supported plates) *non-unilateral* material; and higher the value of h greater FCA and TCA times reduction. The same applies to the safety margin of t_2/ t_1, which decreases with higher contribution of compressive damage (higher value of h). Therefore, if material exhibits *non-unilateral* behaviour (what should be observed by experimentally backed up isochronous failure curves) - this phenomenon has to be taken into account in structural analysis.

Boundary conditions may have considerable influence on structure's behaviour, too: the safety margin for clamped plate can be almost twice as much as for simply supported one. This observation is obvious in the light of cracking networks shown in *Figs* 4 to 7. In all analysed cases the fracture process starts in the most stressed regions. For clamped plates the cracks appear first on the upper surface of the plates along its edges, and then tend to become simply supported ones; in consecutive stages critical cross-cracking is developed. For simply supported plates this critical cracking starts immediately after FCA.

5. References

[1] Garofalo, F., Fundamentals of creep and creep-rupture in metals, MacMillan, New York, 1965.
[2] Palmgren, A., Die Lebensdauer von Kugellagern, Z. Ver. Deutsch Ing., **68**, 1924.
[3] Miner, M.A., Cumulative damage in fatigue, Trans. ASME, J. of Applied Mechanics, **12**, pp A159-A164, 1945
[4] Robinson, L. R., Effect of temperature variation on the creep strength of steels, *Trans. ASME*, **60**, pp 253-259, 1938.
[5] Kachanov, L. M., On the time to rupture in creep conditions (in Russian), *Izv. Ak. Nauk SSSR OTN*, **8**, pp 26-31, 1958.
[6] Rabotnov, Yu.N., Creep problems in structural members, North Holland Publ. Co., Amsterdam, 1969.

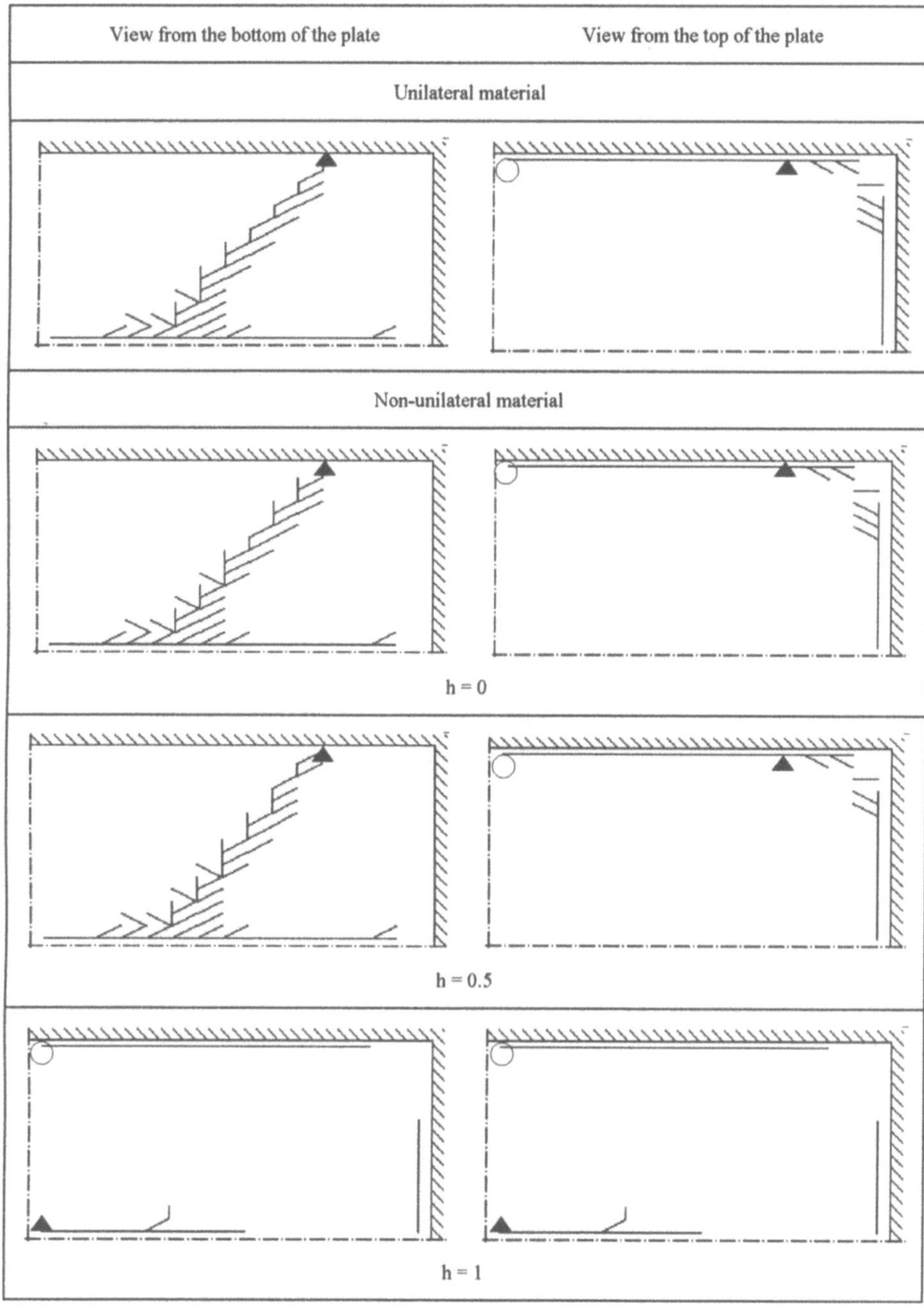

Figure 5. Crack networks at time of TCA in clamped plates; $\alpha = 1$

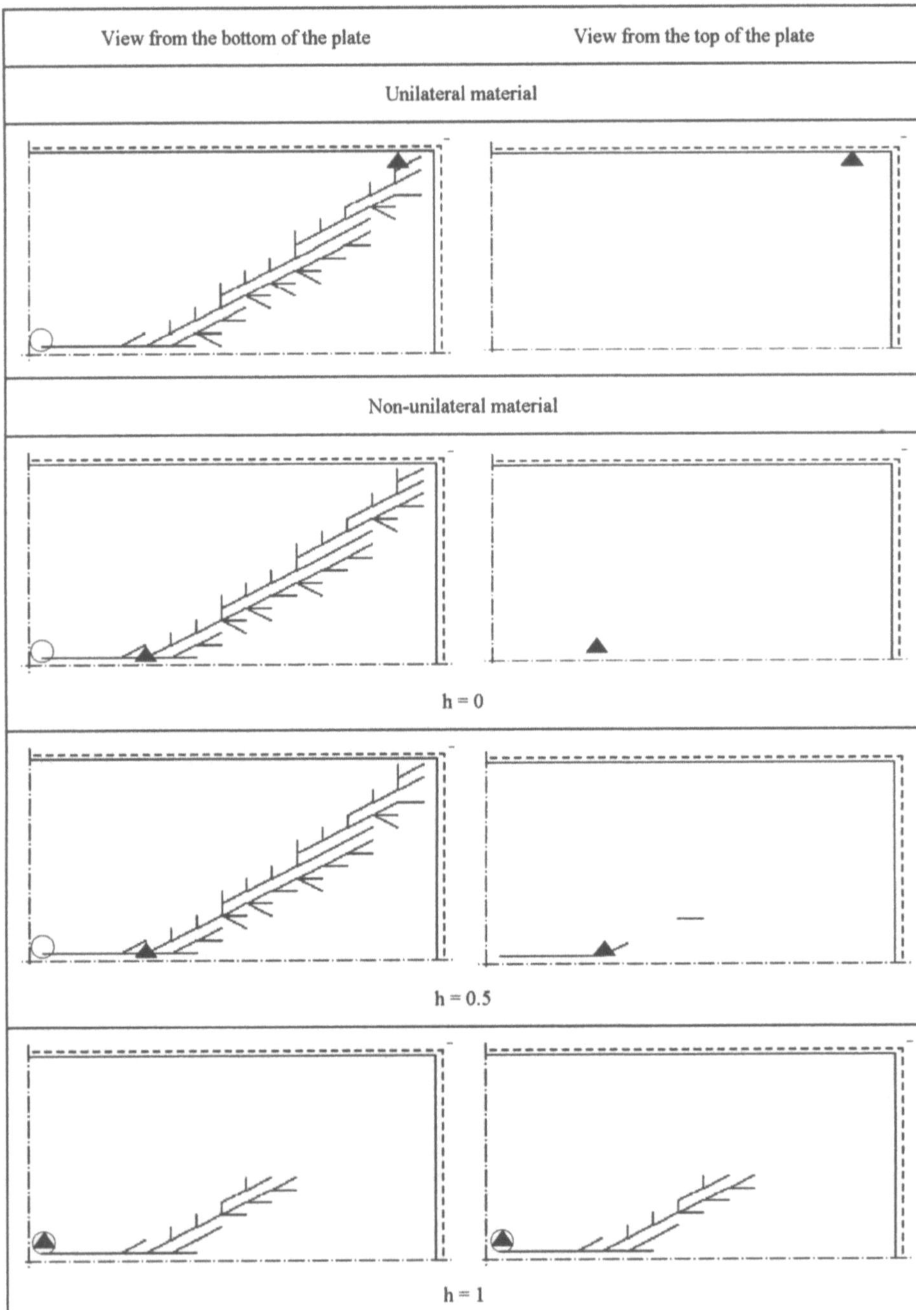

Figure 6. Crack networks at time of TCA in simply supported plates; $\alpha = 0.5$

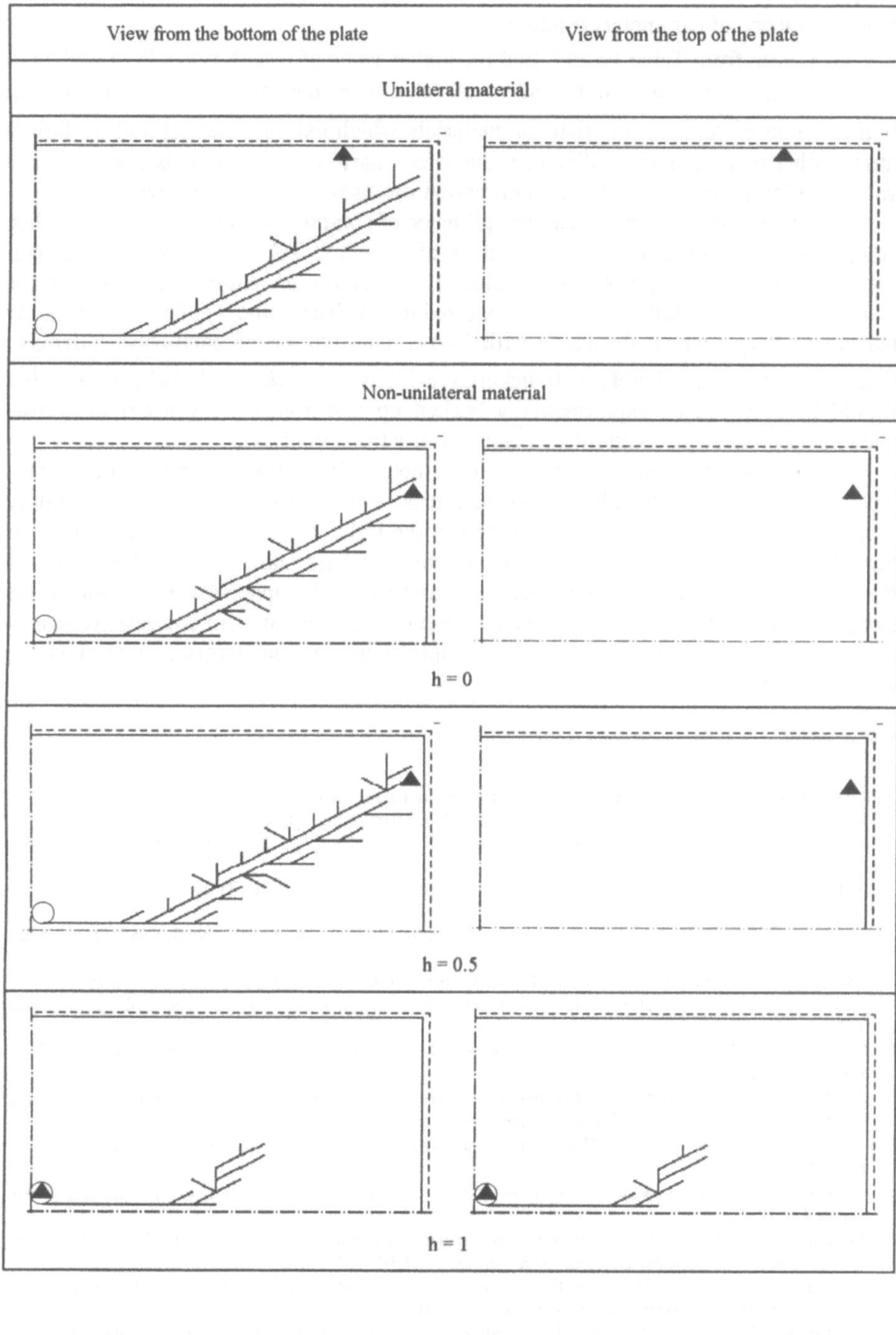

Figure 7. Crack networks at time of TCA in simply supported plates; $\alpha = 1$

simply supported plates differ quantitatively though they exhibit similarity with respect to the influence of parameters α and h.

It is seen from Table 1 and 2 that for higher value of α both times FCA and TCA decrease. But in all cases, independent of value of h, the ratio of t_2/t_1 is higher for higher value of α. It means that for materials which exhibit inter-crystalline failure mode rather then trans-crystalline one, the safety margin is greater. This observation is valid also for materials which are compression insensitive (*unilateral materials*).

For *non-unilateral materials* the influence of compressive damage is sound when compared with *unilateral materials*. Times of FCA and TCA are always shorter (for both, clamped and simply supported plates) *non-unilateral* material; and higher the value of h greater FCA and TCA times reduction. The same applies to the safety margin of t_2/t_1, which decreases with higher contribution of compressive damage (higher value of h). Therefore, if material exhibits *non-unilateral* behaviour (what should be observed by experimentally backed up isochronous failure curves) - this phenomenon has to be taken into account in structural analysis.

Boundary conditions may have considerable influence on structure's behaviour, too: the safety margin for clamped plate can be almost twice as much as for simply supported one. This observation is obvious in the light of cracking networks shown in *Figs* 4 to 7. In all analysed cases the fracture process starts in the most stressed regions. For clamped plates the cracks appear first on the upper surface of the plates along its edges, and then tend to become simply supported ones; in consecutive stages critical cross-cracking is developed. For simply supported plates this critical cracking starts immediately after FCA.

5. References

[1] Garofalo, F., Fundamentals of creep and creep-rupture in metals, MacMillan, New York, 1965.

[2] Palmgren, A., Die Lebensdauer von Kugellagern, Z. Ver. Deutsch Ing., **68**, 1924.

[3] Miner, M.A., Cumulative damage in fatigue, Trans. ASME, J. of Applied Mechanics, **12**, pp A159-A164, 1945

[4] Robinson, L. R., Effect of temperature variation on the creep strength of steels, Trans. ASME, **60**, pp 253-259, 1938.

[5] Kachanov, L. M., On the time to rupture in creep conditions (in Russian), Izv. Ak. Nauk SSSR OTN, **8**, pp 26-31, 1958.

[6] Rabotnov, Yu.N., Creep problems in structural members, North Holland Publ. Co., Amsterdam, 1969.

[7] Hayhurst, D. R., Creep rupture under multi-axial states of stress, J. Mech. Phys. Solids, **20**, pp 381-90,. 1972.

[8] Chaboche, J. L., Continuous damage mechanics - a tool to describe phenomena before crack initiation. *Nucl. Eng. Design*, **64**, pp 309-319, 1981.

[9] Murakami, S., Ohno, N., A continuum theory of creep and creep damage. In Proceedings Creep in Structures, ed. A. R. S. Ponter & D. R. Hayhurst. Springer-Verlag, Berlin, , pp 422-44, 1981.

[10] Krajcinovic, D. , Fonseka, G. U., The continuous damage theory of brittle materials. *J. Appl. Mech.*, 48, pp 809-24, 1981.

[11] Lemaitre, J., Damage constitutive equations, CISM Course on Damage Continuum Mechanics, 13-16 September 1986.

[12] Bodnar, A., Chrzanowski, M., A non-unilateral damage in creeping plates, Trans. Fourth IUTAM Symp. on Creep in Structures, Cracow 1990, Springer Verlag, pp 287-293, 1991.

[13] Bodnar, A., Chrzanowski, M., Latus, P., Lifetime evaluation of creeping structures, Proc. 5th Int. Conf. on Creep of Materials, Grosvenor Resort, Florida, pp 461-469, 1992.

[14] Bodnar, A., Chrzanowski, M., The analysis of life extension for creeping plates, Proc. Inter. Symp. Materials Ageing and Component Life Extension, Milan, **Vol. 2**, pp 1221-1229, 1995.

PLASTIC AND CREEP INSTABILITY OF SHELLS WITH INITIAL IMPERFECTIONS

V.S. GUDRAMOVYCH
National Academy of Sciences of Ukraine
Institute of Technical Mechanics
15, Leshko Popel St., 320600, Dnepropetrovsk, Ukraine

Abstract. - The critical states of cylindrical shells with initial imperfections having plastic and creep deformations are investigated. These states are defined as the load values at which the shell deflections rise sharply. The method of successive loading for which the reaction of the shell to small loading steps is studied and the process of deformation is considered is used. Some experimental results for shells with irregular imperfections are given.

1. Introduction

The actual structures feature different kinds of initial imperfections and inhomogeneity of material properties. These factors are caused by production operations or by diverse operational conditions.

Their effect on the structural behavior under load is governed by the degree of inhomogeneity they give to the structure, and it can be considerable. New effects occur under plastic deformation and creep [1-3]. The inhomogeneity of material properties is determined by residual stresses or the parameters that characterize the deformation anisotropy of the material which emerges in the process of complex loading in the plastic domain. If the flow theory with anisotropic kinematic hardening is used, these parameters are residual microstresses [3, 4].

On the other hand, the geometrical imperfections and residual stresses are included in the basic theoretical models whereby the behavior of nonlinear systems under plastic deformation and creep is studied [2, 5, 6].

In this article the deformation and critical states of shells with shape imperfections under plastic deformation and creep are investigated. Some test results are presented. To be specific, we shall consider cylindrical shells.

2. Critical states of shells

Below are the results of investigation of the deformation and critical state of elastoplastic shells with initial imperfections [3,7]. To be specific, cylindrical shells loaded with the external pressure q and axial compressive force T are considered.

The equations describing the stressed-and-strained state of the shell are obtained using the variational principle of virtual displacements [8].

R. Wang (ed.), IUTAM Symposium on Rheology of Bodies with Defects, 277–289.

The method of successive loading is used. This method is a version of the methods of parameter continuation of solution where the loads are taken as the parameters [3,9-12].

According to this method, the shell loading process is divided into successive loading series. The increments of the loads acting on the shell Δq_i and ΔT_i (i is the step number) on each loading step are small, and the stressed-and-strained state on each loading step is determined. The equations used for investigation of the shell behavior make it possible to determine the increments of the radial displacement and other components of the stressed-and-strained state on each loading step. These increments are added together.

In these investigations the deformation theory of plasticity is used, the relationships for this theory being formulated in terms of the stress increments $\Delta\sigma_{ij}$ and strain increments $\Delta\varepsilon_{ij}$. The increments of the radial displacement Δw_i being determined are added together, and $w=\sum \Delta w_i$.

The calculations for specific initial imperfections are performed.The critical state is defined as the load value at which the shell deflections increase sharply [3,13]. This situation is shown schematically in Fig.1. For one load q or T the critical state is determined by the limiting point A, and for two loads it is determined by the limiting surface.

Below are the results of numerical analysis of the effect of initial imperfections on the deformation and critical states of the shells.

Based on the above relationships, one can calculate the stressed-and-strained state of a shell at the given level of loads. Increasing the loads, one can find such value there of at which the shell deflection rises sharply. This corresponds to the critical state.

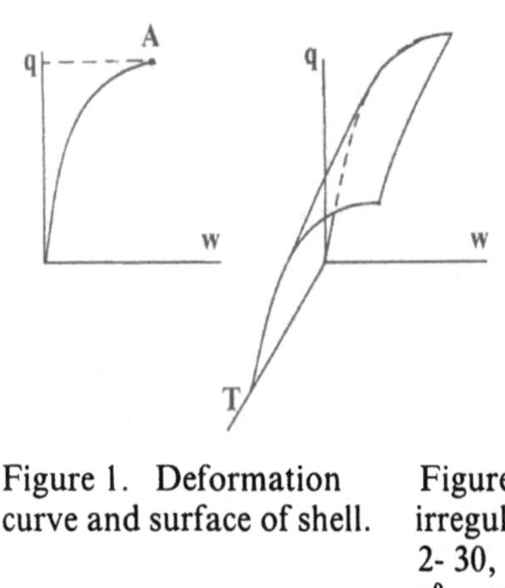

Figure 1. Deformation curve and surface of shell.

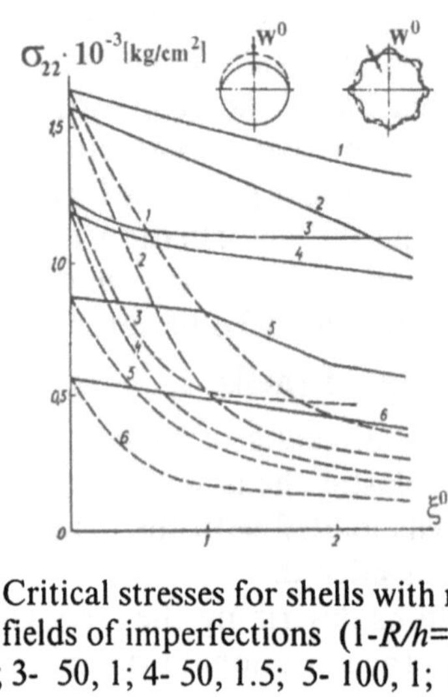

Figure 2. Critical stresses for shells with regular and irregular fields of imperfections (1-R/h=30, L/R=1; 2- 30, 1.5; 3- 50, 1; 4- 50, 1.5; 5- 100, 1; 6- 100, 1.5), $\xi^0=w^0/h$, $\sigma_{22}=qR/h$

Fig.2 shows the critical stresses of the shells under external pressure with regular (dashed curves) and irregular (solid curves) fields of initial shape imperfections. These types of imperfection are shown at the top of the figure. The regular initial imperfections coincide with the shape at which the carrying capacity of shells without imperfections reaches its limit. The irregular initial imperfections impart higher rigidity to the shell and thus the critical loads for such shells are higher.

It may be assumed that it is possible to set up fields of irregular shape imperfections such that the critical loads do not decrease and even increase in comparison with those for shells without imperfections.

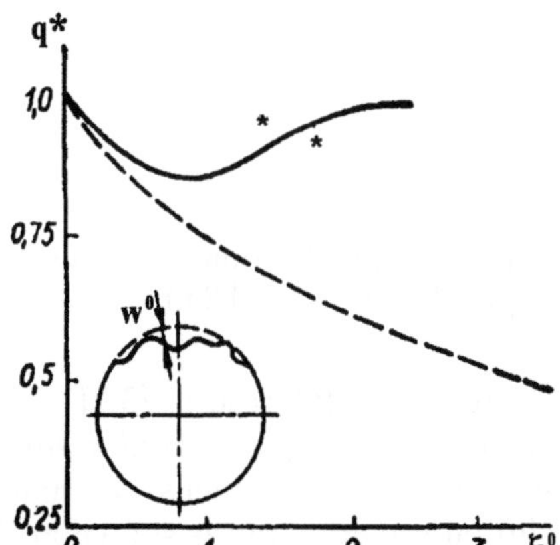

Figure 3. Critical load of shells with respect to hardening of materials ($q^* = q/q_0$, q, q_0 - critical load for shell with and without imperfection, R/h=80,L/R=0.5; * -experimental results)

A shell undergoing initial deflection is accompanied by the hardening of the material. The degree of hardening varies along the cross-sectional circle of the shells as does the change in the initial deflection. The computational methods and algorithms developed allow for these circumstances. Fig. 3 depicts the dependence of dimensionless critical load on $\xi^0 = w^0/h$ for the elastoplastic cylindrical shell with the irregular shape imperfections shown on the lower left. The dashed and solid curves correspond to the analysis of the shell without and with allowance for the uniform hardening of the material, respectively. The symbols * show experimental results. In Fig. 2, 3, 4 captions and below R, h, L are the radius, thickness and length of the shell, respectively.

Fig. 4 depicts the schemes that characterize the development of the deflections of the shell with the shape imperfections shown at the top of the figure. As in Fig. 3, the solid and dashed curves correspond to analyzing the shell with and without allowance for the nonuniform hardening.

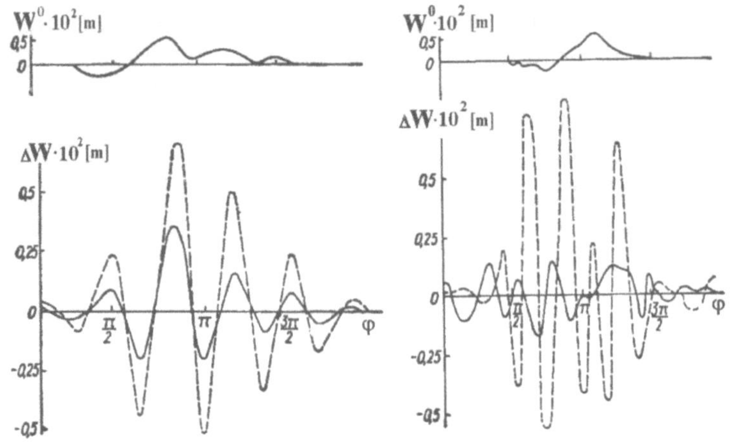

Figure 4. Development of the deflections for shells (to the left: R/h=76, L/R=1.9; to the right: R/h=65, L/R=0.85)

Let us present some experimental results [3,7].

Fig. 5 depicts calculated and experimental results for three types of shells under external pressure. The shells are made of the AMG-6m aluminum alloy. The initial imperfection is shown at the top of the figure. The solid curves show the calculated results, and the dashed curves approximate the experimental data. The shell with R/h=150 exhibited no

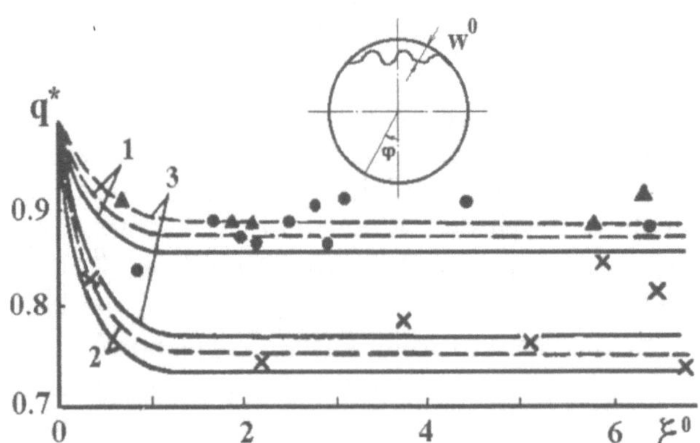

Figure 5. Experimental and calculated results for shells under external pressure (1 - •, R/h=75, L/R=1.5; 2 - *, R/h=100, L/R=1.5; 3 - ▲ , R/h=150, L/R=1.5)

plastic deformation. There is larger difference between the calculation and the experiment (~ 18 %) for the *R/h=150* shell. This may be explained by a significant change in the shape of the shell under loading. In Fig. 5 q^* is the same as in Fig. 3.

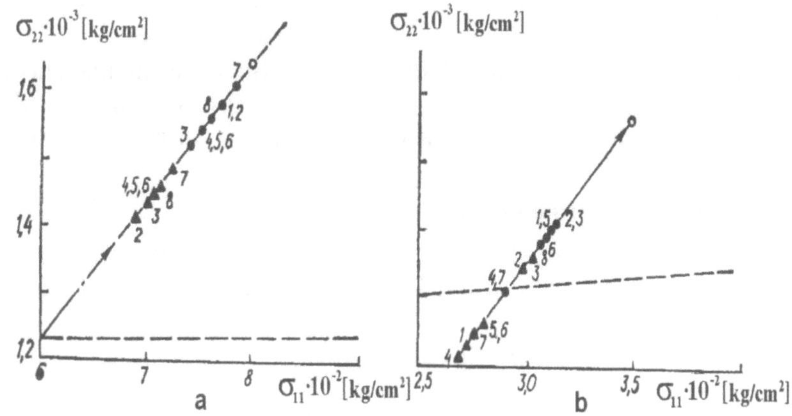

Figure 6. Experimental and calculated results for shell under axial compression and external pressure (a - *R/h*=50, *L/R*=1; b - *R/h*=50, *L/R*=1.5; - ▲- calculated data; • - experimental data; o - shell without imperfections)

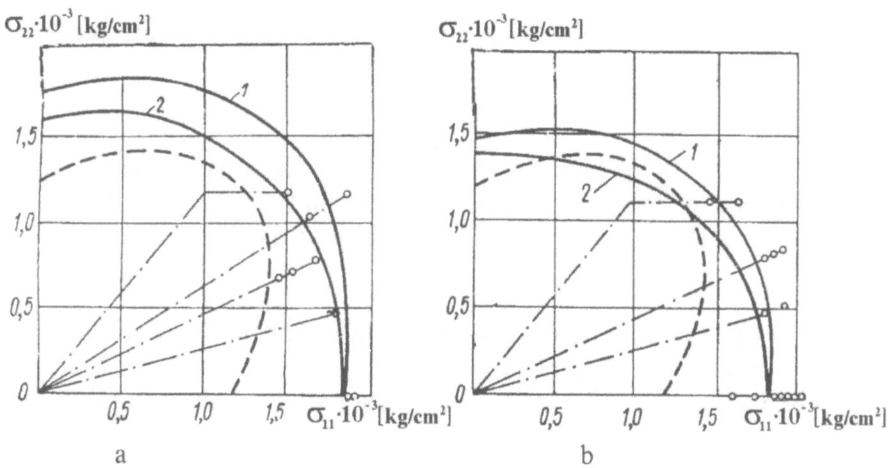

Figure 7. Experimental and calculated results for shells (a - *R/h*=30, *L/R*=1; b - *R/h*=30, *L/R*=1.5; 1-calculated results for ξ^0=0.01, 2- the same for ξ^0 =1.2, o- experimental data)

Fig. 6, 7 show the results of the calculation of the critical load and experimental data for shells with irregular initial defection under axial compression and external pressure. The numbers in Fig. 6 indicate the shell numbers.

The dashed curve in Fig. 6, 7 is a part of the Mises ellipse that separates the elastic of deformation from the plastic one. In Fig. 6 the loads were varied proportionally, and in Fig.7 the loads were varied proportionally too or loading paths were piecewise linear (for two loadings).

A comparison between the calculations and the experiments shows that the computational methods and algorithms developed are suitable for describing the behavior (deformation and critical states) of elastoplastic shells with shape imperfections.

3. Basic equations for structures with creep

To be specific again, we shall consider cylindrical shells under creep. For the material the hypothesis of hardening and the equations of state in the form [1]

$$\dot{p}_i = g(p_i, \sigma_i)\sigma_i \tag{3.1}$$

together with the power law $g=\sigma_i^{n-1}$ are used. Here, $\dot{p}_i, p_i$ and σ_i are creep velocities, creep deformations and stresses intensity.

The plastic deformation is described using the relationships of the flow theory with anisotropic kinematic hardening. This theory is convenient for the analysis of complex loading. The equation for the loading surface is

$$f = (S_{ij}-s_{ij})\ (S_{ij}-s_{ij}) - 2/3\, c^2 = 0\,, \tag{3.2}$$

where S_{ij} are the stress deviators and s_{ij} are the residual microstresses.

c and s_{ij} may be expressed as follows [3,4]

$$c=\sigma_s\,,\quad c= 3/2 S_{ij} S_{ij}[1+\rho(\varepsilon^P_{ij}, \mu)]\,,\quad s_{ij}=3g_0\varepsilon^P_{ij}\,, \tag{3.3}$$

where ε^P_{ij} are the plastic deformations, ρ is the function that allows for Baushinger's effect, μ is Lode's parameter and σ_s is the yield point.

With a sufficient degree of accuracy we may assume that

$$\rho = \sigma_s^- / \sigma_s^+\,. \tag{3.4}$$

The superscripts + and - denote the values of the yield point in tension and in compression, respectively.

The values of s_{ij} for different loading paths are given in [3,4]. Finally we have

$$s_{11}=A_1(\sigma_{11})\sigma_{11}\,,\quad s_{22}=A_2(\sigma_{22})\sigma_{22}\,, \tag{3.5}$$

where $A_1(\sigma_{11}), A_2(\sigma_{22})$ are the specific functions for different complex loading paths.

Applying for the description of plastic deformation the relationships

$s_{ij}^p = s_{ij}/3g_0$ for the intensity of creep deformation we have [1,14]

$$p_i = e_i - s_i/3g_0 - \sigma_i/E, \tag{3.6}$$

where $p_i=(2/3\, p_{ij}p_{ij})^{1/2}$, $e_i=(2/3\, e_{ij}e_{ij})^{1/2}$, $s_i=(2/3\, s_{ij}s_{ij})^{1/2}$, $\sigma_i=(2/3\, S_{ij}S_{ij})^{1/2}$, and e_{ij} are the strain deviators.

According to the deformation theory for e_{ij} we have $e_{ij} = 3e_i S_{ij}/2\sigma_i$ [1]. The variation δe_{ij} is

$$\delta e_{ij} = 3/2\, [e_i(\delta S_{ij} - S^0_{ij}\delta\sigma_i)/\sigma_i + S^0_{ij}\delta e_{ij}], \quad S^0_{ij} = S_{ij}/\sigma_i. \tag{3.7}$$

Linearization of (3.6) and allowing for the fact that

$$\delta p_i = g\int_0^{p_i}\frac{1+b}{g\sigma_i}\delta\sigma_i\, dp_i, \quad b = \frac{\sigma_i}{g}\partial g/\partial\sigma_i \tag{3.8}$$

gives

$$\delta e_i = \delta\sigma_i/E + \delta S_i/3g_0 + g\int_0^{p_i}\frac{1+b}{g\sigma_i}\delta\sigma_i\,\delta p_i. \tag{3.9}$$

Rearrangement of (3.7) with allowance for (3.9) gives

$$\delta S_{ij} = [2E\delta e_{ij}/3 + \sigma^0_{ij}(\chi\delta\sigma - g\int_0^{\xi}\frac{1+b}{g}\,\delta\sigma_i\, d\xi - E\delta s_i/3g_0)]/(1+\chi), \tag{3.10}$$

where

$$\chi = \xi + Es_{ij}/3g_0\sigma_i, \quad \xi = Ep_i/\sigma_i.$$

ξ is the dimensionless time obtained as a result of going over from dt to dp_i by (3.1) $\left(\frac{d}{dt} = \frac{d}{dp_i}\frac{dp_i}{dt} = \frac{d}{dp_i}g\sigma_i\right)$ [1,14].

The generalized moments and forces in the middle surface of the shell are

$$M_{11} = \int(2\,\delta S_{11} + \delta S_{22})zdz, \quad M_{12} = \int\delta S_{12}zdz$$

$$T_{11} = \int(2\,\delta S_{11} + \delta S_{22})zdz, \quad T_{12} = \int\delta S_{12}dz, \quad 11 \rightleftarrows 22. \tag{3.11}$$

Substitution of (3.10) in (3.11) with allowance for the geometrically nonlinear relationships for δe_{ij} gives

$$M_{11} = [-1/2D(2\tilde{w}_{,xx} + \tilde{w}_{,yy}) + \sigma_{11}{}^0\tilde{M}_i\,]/(1+\chi)$$

$$M_{22} = [-1/2D(2\tilde{w}_{,yy} + \tilde{w}_{,xx}) + \sigma_{22}{}^0\tilde{M}_i\,]/(1+\chi)$$

$$M_{12}=(-1/2D\widetilde{w}_{,xx}+\sigma_{12}^{\ 0})/(1+\chi) \tag{3.12}$$

$$T_{11}=\{[2/3\,B(2u_{,x}+v_{,y}+\widetilde{w}/R)+w^2_{,x}-w^{02}_{,x}+(w^2_{,y}-w^{02}_{,y})/2]$$
$$+\sigma_{11}^{\ 0}\widetilde{T}_i\}/(1+\chi)\,,$$

$$T_{22}=\{[2/3\,B(2v_{,y}+u_{,x}+2\widetilde{w}/R)+w^2_{,y}-w^{02}_{,y}+(w^2_{,x}-w^{02}_{,x})/2]$$
$$+\sigma_{22}^{\ 0}\widetilde{T}_i\}/(1+\chi)\,,$$

$$T_{12}=\{[2/3\,B(1/2\,u_{,y}+v_{,x})+1/2(w_{,x}w_{,y}-w_{,x}^{\ 0}w_{,y}^{\ 0})]$$
$$+\sigma_{12}^{\ 0}\widetilde{T}_i\}/(1+\chi),$$

where $\quad B=Eh\,,\quad D=Eh^3/3\,,\quad \widetilde{w}=w-w^0,\quad \sigma_{ij}^{\ 0}=\sigma_{ij}/\sigma_i$

$$\widetilde{M}_i=\chi M_i-E\,M_i^*/3g_0-g\int_0^{\xi}\frac{1+b}{g}\,M_i\,d\xi$$

$$\widetilde{T}_i=\chi\,T_i-E\,T_i^*/3\,g_0-g\int_0^{\xi}\frac{1+b}{g}\,T_i\,d\xi$$

$$T_i=\int\delta\sigma_i\,dz,\quad M_i=\int\delta\sigma_i\,zdz,\quad T_i^*=\int\delta s_i\,dz,\quad M_i^*=\int\delta s_i\,zdz$$

w^0 is the initial displacement, $(\cdot)_{,x}=\partial(\cdot)/\partial x\,,(\cdot)_{,xx}=\partial^2(\cdot)/\partial x^2,\ x\rightleftarrows y$.

The variations $\delta\sigma_i$ and δs_i are

$$\delta\sigma_i=\overline{\sigma}_1\,\delta\sigma_{11}+\overline{\sigma}_2\,\delta\sigma_{22}+3\sigma_{12}\delta_{12} \tag{3.13}$$

$$\delta s_i=3(\,\overline{s}_1\,\delta s_{11}+\,\overline{s}_2\,\delta s_{22}), \tag{3.14}$$

where $\quad \overline{\sigma}_1=\sigma_{11}^{\ 0}-\sigma_{22}^{\ 0}/2,\quad \overline{\sigma}_2=\sigma_{22}^{\ 0}-\sigma_{11}^{\ 0}/2\,,\quad \overline{s}_1=(s_{11}+s_{22}/2)/s_i\,,$

$$\overline{s}_2=(s_{22}+s_{11}/2)/\,s_i\,.$$

This determines T_i and M_i

$$T_i=\overline{\sigma}_1\,T_{11}+\overline{\sigma}_2\,T_{22}+3\sigma_{12}^{\ 0}T_{12}$$

$$M_i=\overline{\sigma}_1\,M_{11}+\overline{\sigma}_2\,M_{22}+3\sigma_{12}^{\ 0}M_{12}. \tag{3.15}$$

Taking into account (3.5), we have from (3.14)

$$\delta s_i=A(\sigma_{11})\,\delta\sigma_{11}, \tag{3.16}$$

where $\qquad A(\sigma_{11})=3[\,\overline{s}_1A_1(\sigma_{11})+\overline{s}_2\,A_2(\sigma_{11})]\,,\qquad$ and

$$M_i^* = A(\sigma_{11})M_{11}\,,\ T_i^* = A(\sigma_{11})T_{11}\,. \tag{3.17}$$

Let us use the equilibrium equations of a shallow shell

$$T_{11,x} + T_{12,y} = 0, \qquad T_{22,y} = T_{12,x} = 0$$

$$M_{11,xx} + M_{22,yy} + 2M_{12,xy} - T_{22}/R + T_{11}w_{,xx} + T_{22}w_{,yy} + 2T_{12}w_{,xy} + q = 0. \tag{3.18}$$

Introducing the force function and allowing for (3.15), (3.17), we obtain from (3.18) the equation of strain consistency [14]

$$(1+\chi)\,\nabla\nabla\,\Phi = B\Gamma(w,w^0) + \chi\Lambda_1\Lambda_1\Phi - g\int_0^\xi \frac{1+b}{g}\,\Lambda_1\Lambda_1\Phi\,d\xi$$

$$-\,E\,A_1(\sigma_{11})\Lambda_1\Phi_{,yy}/3g_0 \tag{3.19}$$

$$T_{11} = \Phi_{,yy}\,,\ T_{22} = \Phi_{,xx}\,,\ T_{12} == -\,\Phi_{,xy}\,.$$

Allowing for (3.12), the third equation in (3.18) gives

$$(1+\chi)U(w,w^0,\Phi) + \chi D\,\nabla\nabla\,\tilde{w} + \ \chi\,(M_i) - g\int_0^\xi \frac{1+b}{g}\,\Lambda(M_i)\,d\xi$$

$$-\,E\Lambda(M_i^*)\,/3g_0 = 0\,. \tag{3.20}$$

Next M_i is eliminated from (3.20) . Finally, instead of (3.20) we obtain

$$U(w,w^0,\Phi) + \chi\,(1+\chi)^{-1}\,D(\,\nabla\nabla\text{-}3/4\Lambda\Lambda)\,\tilde{w} + g\int_0^\xi \frac{1+b}{g}\,\tilde{U}\,d\xi$$

$$-\ E\Lambda(M_i^*)\,/3g_0 = 0\,, \tag{3.21}$$

where

$$\tilde{U} = U(w,w^0,\Phi) + D\,(3/4\Lambda\Lambda + \chi\nabla\nabla)\,\tilde{w}\,/(1+\chi).$$

Allowing for (3.17) we find $\Lambda(M^*{}_i)$ and obtain the equilibrium equation in the form as follows [14]

$$U(w,w^0,\Phi) + \chi\,(1+\chi)^{-1}\,D(\,\nabla\nabla\text{-}3/4\Lambda\Lambda)\,\tilde{w} + g\int_0^\xi \frac{1+b}{g}\,\tilde{U}\,d\xi$$

$$+\,E\Lambda(\sigma_{11})/[3g_0\,(1+\chi) + \sigma_{11}{}^0\,E\Lambda(\sigma_{11})\,]\,[D\Lambda(\,\tilde{w}_{,xx} + 1/2\,\tilde{w}_{,yy})$$

$$+\sigma_{11}{}^0\chi\tilde{U} - g\sigma_{11}{}^0 \int_0^\xi \frac{1+b}{g}\ \tilde{U}\,d\xi] = 0\,. \tag{3.22}$$

In the above equations the following notations are used

$$U(w,w^0,\Phi) = -D\,\nabla\nabla\tilde{w} - \Phi_{,xx}/R + \Phi_{,yy}w_{,xx} + \Phi_{,xx}w_{,yy}$$

$$-\,\Phi_{,xy}w_{,xy} + h\sigma_i\Lambda\,w + q$$

$$\Gamma(w,w^0\,) = w_{,xy}{}^2 - w_{,xy}{}^{0\,2} - w_{,xx}w_{,xy} + w_{,xx}{}^0\,w_{yy}{}^0 + \tilde{w}_{,xx}/R \tag{3.23}$$

$$\nabla(\cdot)=(\cdot)_{,xx}+(\cdot)_{,yy}\,,\ \Lambda(\cdot)=\sigma^0_{11}(\cdot)_{,xx}+2\sigma^0_{12}(\cdot)_{,xy}+\sigma^0_{22}(\cdot)_{,yy}$$

$$\Lambda_1(\cdot)=\sigma^0_{11}[(\cdot)_{,yy},-1/2(\cdot)_{,xx}]+\sigma^0_{22}[(\cdot)_{,xx}-1/2(\cdot)_{,yy}]-3\,\sigma^0_{12}\,(\cdot)_{,xy}\,.$$

If in (3.19), (3.22) we put the plastic strains equal to zero ($A(\sigma_{11})=0$), we shall obtain the equations given in [15].

4. Buckling of shells under creep

The critical state of structures under creep is determined by critical time it takes for the structure to fail.

The approach to studying the time evolution of initial shape imperfections is well defined and has received wide acceptance in studying the creep buckling of such structural elements as shells, plates, and beams [1,6,14,16]. When studying the creep, the loads are constant. The time at which the deflection exhibits a sharp increase is the critical time. Each load level is characterized by its own critical time.

Let us present some results of the analysis of a shell under internal pressure and compression based on the equations given in section 3.

We shall use the power creep law at $n=3$. The shell with initial deflection w^o_{max} is made of the AMG-6m aluminium alloy. Fig. 8 shows the curves that determine the time evolution of the deflection at different values of the internal pressure q. At the moment of buckling the velocity of radial displacement tends to infinity. If the axial load is constant, the internal pressure has a reinforcing effect.

Curves 1, 2, 3 correspond to the values of q equal to zero, 1 kg/cm², 2 kg/ cm².

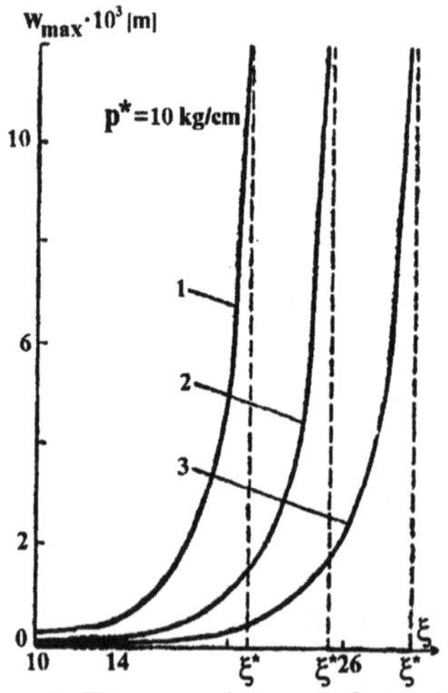

Figure 8. Time evolution of the defections and critical time for shell ($w^0_{max}=4h$, $R/h=176$, $L/R=4.83$, $T^0=250$°C)

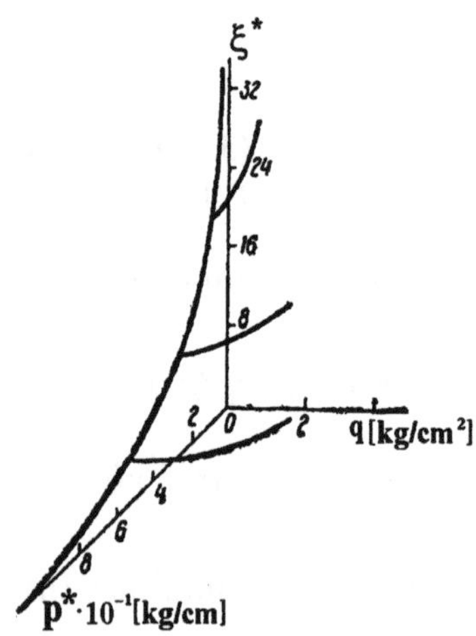

Figure 9. Surface determining the critical state for shell under creep ($R/h=176$, $L/R=4.83$, $T^0=150$°C)

Shown in Fig. 9 in the three-dimensional coordinate system $p^*=T/2\pi R$, q, ξ^* is the surface determining the critical state of the shell under creep for the given initial deflection [17].

In the calculations the deflection was specified as

$$w = f_1 sin\lambda x sin\eta y + f_2 sin^2\lambda x, \;\; \lambda = m\pi/L, \;\; \eta = n/R, \qquad (4.1)$$

where m and n are the numbers of waves under buckling in the axial and lateral directions. During the calculations the convergence with respect to time step was examined. At $d\xi = 0.01$ and $d\xi = 0.001$ the corresponding values of deflection differ by less than 5 percent [14].

If the plastic strain is allowed for, the effect of internal pressure undergoes a qualitative change: with increasing pressure and constant axial load the critical time shortens. When calculating the critical time, in addition to the proportional loading where $k = \sigma_{22}/\sigma_{11} = -0.33$ (path 1), complex step loading paths were realized as well (for example, increasing the compressive force beyond the yield point followed by varying the loads proportionally (σ_{110} = 1440 kg/cm², k = -2) in the plastic domain of deformation (path 2)). To determine the instantaneous plastic deformations for different complex loading paths, the flow theory with anisotropic kinematic hardening [3,4] is used.

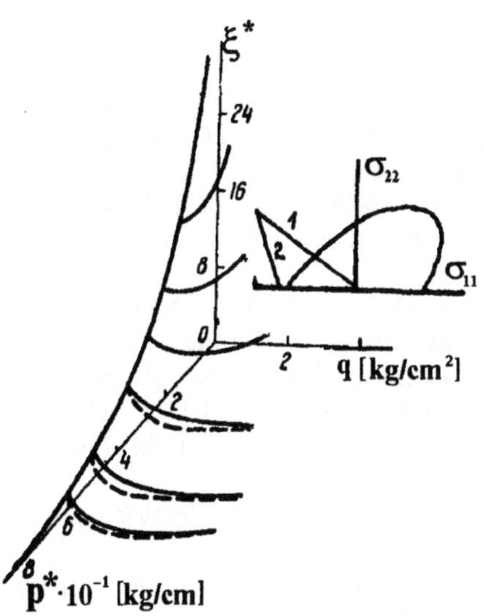

Figure 10. Surface for determining the critical state for shell under creep and plastic deformations (R/h=176, L/R=4.83, T^0=150°C)

In Fig. 10 the loading paths and Mises ellipse are shown at the top right of the figure. The buckling surface is also shown. The solid lines correspond to the calculation in the case of elastic deformation, the dashed lines correspond to allowing for plastic deformation under complex loading for path no 1. The shell is made of the AMG-6m aluminium alloy.

Loading path no 1 gave a lower value of the critical time in comparison with loading path no 2. Allowing for plastic deformation has an insignificant effect on the value of the critical time for the specific shells considered above. Similar results were obtained when studying the creep buckling of beams [18].

5. Conclusions

The results of the investigation described above show a great diversity of phenomena that are due to the effect of initial imperfections on structural behavior. This effect can be significant.

Of great importance is the development of a database containing data on different structural initial imperfections which are used in the design of a variety of structures of modern engineering [19,20].

Such studies are of importance for the development of optimum designs and choice of optimum technologies. It is possible to set up special shape imperfections deliberately in order to increase the carrying capacity and service life of structures.

6. References

1. Rabotnov, J.N. (1969) *Creep problems in structural members*, North-Holland Publ. Co, Amsterdam.
2. Hutchinson, J.W. (1974) Plastic buckling, *Adv. Appl. Mech.* **14**, 67-114.
3. Gudramovych, V.S. (1987) *Stability of elastoplastic shells*, Academic Publ. Naukova Dumka, Kiev (in Russian).
4. Kadashevich, Y.I. and Novozhilov, V.V. (1981) The theory of plasticity and creep allowing for microstresses, *Izv. AN SSSR. Mech. Tverdogo Tela* **5**, 99-110 (in Russian).
5. Onat, E. T. and Drucker, D. C. (1953) Inelastic instability and incremental theories of plasticity, *J. Aeronaut. Sci.* **20**, 181-186.
6. Hoff, N. I. (1956) Creep buckling, *J. Aeronaut. Quart.***7**, no 1.
7. Gudramovych,V.S. and Demenkov, A.F. (1991) *Elastoplastic structures with imperfections and residual stresses,* Academic Publ. Naukova Dumka, Kiev (in Russian).
8. Washizu, K. (1982) *Variational methods in elasticity and plasticity,* Pergamon Press Publ., Oxford.
9. Lahaye, M.E. (1948) Solution of system of transcendental equations, *Acad.Roy. Belg. Bull Cl. Sci.* **5** , 805-822.
10. Lin, T.H. (1950) Buckling of the nonelasticity column, *J. Aeronaut. Sci.* **25**, n.3, 159-172.
11. Oden, I.T. (1972) *Finite elements of nonlinear continua*, Mc Graw-Hill Book Co.
12. Grigoljuk, E.I. and Shalashilin, V.I. (1988) *Nonlinear problems of deformation. The method of parameter continuation of solution in nonlinear problems of mechanics of deformable bodies.* Publ. Nauka, Moscow (in Russian).

13. Bushnell, D. (1981) Buckling of shell-pitfall for designers, *AIAA Journ.* **19** ,n.9, 1183-1226.
14. Gudramovych, V. S., Gerasimov, V. P., Konovalenkov, V. V., and Poshivalov, V.P. (1984) *Critical states of shells under complex loading and creep*, Academic Publ. Naukova Dumka, Kiev (in Russian).
15. Kurshin, L.M. (1965) On statement of problem of shell buckling under creep, *Dokl. Acad. Nauk SSSR* **163**, n. 1, 46-49 (in Russian).
16. Hoff, N.I., Jahsman, W.E., and Nachbar, W.A. (1959) Study of creep collapse of a long circular cylindrical shell under uniform external pressure, *J.Aeronaut. Sci.* **25**, n. 10, 663-669.
17. Gudramovych, V.S. and Poshivalov, V.P. (1979) On investigation of shell buckling under creep, *Dokl.Acad.Nauk SSSR* **245**, n.6, 1329-1332 (in Russian).
18. Shesterikov, S. A. (1963) Buckling under creep allowing for instantaneous plastics deformation, *Prikl. Mech. and Tech. Phys.* **2**, 124-129 (in Russian).
19. Arbocz, I. (1983) Shell stability analysis: theory and practice, *Collapse. The buckling of structures in theory and practice*, Cambridge. Univ. Press, 43-74.
20. Lutteroth, A. (1983) Neubearleitung des Standarts TGL 13503 Stabilitat von Stahltragwerken, *Bauplan.-Bautechn.* **37**, n. 5, 212-215.

13. Bushnell, D. (1981) Buckling of shells-pitfall for designers, AIAA Journal 19, 1183–1226.
14. Gudramovich, V.S., Gerasimov, V.P., [illegible], V.V. and Pochtman, V.R. (19[illegible]) Limit states of shells under complex loading and creep, Academic Publ. Naukova Dumka, Kiev (in Russian).
15. Kurshin, L.M. (1965) On statement of problem of shell buckling under creep, Dokl. Akad. Nauk SSSR 163, [illegible] (in Russian).
16. Hoff, N.J., Jahsman, W.E. and Nachbar, W.A. (1959) Study of creep collapse of a long circular cylindrical shell under uniform external pressure, [illegible]
17. [illegible]
[illegible]

AUTHOR'S INDEX

Mechanics

SOLID MECHANICS AND ITS APPLICATIONS

Series Editor: G.M.L. Gladwell

Aims and Scope of the Series

The fundamental questions arising in mechanics are: *Why?, How?,* and *How much?* The aim of this series is to provide lucid accounts written by authoritative researchers giving vision and insight in answering these questions on the subject of mechanics as it relates to solids. The scope of the series covers the entire spectrum of solid mechanics. Thus it includes the foundation of mechanics; variational formulations; computational mechanics; statics, kinematics and dynamics of rigid and elastic bodies; vibrations of solids and structures; dynamical systems and chaos; the theories of elasticity, plasticity and viscoelasticity; composite materials; rods, beams, shells and membranes; structural control and stability; soils, rocks and geomechanics; fracture; tribology; experimental mechanics; biomechanics and machine design.

1. R.T. Haftka, Z. Gürdal and M.P. Kamat: *Elements of Structural Optimization.* 2nd rev.ed., 1990 ISBN 0-7923-0608-2
2. J.J. Kalker: *Three-Dimensional Elastic Bodies in Rolling Contact.* 1990 ISBN 0-7923-0712-7
3. P. Karasudhi: *Foundations of Solid Mechanics.* 1991 ISBN 0-7923-0772-0
4. *Not published*
5. *Not published.*
6. J.F. Doyle: *Static and Dynamic Analysis of Structures.* With an Emphasis on Mechanics and Computer Matrix Methods. 1991 ISBN 0-7923-1124-8; Pb 0-7923-1208-2
7. O.O. Ochoa and J.N. Reddy: *Finite Element Analysis of Composite Laminates.* ISBN 0-7923-1125-6
8. M.H. Aliabadi and D.P. Rooke: *Numerical Fracture Mechanics.* ISBN 0-7923-1175-2
9. J. Angeles and C.S. López-Cajún: *Optimization of Cam Mechanisms.* 1991 ISBN 0-7923-1355-0
10. D.E. Grierson, A. Franchi and P. Riva (eds.): *Progress in Structural Engineering.* 1991 ISBN 0-7923-1396-8
11. R.T. Haftka and Z. Gürdal: *Elements of Structural Optimization.* 3rd rev. and exp. ed. 1992 ISBN 0-7923-1504-9; Pb 0-7923-1505-7
12. J.R. Barber: *Elasticity.* 1992 ISBN 0-7923-1609-6; Pb 0-7923-1610-X
13. H.S. Tzou and G.L. Anderson (eds.): *Intelligent Structural Systems.* 1992 ISBN 0-7923-1920-6
14. E.E. Gdoutos: *Fracture Mechanics.* An Introduction. 1993 ISBN 0-7923-1932-X
15. J.P. Ward: *Solid Mechanics.* An Introduction. 1992 ISBN 0-7923-1949-4
16. M. Farshad: *Design and Analysis of Shell Structures.* 1992 ISBN 0-7923-1950-8
17. H.S. Tzou and T. Fukuda (eds.): *Precision Sensors, Actuators and Systems.* 1992 ISBN 0-7923-2015-8
18. J.R. Vinson: *The Behavior of Shells Composed of Isotropic and Composite Materials.* 1993 ISBN 0-7923-2113-8
19. H.S. Tzou: *Piezoelectric Shells.* Distributed Sensing and Control of Continua. 1993 ISBN 0-7923-2186-3

Kluwer Academic Publishers – Dordrecht / Boston / London

Mechanics

SOLID MECHANICS AND ITS APPLICATIONS

Series Editor: G.M.L. Gladwell

20. W. Schiehlen (ed.): *Advanced Multibody System Dynamics*. Simulation and Software Tools. 1993 ISBN 0-7923-2192-8
21. C.-W. Lee: *Vibration Analysis of Rotors*. 1993 ISBN 0-7923-2300-9
22. D.R. Smith: *An Introduction to Continuum Mechanics*. 1993 ISBN 0-7923-2454-4
23. G.M.L. Gladwell: *Inverse Problems in Scattering*. An Introduction. 1993 ISBN 0-7923-2478-1
24. G. Prathap: *The Finite Element Method in Structural Mechanics*. 1993 ISBN 0-7923-2492-7
25. J. Herskovits (ed.): *Advances in Structural Optimization*. 1995 ISBN 0-7923-2510-9
26. M.A. González-Palacios and J. Angeles: *Cam Synthesis*. 1993 ISBN 0-7923-2536-2
27. W.S. Hall: *The Boundary Element Method*. 1993 ISBN 0-7923-2580-X
28. J. Angeles, G. Hommel and P. Kovács (eds.): *Computational Kinematics*. 1993 ISBN 0-7923-2585-0
29. A. Curnier: *Computational Methods in Solid Mechanics*. 1994 ISBN 0-7923-2761-6
30. D.A. Hills and D. Nowell: *Mechanics of Fretting Fatigue*. 1994 ISBN 0-7923-2866-3
31. B. Tabarrok and F.P.J. Rimrott: *Variational Methods and Complementary Formulations in Dynamics*. 1994 ISBN 0-7923-2923-6
32. E.H. Dowell (ed.), E.F. Crawley, H.C. Curtiss Jr., D.A. Peters, R. H. Scanlan and F. Sisto: *A Modern Course in Aeroelasticity*. Third Revised and Enlarged Edition. 1995 ISBN 0-7923-2788-8; Pb: 0-7923-2789-6
33. A. Preumont: *Random Vibration and Spectral Analysis*. 1994 ISBN 0-7923-3036-6
34. J.N. Reddy (ed.): *Mechanics of Composite Materials*. Selected works of Nicholas J. Pagano. 1994 ISBN 0-7923-3041-2
35. A.P.S. Selvadurai (ed.): *Mechanics of Poroelastic Media*. 1996 ISBN 0-7923-3329-2
36. Z. Mróz, D. Weichert, S. Dorosz (eds.): *Inelastic Behaviour of Structures under Variable Loads*. 1995 ISBN 0-7923-3397-7
37. R. Pyrz (ed.): *IUTAM Symposium on Microstructure-Property Interactions in Composite Materials*. Proceedings of the IUTAM Symposium held in Aalborg, Denmark. 1995 ISBN 0-7923-3427-2
38. M.I. Friswell and J.E. Mottershead: *Finite Element Model Updating in Structural Dynamics*. 1995 ISBN 0-7923-3431-0
39. D.F. Parker and A.H. England (eds.): *IUTAM Symposium on Anisotropy, Inhomogeneity and Nonlinearity in Solid Mechanics*. Proceedings of the IUTAM Symposium held in Nottingham, U.K. 1995 ISBN 0-7923-3594-5
40. J.-P. Merlet and B. Ravani (eds.): *Computational Kinematics '95*. 1995 ISBN 0-7923-3673-9
41. L.P. Lebedev, I.I. Vorovich and G.M.L. Gladwell: *Functional Analysis*. Applications in Mechanics and Inverse Problems. 1996 ISBN 0-7923-3849-9
42. J. Menčik: *Mechanics of Components with Treated or Coated Surfaces*. 1996 ISBN 0-7923-3700-X
43. D. Bestle and W. Schiehlen (eds.): *IUTAM Symposium on Optimization of Mechanical Systems*. Proceedings of the IUTAM Symposium held in Stuttgart, Germany. 1996 ISBN 0-7923-3830-8

Kluwer Academic Publishers – Dordrecht / Boston / London

Mechanics

SOLID MECHANICS AND ITS APPLICATIONS

Series Editor: G.M.L. Gladwell

44. D.A. Hills, P.A. Kelly, D.N. Dai and A.M. Korsunsky: *Solution of Crack Problems*. The Distributed Dislocation Technique. 1996 ISBN 0-7923-3848-0
45. V.A. Squire, R.J. Hosking, A.D. Kerr and P.J. Langhorne: *Moving Loads on Ice Plates*. 1996 ISBN 0-7923-3953-3
46. A. Pineau and A. Zaoui (eds.): *IUTAM Symposium on Micromechanics of Plasticity and Damage of Multiphase Materials*. Proceedings of the IUTAM Symposium held in Sèvres, Paris, France. 1996 ISBN 0-7923-4188-0
47. A. Naess and S. Krenk (eds.): *IUTAM Symposium on Advances in Nonlinear Stochastic Mechanics*. Proceedings of the IUTAM Symposium held in Trondheim, Norway. 1996 ISBN 0-7923-4193-7
48. D. Ieşan and A. Scalia: *Thermoelastic Deformations*. 1996 ISBN 0-7923-4230-5
49. J. R. Willis (ed.): *IUTAM Symposium on Nonlinear Analysis of Fracture*. Proceedings of the IUTAM Symposium held in Cambridge, U.K. 1997 ISBN 0-7923-4378-6
50. A. Preumont: *Vibration Control of Active Structures*. An Introduction. 1997 ISBN 0-7923-4392-1
51. G.P. Cherepanov: *Methods of Fracture Mechanics: Solid Matter Physics*. 1997 ISBN 0-7923-4408-1
52. D.H. van Campen (ed.): *IUTAM Symposium on Interaction between Dynamics and Control in Advanced Mechanical Systems*. Proceedings of the IUTAM Symposium held in Eindhoven, The Netherlands. 1997 ISBN 0-7923-4429-4
53. N.A. Fleck and A.C.F. Cocks (eds.): *IUTAM Symposium on Mechanics of Granular and Porous Materials*. Proceedings of the IUTAM Symposium held in Cambridge, U.K. 1997 ISBN 0-7923-4553-3
54. J. Roorda and N.K. Srivastava (eds.): *Trends in Structural Mechanics*. Theory, Practice, Education. 1997 ISBN 0-7923-4603-3
55. Yu. A. Mitropolskii and N. Van Dao: *Applied Asymptotic Methods in Nonlinear Oscillations*. 1997 ISBN 0-7923-4605-X
56. C. Guedes Soares (ed.): *Probabilistic Methods for Structural Design*. 1997 ISBN 0-7923-4670-X
57. D. François, A. Pineau and A. Zaoui: *Mechanical Behaviour of Materials*. Volume I: Elasticity and Plasticity. 1998 ISBN 0-7923-4894-X
58. D. François, A. Pineau and A. Zaoui: *Mechanical Behaviour of Materials*. Volume II: Viscoplasticity, Damage, Fracture and Contact Mechanics. 1998 ISBN 0-7923-4895-8
59. L. T. Tenek and J. Argyris: *Finite Element Analysis for Composite Structures*. 1998 ISBN 0-7923-4899-0
60. Y.A. Bahei-El-Din and G.J. Dvorak (eds.): *IUTAM Symposium on Transformation Problems in Composite and Active Materials*. Proceedings of the IUTAM Symposium held in Cairo, Egypt. 1998 ISBN 0-7923-5122-3
61. I.G. Goryacheva: *Contact Mechanics in Tribology*. 1998 ISBN 0-7923-5257-2

Kluwer Academic Publishers – Dordrecht / Boston / London

ICASE/LaRC Interdisciplinary Series in Science and Engineering

1. J. Buckmaster, T.L. Jackson and A. Kumar (eds.): *Combustion in High-Speed Flows.* 1994 ISBN 0-7923-2086-X
2. M.Y. Hussaini, T.B. Gatski and T.L. Jackson (eds.): *Transition, Turbulence and Combustion.* Volume I: Transition. 1994 ISBN 0-7923-3084-6; set 0-7923-3086-2
3. M.Y. Hussaini, T.B. Gatski and T.L. Jackson (eds.): *Transition, Turbulence and Combustion.* Volume II: Turbulence and Combustion. 1994 ISBN 0-7923-3085-4; set 0-7923-3086-2
4. D.E. Keyes, A. Sameh and V. Venkatakrishnan (eds): *Parallel Numerical Algorithms.* 1997 ISBN 0-7923-4282-8
5. T.G. Campbell, R.A. Nicolaides and M.D. Salas (eds.): *Computational Electromagnetics and Its Applications.* 1997 ISBN 0-7923-4733-1
6. V. Venkatakrishnan, M.D. Salas and S.R. Chakravarthy (eds.): *Barriers and Challenges in Computational Fluid Dynamics.* 1998 ISBN 0-7923-4855-9

KLUWER ACADEMIC PUBLISHERS – DORDRECHT / BOSTON / LONDON

y you have any concerns about our products,
you can contact us on
roductsafety@springernature.com

e Publisher is established outside the EU,
the EU authorized representative is:
Springer Nature Customer Service Center GmbH
Europaplatz 3, 69115 Heidelberg, Germany

Printed by Libri Plureos GmbH
in Hamburg, Germany

MIX
Papier aus verantwortungsvollen Quellen
Paper from responsible sources
FSC® C105338

If you have any concerns about our products,
you can contact us on
ProductSafety@springernature.com

In case Publisher is established outside the EU,
the EU authorized representative is:
Springer Nature Customer Service Center GmbH
Europaplatz 3, 69115 Heidelberg, Germany

Printed by Libri Plureos GmbH
in Hamburg, Germany